T0202733

Springer Biographies

The books published in the Springer Biographies tell of the life and work of scholars, innovators, and pioneers in all fields of learning and throughout the ages. Prominent scientists and philosophers will feature, but so too will lesser known personalities whose significant contributions deserve greater recognition and whose remarkable life stories will stir and motivate readers. Authored by historians and other academic writers, the volumes describe and analyse the main achievements of their subjects in manner accessible to nonspecialists, interweaving these with salient aspects of the protagonists' personal lives. Autobiographies and memoirs also fall into the scope of the series.

More information about this series at http://www.springer.com/series/13617

Rob Herber

Nico Bloembergen

Master of Light

Rob Herber
Utrecht University
Utrecht, The Netherlands

Netherlands Organisation
for Scientific Research

This book could be achieved by support of the Netherlands Organisation for Scientific Research.

ISSN 2365-0613   ISSN 2365-0621   (electronic)
Springer Biographies
ISBN 978-3-030-25739-2   ISBN 978-3-030-25737-8   (eBook)
https://doi.org/10.1007/978-3-030-25737-8

This Springer imprint is published by the registered company Springer Nature Switzerland AG
The registered company address is: Gewerbestrasse 11, 6330 Cham, Switzerland

To the memory of my late wife Marga. Night and day, you were the one.

Preface

For the Dutch Publication

Around 1950, my dad Frans took me in Bilthoven to a field hockey match of the local hockey club Stichtsche Cricket en Hockey Club. I went to a field hockey game for the first time and did not know much about it. My father showed me the blue-red-dressed players and said that especially the two brothers Bloembergen were very good. The brothers were Auke and Herbert. It struck me that Auke, who was not big, screamed loudly in an unadulterated accent over the field. That was my first acquaintance with the Bloembergen.

Ten years later, I worked as a laboratory technician in the Chemical Laboratory of the Royal Tropical Institute in Amsterdam. The Chemical Laboratory belonged to the Tropical Products department. Director of the department was Dr. Rinse Brink. This man was the father of Deli Brink, the wife of Nico Bloembergen. That was my first acquaintance with the Brink family.

After my retirement from the AMC in 2000, I was able to turn my old interest in history into something useful by becoming editor-in-chief of the journal of local history of De Bilt. The magazine is called *De Biltse Grift,* and I conceived the plan to generate interest for the many scientists who had lived in thanks to the proximity of Utrecht University, the National Institute for Public Health and the Environment and the Royal Meteorological Institute in De Bilt or Bilthoven.

First, I thought of Bloembergen, because he was relatively unknown. *Nico Bloembergen, a physicist in light, appeared in 2006* [1]. Because I had immersed myself in the work of Bloembergen, I wondered whether a biography was written about him. That turned out not to be the case.

Bloembergen's life and career took place in the Netherlands but mostly in the USA: Perhaps that is why no biography was published.

I submitted my idea to Bloembergen in autumn 2006 by e-mail, but at first, he did not like it at all. He himself had written a number of pieces that could be considered autobiography and that seemed sufficient to him. Moreover, he doubted whether the *"project was feasible."* However, I was able to convince him quite quickly that a biography was useful and that I was indeed equipped for such an undertaking. Firstly, after consulting the family chronicle of Auke Reitze, the youngest brother of Nico, I decided to interview the "nicest" brother Herbert in Gorinchem. After Herbert, I visited the oldest brother Evert in Zeist. Evert had poor health, was in a wheelchair, was almost blind, but could still tell about the past. Shortly after visiting Evert, I visited Jan Caro in a nursing home in Bilthoven. Nico Bloembergen had been friends with him since the gymnasium. Caro was approachable despite his illness. On 5 December 2006, I visited Nico and his wife Deli Bloembergen in Tucson, Arizona. Nico and I saw each other for the first time. I was received very warmly, and both were willing to answer questions from me. Deli, after I had made known in advance that I am a vegetarian, prepared an extensive vegetarian lunch.

In the following years, I met Nico and Deli twice in Amsterdam, where they stayed in transit from Lindau. The Bloembergen also met my wife Marga Verschoor the second time in Amsterdam. Many e-mails and telephone conversations show intensive contact between biographer and the Bloembergen. I have come to know both as very warm; they sent me numerous photographs, letters and other documents asked and unsolicited. The last time I have met Nico was during a congress in Barcelona, Spain, in 2012.

In early 2007, I interviewed the youngest brother Auke Reitze. I am extremely grateful that he is in 2003 recorded the family life of father and mother Bloembergen and their children. This has laid a solid foundation for my biography.

The youngest daughter of Deli and Nico, Juliana Dalton-Bloembergen, I could interview in Lexington in 2009. She gave me a picture of the family life in Lexington and treated Marga, her daughter and me for lunch. What a warm woman!

My thanks go to Oswald de Bruin, brother-in-law of Clara Shoemaker-Brink.

I have interviewed many fellow students, colleagues and former colleagues of Nico. Some were old and passed away during the writing of the biography. My thanks go to the following persons:

Maarten Bouwman (Emeritus Professor of Physics and Medicine at Utrecht University) †, Jan Caro (Lawyer) †, Gerald Holton (Research Professor of Physics and Research at the University of Harvard University), Cees de Jager (Emeritus Professor of Astronomy, Utrecht University), Nico van Kampen (Emeritus Professor of Physics, Utrecht University) †, Bob Lansberg (Aerospace Engineer), Eric Mazur (Professor of Physics and Applied Physics, Harvard University), Peter Pershan (Professor of Science, Harvard University), Robert Pound (Professor of Physics Emeritus, Harvard University) †, Norman Ramsey (Professor of Applied Physics, Harvard University) †, Barend Streefland (retired Senior Academic, Utrecht University), Tony Siegman (Emeritus Professor of Engineering, Stanford University) †, Charles Townes (University Professor, University of California) †, Martinus Veltman (Emeritus Professor of Theoretical Physics at Utrecht University) and Jan Visser (retired Lecturer in Classical Languages, Utrecht Urban Gymnasium).

I am indebted to the following persons for their cooperation:

Joe Anderson (Niels Bohr Library Director, American Institute of Physics), Lex Boumans (former Director of FOM, Amsterdam) †, Dirk van Delft (Director, Boerhaave Museum, Leiden), Wiel Hoekstra (Emeritus Professor of Microbiology, Utrecht University), Jeroen Jekel (Assistant, University Museum, Utrecht University), Jo Nelissen (Researcher, Freudenthal Institute, Utrecht University), Leendert Plaisier (ABN-AMRO Bank), Maarten Tromp (University Administration, Utrecht University) and Jelle van der Zee (hockey club SCHC).

Finally, I would like to thank my fellow readers who have read my epistles without any compensation or without having the least interest in them. The commentary of the co-readers was occasionally not wrong: *"That does not read anybody,"* or *"Nothing fits that."* With this, I hope to have illustrated the independent character of my fellow readers.

Pieter Span is the one who motivated me, encouraged me and followed critically. He struggled through the physics as Professor Emeritus in the psychology of Utrecht University. What patience and what have I learned a lot from giftedness! With the progress of the biography, Pieter also found it increasingly exciting and of course I liked that in my turn.

Frans van Lunteren had, as he wrote it in an e-mail, occasionally horribly busy. Nevertheless, he has seen a chance between his research, education and guidance of countless Ph.D. students as professor at the Free University in Amsterdam and the Leiden University in the history of science, to assist me and to suggest numerous improvements in physics. The nice thing is that, as he pointed out, we complemented each other quite nicely: he as the theoretical man in the history of science and I as an experimentalist. Frans has ensured that I received an appointment as a guest employee at the Utrecht University. Furthermore, I was asked by Frans to write a piece about Bloembergen in the Dutch Journal of Physics in the commemorative issue about 50 years of laser [2].

Anne Doedens spontaneously stated, when he heard about my biography, that he wanted to assess the biography on historical merits as a critical reader. As a retired lecturer at the high school Inholland in Diemen, he did have some time, he said. Now, I could have told him that a retired teacher has no time at all. Then, there is time for all the fun things you would like to do. But it did turn out anyway. Anne has helped me with the first chapters, in which a lot of history occurs.

In the biography, I could combine my knowledge of spectrometry with my interest in history, a good starting point for writing. The biography had been written without the interference of Nico Bloembergen. When I started with the biography, I tried to get a grant. I foresaw that I should visit Nico, Harvard University and the American Institute of Physics. The subsidy did not work out, so I asked myself the question: do I have to continue? My life partner Dr. Marga Verschoor convinced me to continue without a subsidy. The support she gave me, even during difficult times since 2012, has been indispensable for the success of the biography. To my great sadness, Marga could no longer see the publication of the Dutch biography in 2016. She passed away in 2014. We had been happy together for 45 years.

For the English Publication

After the publication of the Dutch book [3], Nico Bloembergen suggested that the manuscript should be published in English. He rated the Dutch book as *"very good"* and that after editing it could be published. Nico contacted his former student Dr. Michael Salour, now his friend and CEO of IPITEK, Carlsbad, California. Michael has searched intensively and approached Springer Verlag in Germany. I have to thank Michael very much for his effort.

Springer agreed to publish the biography, but translation was too expensive. We then choose to try the DeepL translation program. I want to thank Dr. Liesbeth Mol, at that time Editorial Director of the Physics and Astronomy Department of Springer, as she was very enthusiastic about publishing the manuscript in English. Machine translation has still its limits to write it friendly. Dr. Stephen Lyle, Copy-Editor of Springer, had the task to iron away the many flaws in the text. It was a very nice cooperation.

I redesigned the Dutch manuscript to focus on the international reader; shortened the first chapters and extended some other chapters, added new facts, etc. Many interviews have been processed in the biography. The quotations are printed in italics. The interviews taken by myself are placed in quotation marks.

Utrecht, The Netherlands   Rob Herber

References

1. Herber R. Nico Bloembergen, fysicus in licht. (Physicist in Light) De Biltse Grift 15:34–49 (2006)
2. Herber R. Nico Bloembergen: laserpionier en Nobelprijswinnaar (Laser pioneer and Nobel Price winner). Ned Tijdschr Natuurk 76: 228–231 (2010)
3. Herber R. Nico Bloembergen, Meester van het licht. (Master of the Light) Eburon, Delft (2016)

Introduction

It is Saturday 19 February 1887. A typical Dutch winter day; last night, there was frost, and it is now a few degrees above zero and almost clear sky [1]. In the center of the Frisian city of Leeuwarden are everywhere the Dutch three color and the Frisian flag with pump blades. On the church for the Large Jacobijnerkerkhof (cemetery) numerous carriages, predominantly brigs in black and red. A single coach is an expensive Spijker, of the manufacturer who in 1898 The Golden Carriage for Queen Wilhelmina will make.

Groups of coachmen in their black costumes, some in green or gray broadcloth, talk between the carriages. The church square is too small to all coaches to park, and the side streets are also full. Opposite the Large Church is the in 1866 opened Jewish school. Everywhere people walk on the church square. Inside the church is the in gold and brown tones painted Müller Organ from the years 1724–1727 [2]. On the left-hand side is the seventeenth-century pulpit. On the other right-hand is the balcony where the Frisian Nassaus sat—which had a separate entrance. The Frisian stewards are the direct ancestors of the members of the Royal House and are buried for centuries in the Large Church. The church is packed full, all five hundred seats are occupied, and there are even people to stand. There is clearly a party going on [3]. The mostly bearded and mustachioed gentlemen wear three-piece black, with or without a watch chain on the belly; the ladies wear crinolines in pastel shades and hats. Most have an orange cockade or another orange decoration on their hat or dress. All belong clearly to the elite of Leeuwarden. In anticipation of the speaker, one maintains with one another the full awareness of one's own importance. The meeting in the Large Church is the capstone of a series of festivities held in honor of the seventieth birthday of King William III on 19 February.

When the only speaker of that afternoon comes in, everyone falls silent. The speaker steps to the lectern. He looks at the colorful stained-glass windows to the right of him where, the sun lights up the colors of the depicted Orange, and starts. He is one of the most important inhabitants of the city of Leeuwarden, the member of the Provincial States and banker Auke Bloembergen.

94 years later, Tuesday 10 December 1981, Council House of Stockholm. On 7 December there was a blizzard; now, it is sunny and it freezes. The Stockholms

Stadshus (Council House) is beautifully situated on the banks of two waterways: the Riddarfjärden (Knight Fjord and the Channel Klara Sjö (Klara More) [4]. From the balcony of the Blue Hall there is a beautiful view of the room, where during the day the light by high windows [5]. The walls of the hall are brick; in front of the walls are high pillars of several meters that carry the balconies. There is also an organ in the Blue Room: E. F. Walcker and Cie from Ludwigsburg constructed it in 1925 with more than 10,000 pipes. On this dark winter evening, the hall is festively illuminated and filled with 1200 invited guests, including several hundred students sitting at a banquet. Accompanied by trumpet sound, King Carl XVI Gustav and Queen Silvia of Sweden, the Nobel laureates of 1981, the guests and the government officials descend the stairs to the covered main table in the middle of the hall. Everyone stands up when the procession—including Nobel Prize winner Nico Bloembergen—enters the hall.

He speaks to the guests with a word of thanks during the banquet as one of the Nobel Prize winners.

In the afternoon, the official presentation took place of the Nobel Prize Medal in the Konserthuset (the Stockholm Concert Hall) by the king. On that occasion, Bloembergen was awarded the Nobel Prize Diploma. He was awarded the prize for the application of spectroscopy, a physical technique in which colors play an important role.

Nico Bloembergen is the great-grandson of the Auke Bloembergen who addressed the guests in Leeuwarden in 1887.

In this biography not only the scientific career of Bloembergen is highlighted, but there is also attention for his origins, the Bloembergen family and the environment in which he grew up, the village of Bilthoven.

Physicists in Western Europe have been relatively unknown in recent decades. That used to be different. When in 1928 Dutch national physicist Hendrik Antoon Lorentz was buried in Haarlem, there was national mourning. Flags hung half-mast, and street lamps were wrapped in black cloth. Telephone and telegraph services were shut down for three minutes [6]. Thousands stood along the funeral procession; men took off their hats and hats. The funeral was attended by, among others, Albert Einstein, Ernest Rutherford, members of the cabinet and the Royal House [7].

From 1986–1994, the German television channel Zweites Deutsches Fernsehen (ZDF) broadcasted a television series *Die Stillen Stars* (The silent stars). The subtitle is significant: *Nobelpreisträger privat gesehen* (Nobel Prize winners seen privately). In this series of no less than 113 episodes, the Austrian Frank Elster interviewed a number of Nobel Prize winners in physics at home or at their workplace. Bloembergen was also interviewed for this series [8]. The title of the TV series clearly indicates what is intended: to bring important people in the spotlight for the development of science.

These scientific developments, in particular developments in physics, have in turn led to important technological and social developments. As an example, I mention the proposal by the (English) computer scientist Timothy John Berners-Lee and (Belgian) computer scientist Robert Callieu in March 1989 to develop a computer and information management system for communication within

CERN (originally: Conseil Européen pour la Recherche Nucléaire), the research center in Geneva of the European organization for research into elementary particles [9]. This computer system originally intended for the exchange of data in the field of elementary particle physics has grown into the World Wide Web with billions of users and with great implications for society.

Bloembergen has also contributed to major technological and social developments through its research. The research that Bloembergen did, however, was not intended by him to generate applications. Bloembergen was interested in the physics research itself and did not think about applications. Yet the work of Bloembergen has had a major influence on our current way of life.

The first major research that Bloembergen was involved in led to developments in the structural analysis of chemical compounds and is still used in chemistry and biology. This same research, together with developments in image technology and computers, led to MRI, magnetic resonance imaging. MRI is used in hospitals to form an image of soft tissues.

There is a laser in every CD or DVD player. Lasers are used at cash registers in the shop, in medicine, in industry, in surveying and in defense. The number of applications runs in the thousands. Bloembergen devised an energy pump system, making the maser, the predecessor of the laser, practically usable. The laser works according to the same principle as the maser and Bloembergen's energy pump system is used in all these lasers.

For this, the World Wide Web is briefly mentioned. The World Wide Web has only been used worldwide since the introduction of the optical fiber connection. The optical fiber is another application of Bloembergen research, the nonlinear optics. The glass fiber, together with the aforementioned laser, has ensured that a large amount of data can be transported very quickly over large distances.

Bloembergen studied at the then Utrecht University from 1938 to 1943. The period of physics in Utrecht between 1896 and 1940 was described in 1994 by Han Heijmans. In this study, an impression of Leonard Ornstein is also outlined. Ornstein was a major research and education innovator, and he has given the physics laboratory a great international reputation. Bloembergen has been a lecturer at Ornstein, but much more important is that Bloembergen could see at Ornstein how the research was done. Bloembergen was immediately sold and from that moment put everything in the service of research.

Bloembergen obtained his Ph.D. in 1948 at the Leiden University. He met Hendrik (Hans) Kramers there. The latter was Professor from 1926 to 1934 at the Utrecht University. The daughter of Kramers, Suusje, was in the first class of the Utrecht City Gymnasium (1932–1933), just like Bloembergen. In the Hunger Winter (1944–1945), Bloembergen studied Kramers' standard work on quantum mechanics. In his Leiden period, Bloembergen had many discussions with Kramers. In 1987, a biography of Hendrik Kramers by Max Dresden was published [10]. Kramers was an important but less known theoretical physicist. He was assistant to Niels Bohr and has worked with Werner Heisenberg and Ralph Kronig. Kramers was Founder of the Center for Mathematics and Computer Science in Amsterdam. In 1947, Kramers received the Lorentz medal for theoretical physics.

A number of biographies have appeared about American physicists, which Bloembergen had to deal with. The life and work of Isadore Rabi, founder of nuclear magnetic resonance, was described in 2001 by John S. Rigden [11]. Robert Pound has its long-term cooperation with Edward Purcell in biographical memories recorded [12]. The Julian Schwinger biography appeared in 2003 by Kimball Milton [13].

Dirk Van Delft writes in his biography about Kamerlingh Onnes: *And the more of those calls I performed, the stronger the awareness that insisted biographer cannot be careful enough with old memories, that he will be better to use the archives* [14]. I did not have that choice. There is no family correspondence, and no significant correspondence with colleagues from Bloembergen has been preserved. However, the youngest brother of Nico Bloembergen, Auke Reitze, published a family chronicle some years ago [15]. This chronicle is also largely based on conversations with the other brothers and on the memories of the author himself. I (RH) am very grateful that the booklet was there and I made grateful use of it. A limited number of letters have been retained from Bloembergen's correspondence with his supervisor, Cor Gorter from the Leiden University and with Pim Milatz from the Utrecht University. In addition, Bloembergen himself has written a number of autobiographical documents in a few collected works that he has released [16]. Its largest competitor Charles Townes has published an autobiography [17]. My intention to write a biography about Nico Bloembergen came just in time to interview some of his (sometimes very elderly) friends, fellow students and colleagues. Strikingly, everyone was happy to cooperate with the interviews. This already indicates that Bloembergen is someone who raised predominantly positive feelings.

References

1. www.knmi.nl
2. www.grotekerkleeuwarden.nl
3. Plaisir L. Bloembergen: Een Leeuwarder familie in bankzaken (A Leeuwarder family in banking). Fryslan 3–7, J07 (2001)
4. www.stockholm.se/
5. Nobelprize.org/award_ceremonies/banquet/vr-cityhall/linked-java/index.html
6. nl.wikipedia.org
7. www.geschiedenis24.nl/speler.program.7063332.html
8. Frank Elstner Interview with N Bloembergen. Die Stillen Stars (The Silent Stars). Nobelpreisträger privat gesehen (Nobel Prize winners seen privately.) Heute (Now): Prof. Nicolaas Bloembergen. ZDF-documentary (1987)
9. en.wikipedia.org
10. Dresden M.H.A. Kramers - Between tradition and Revolution. Springer, Heidelberg (1987)
11. Rigden JS. Rabi: Scientist and citizen. Harvard University Press, Cambridge (2001)
12. Pound R. Edward Mills Purcell, 1912–1997: A biographical memoir. National Academy Press, Washington (2000)

13. Milton KA. Climbing the Mountain: The Scientific Biography of Julian Schwinger. Oxford University Press, New York (2003)
14. Delft D. De man van het absolute nulpunt. (The man of the absolute zero). Bert Bakker, Amsterdam (2005)
15. Bloembergen AR. Het gezin van Rie en Auke Bloembergen 1917–1956 (The family of Rie and Auke Bloembergen 1917–1956). Private Management, Wassenaar, 2003
16. Bloembergen N. Encounters in magnetic resonances. World Scientific, Singapore (1996)
17. Townes CH. How the Laser happened. Oxford University Press, New York (1999)

Contents

Part I
Family and Upbringing

Chapter 1
The Bloembergen Dynasty: High Peaks and Deep Valleys

1.1 The Bloembergens in Friesland

Ninety-four years, from 1887 to 1981, separate Auke and Nico Bloembergen as I have described them in the Introduction. Certain themes connect the great-grandfather Auke, the banker and politician who addressed the local elite in a Protestant church, to the great-grandchild Nico, the scientist who addressed an international elite in a town hall. These themes are clear from the history of the Bloembergen dynasty, that is, the history of the Bloembergen's from their arrival in Leeuwarden to the establishment of Evert Bloembergen in the west.

The Bloembergen's were got where they were through their social activities and it was this training that they passed on to their descendants, including Nico and his brothers and sisters. This is what we call 'nurture'. On the other hand, there is another important factor in our construction, called 'nature'. The representation of the family's history is important, because it may be that clues can be found in the family about this predisposition.

In this part of my story, there are two Bloembergen's that especially demand attention. Auke Bloembergen (1838–1901) was a politician and banker, in that order of importance. His younger brother Reitze (1844–1914) was also a banker and later a politician. How did these Bloembergen brothers reach these positions?

Cashiers and bankers
Auke Bloembergen's ancestors had founded a small cashier's office in the Nieuwestad in Leeuwarden. This building in the Nieuwestad was very centrally located opposite the Stadswaag (the town weighbridge) and would remain family property until the twentieth century—on the site of the demolished building was the former V&D department store.

The bank was one of the largest in Friesland, measured by the amount of debt and deposit that appeared on its balance sheets [1].

The extent of the activities the Bloembergen's developed in Leeuwarden society was astonishing. They held important positions in almost every field: in politics,

© Springer Nature Switzerland AG 2019
R. Herber, *Nico Bloembergen*, Springer Biographies,
https://doi.org/10.1007/978-3-030-25737-8_1

business, and economic development, in the church, in freemasonry, in sport, in care for the poor, and in the provision of work [2]. The Bloembergen bank was, according to the description in the *Nederland's Patriciaat* (1993), "*banking for the rich and the super-rich*" [3].

Auke was the first in the Bloembergen family to obtain a Ph.D. After completing his studies, he settled as a lawyer in Leeuwarden and continued in this line until 1864. Later he became chairman of the Masonic Lodge, the *Friesche Trouw*. Auke was active both in freemasonry and in the Protestant church. He became chairman of the gentlemen's club *Amicitia* and of the ecclesiastical electoral association *De moderne richting* (The modern direction). In 1878 Auke became a deputy judge and in 1887 a member of the Provincial Council of Friesland (Fig. 1.1).

The title of the speech which Auke gave on 19 February 1887 in the Grote Kerk (Great Church) in Leeuwarden, was *The king's reign supreme.* The local newspaper, the *Leeuwarder Courant* of 21 February 1887, reported the meeting as follows:

Soon the time had come for the festive gathering in the building of the Groote Kerk. A more appropriate place than the main Hervormden [Reformed] church could not have been chosen - the church where the ancestors of the King had so often come in former times, where so many of them had found their last resting place, and where so much more reminds us of the Nassau heroes and statesmen, to whom Friesland

Fig. 1.1 Auke Ezn Bloembergen (1838–1901), great-grandfather of Nico Bloembergen. *Source* Auke Bloembergen (Nico's nephew)

owes such a great debt, not to mention the King himself, who, whenever he was in Leeuwarden, also made his way there and took his place on the throne still preserved for him after several centuries? All the seats were occupied long before the allotted time; many had to make do with standing room. Under the deep silence of all those hundreds present, Mr. A. Bloembergen Esq addressed the assembly [4].

On 31 August 1898 Auke became Knight of the Order of the Dutch Lion. In April 1901, he said goodbye to the Provincial Councils, because he became a Liberal Member of the Senate on 22 April [5]. He died from a heart attack in The Hague on 20 May 1901.

The Demise of the Bank and the Bloembergen's

The decline came after the long social rise of the Bloembergen's, which had reached its peak with the successes of the banker Reitze and the politician Auke.

In 1911, one of the bank's major debtors, a grain merchant and owner of a cattle feed factory went bankrupt. The bank's agent demanded access to the books, but was not given access to them, and new loans were refused. The bank threatened to go into liquidation and in the deepest secrecy an acquisition was prepared by the *Rotterdamsche Bankvereeniging* [6]. On October 14, 1912, an announcement appeared in the *Leeuwarder Courant* that the articles of association had been amended and that E. Th Geesink had been appointed director. Reitze was dismissed as director [7].

On 1 January 1913, Reitze moved away from Leeuwarden and travelled to The Hague in a bitter mood.

The name Bloembergen Bank was retained until the bank was incorporated into the National Banking Association on 1 July 1916. Of the once so influential and powerful Bloembergen family, there has never been any representative in Leeuwarden since then.

In 1895, however, a son of Auke, Evert, brought about a further decline.

Evert Bloembergen (1865–1925)—Not a banker, but a manufacturer—Nico Bloembergen's grandfather

Evert was born in 1865, the eldest son of Auke, just three months after his father's marriage with his mother Johanna Singels. Auke and Johanna were married when Johanna was already six months pregnant.

Evert followed the family tradition and started working in the bank Bloembergen and sons. He completed business studies in Amsterdam in 1888 and was appointed as the bank's procurator. On 4 June 4 1891, Evert married Antonia Jacoba (Toon) Burger. Her father was Combertus Pieter Burger.

Combertus Pieter Burger (1825–1908), professor and great-grandfather of Nico Bloembergen

Combertus Pieter Burger was born on 25 April 1825 in Rotterdam. He was the first of his family to study mathematics and physics in Leiden.

In 1851, Burger obtained his Ph.D. at Leiden University with a thesis written in Latin and entitled *Specimen academicum inaugurale de solutione problematis Keppleriani* [8]. At the front of the dissertation is a loose-leaf sheet of paper measuring

15 by 15 cm, in which the author explains a few things briefly in Dutch in the most elegant handwriting. The Dutch title then read: *Simple proof of Kepler's theorem.*

The dissertation dealt with the different mathematical approaches to Kepler's theorem, nowadays usually referred to as Kepler's laws. Johannes Kepler (Weil der Stadt, 27 December 1571—Regensburg, 15 November 1630) was a German astronomer and mathematician, who is remembered for the determination of planetary motions. Isaac Newton (1643–1727) would later explain these discoveries physically by his universal law of gravity.

Burger was a lecturer at the Kinsbergen Institute in Elburg (1849), a grammar school teacher in Gouda (1853), a grammar school teacher in Zutphen (1858), a teacher at the Delft Academy (1859), a professor at the Polytechnic School in Delft (1864), and director of the State High School in Leeuwarden (1867–1892). In 1867, he became Titular Professor. He wrote a classic textbook called *Foundations of Mathematical Geography* [9]. This book was reprinted many times to teach what used to be called cosmography (astronomy). He also published a booklet called *On irrational numbers and efficient methods for calculations with decimal fractions* [10].

Burger was appointed Knight of the Order of the Dutch Lion in 1899.

He got married for the first time in Elburg on 19 July 1852, to Haasloop Werner, who was born in Kampen on 21 November 1825 and died in Leeuwarden on 13 October 1881. The couple had twelve children, four of whom died young.

Their daughter Antonia Jacoba Burger, grandmother of Nico Bloembergen, was born in Delft on 25 August 1867 and died in Utrecht on 28 July 1933. All six of her brothers had studied; five obtained Ph.D's. The most famous was the physicist Hendrik (Henk), whom we have met before as Fig. 1.2 the father of Suusje, Nico's class mate. He became professor at the University of Amsterdam.

Until Combertus Pieter came along, the Burgers were a family of merchants, just like the Bloembergens. With Combertus Pieter, Nico's great-grandfather, the brain boxes made their entrance. The brothers of Antonia Jacoba, Nico's grandmother, had all undertaken academic study. It seems likely that something seeped through to the family of Antonia Jacoba and Evert, Nico's parents, from this desire for academic study.

Grandfather Evert Bloembergen: sequel in sin

When Evert Bloembergen and Toos Burger got married in 1891, the couple settled in Leeuwarden. As mentioned earlier, Evert was the bank's authorized representative. According to the genealogy by Van Lennep from 1958, Evert left the bank on 20 July 1895 and went to Groningen. Then in March 1897 he moved on to Rotterdam [11]. However, the Bloembergen bank had no branch in Groningen. In the *Leeuwarder Courant*, there is nothing about a departure, nor anything else about Evert. But Plaisier discusses what was going on in *A Leeuwarder family in banking*, in 2001 [12]. In June 1895 Evert was arrested because he had committed sexual acts on several occasions with a thirteen-year-old girl. The public prosecutor demanded three years' imprisonment, which the court reduced to one year (Fig. 1.3).

With the codification of the Civil Code in 1838, the paternalistic element, already strong in the eighteenth century, was further strengthened. The man, as 'head of the marriage association'

Fig. 1.2 Antonia
Bloembergen-Burger
(1867–1933), Nico's paternal
grandmother. *Source* Auke
Bloembergen (Nico's
nephew)

Fig. 1.3 Evert Bloembergen
(1865–1925), Nico's paternal
grandfather. *Source* Auke
Bloembergen

and as father and guardian of his children, legally gained absolute control over his family. Children remained immature for a long time, while women owed their husbands 'obedience'. In the eighteenth century the mother cult came into being, providing a role for the woman as the highlight of the enlightened bourgeoisie, especially in the nineteenth century. Key concepts here were virtue and passivity.

In bourgeois circles, the various stages of life were characterized by transitional rites. In the promotion to kindergarten, boys were given trousers, and in the promotion to schoolboy, a school uniform. After 1870, schoolboys from the well-to-do bourgeoisie were known to wear knee-length trousers, and on Sundays they wore a sailor's suit until the third grade (age 9–10 years). From the fourth grade on, they wore a bowler hat and a walking stick. In the last quarter of the nineteenth century, as high school girls or 'flappers', daughters of civilians first got half-length dresses and only later full-length dresses. Outgoing young ladies were supposed to put the hair in a bun. For civilian children, the transition to young gentleman or young lady was the most significant transition.

One could only become a fully mature member of civil society through marriage and by setting up one's own household. At a more mature age, over forty, people already belonged to the elderly and had to dress and behave in accordance with that status.

The great prudery which spread in bourgeois and petty-bourgeois circles after 1830 left its mark on community life, public life, and social conventions. In the nineteenth century, a world of men and women emerged. The forbidden fruit, which was shrouded in mystery and considered so dangerous, kept old and young under its spell. Many Victorian citizens (especially women) showed 'neurotic' behaviour. See for example *Eline Vere* by the famous writer Louis Couperus, published in 1889 [13]. The following review of *Eline Vere* appeared in the *De Gids* (The Guide) of 1889:

This victim of our nervous age, with her need for love in all kinds of forms, a good little creature at heart, but spoilt and softened by a life of luxury, has become weak-willed and overwrought; this union of affection and sincerity, from which we cannot withhold our sympathy, even though she herself is the main cause of all the suffering that afflicts her, has been described by Mr. Couperus with a sense for character, psychological analysis, color, and passion [14].

Middle-class girls met the social conventions and presented themselves as innocent and ignorant with so much conviction that contemporaries had difficulty keeping appearances and reality apart. Lower class girls (not middle-class girls) and married (middle-class) women were aware that they had to forgive sexual favours. Civic officials took advantage of this, because it was a safer and more adventurous alternative to visiting the brothel. Incidentally, in middle-class circles, until the turn of the century the latter was still preferred to self-gratification, which was considered to undermine mental and physical health.

Mathijsen uses the umbrella term 'masquerade' for the tangle of progressive and conservative, social and elitist, liberating and oppressive movements, which could be defended by the same person [15]. The official sexual morality is spelt out in public literature. For what was considered normal and what was written about, one has to rely on scarce accounts by the elderly and especially on the undercurrent of literature such as pamphlets, pornography, and gossip columns.

In bourgeois Holland, the conditions a spouse had to meet were fairly clear. In addition to sufficient income from their profession or assets, these were a healthy constitution, a representative appearance, and of course also compatibility of character and preferably an affinity of souls. In the community, during children's balls and at parties, young people were given the opportunity to meet each other under the watchful eye of family and acquaintances. Since the last quarter of the nineteenth century, young people from liberal and Protestant circles have also met in mixed associations and, since the turn of the century, in schools and lecture rooms. In that context, there was no longer direct supervision by family and acquaintances, only mutual control by peers. In practice, these conditions and rules led to unions between sons and daughters of friendly families, fellow students, and professional colleagues, and sometimes to marriages between nephews and nieces, while in the event of remarriage, it could be with a sister-in-law, for example, and subsequent marriages would often remain within the same household. From the eighteenth century onwards, the meaning of 'family' shifted from a neutral term for housemates, including servants under the authority of the owner of the house, to that of an affective and exclusive group of father, mother, and children. The cult of the family meal without outsiders indicates an increasing need for family members to spend time together. Social intercourse or 'community life' was mainly limited to the circle of fellow citizens who knew the courtesy rituals and with whom they could go to the town hall without problems.

The relationship between man and woman was ideally characterized by a 'harmonious order' between breadwinner and housewife. Apart from an affective relationship, marriage was mainly a business agreement to manage one's household and one's sex drive. In the first years of marriage, spouses were allowed to respond moderately to their sex drive, and at a more mature age, after their fortieth year of life, they should strive to spiritualize their love. By curbing his lust in the form of withdrawal or abstinence, the man could show that his wife's health was close to his heart. In practice, this striving for spiritualization was often accompanied by frustrations and irritations.

Testimonies about experiences as a child generally give the image of a strict father and a loving mother. Despite this rigor, fathers in free or learned professions were often very closely involved in family life and education. But involvement is not in contradiction with rigor. The father was concerned about the future of his children: sons had to study and find a good job, daughters had to meet standards of decency and marry a good husband. In modest households with no more than two servants, the relationship between parents and children was rather intensive and often accompanied by profound emotions. In the course of the nineteenth century, this approach shifted from a more distant, formal enforcement of authority to an authority based on mutual involvement.

After returning to society, Evert probably no longer had employment in the bank, although he remained there until 1905. In March 1897, he and his family left for the west. This departure meant that he left the small Leeuwarden elite. In other words, he was thrown out. Evert's father Auke and his uncle Reitze achieved their social status because of their important position as bankers. Through this status, they had also ended up in politics. But none of this was now possible for Evert in Leeuwarden. If he wanted to pursue such a position in the West, it would be much more difficult, because the group of wealthy citizens was much larger and had a much more diffuse character. Moreover, Evert's past would be held against him. He could not therefore fulfil any important social or political functions.

1.2 The Trek to the West

Grandfather Evert and grandmother Toon—From north to west and from paper-pushing to (artificial) manure
Evert had made life impossible for himself in Leeuwarden and left for Rotterdam in 1897. Meanwhile, in Leeuwarden Evert and Toon had become father and mother of

three children: Auke (father of Nico) on 8 April 1892, Maria Agnes (Mity) on 31 March 1893, and Johanna Elisabeth (Annabet) on 7 July 1895. Their fourth child, Wilhelmina Louise Petronella (Mien) was born on 2 August 1897 in Rotterdam.

How Evert ended up in Rotterdam is unknown. In any case, he turned up in Rotterdam in March 1897 as secretary to the commissioners of the International Guano and Superphosphate Works (IGS) in the village of Zwijndrecht. He may have ended up there through his connections. Manure would play a major role in the family; not only Evert, but also his son Auke and grandson Evert would become director of the IGS or one of its successor companies.

Evert's family had moved to Rotterdam, because that was where the headquarters of the Zwijndrecht factory was located. Evert became director of the International Guano and Superphosphate Works (IGSW) in 1901. In the years following its foundation in 1895, the company did not do well and no dividend was paid until 1901. After that things got better, according to family chronicler Auke Reitze, probably thanks to the new director Evert, and every year 10% or more dividend was paid out. In 1901 the family moved from Rotterdam to Zwijndrecht, then a village across the river from Dordrecht [16].

For a while, the production of fertilizer went well in the Netherlands, but then in 1914, the First World War broke out.

On 28 June 1918 the War to End All Wars broke out. Germany declared war on France on 3 August, after French troops had been mobilized on 1 August. Belgium and the Netherlands were neutral, but Germany nevertheless invaded Belgium on 4 August. The German army swept through Belgium, executing civilians and razing villages. The application of military action against a civilian population further galvanized the allies. Newspapers condemned the German invasion, violence against civilians, and destruction of property, which became known as the 'Rape of Belgium' [17].

Although the Netherlands remained neutral and no acts of war took place on Dutch territory, this war did affect the daily life of the population.

On 1 August 1914, the Netherlands mobilized its armed forces. The Dutch government closed the Scheldt estuary to British ships on 3 August and declared itself neutral. In total, the Netherlands had 200,000 men under its arms [18]. After the invasion of Belgium on 3 August, a state of war was declared in Zeeland, Noord-Brabant, Limburg, and part of Gelderland [19].

At the beginning of October, the Belgian army evacuated the fortress of Antwerp. Over a million people fled to the Netherlands. Never before had the Netherlands received such a large influx of refugees in just a few weeks. And never afterwards either.

Municipalities in the provinces bordering Belgium were particularly hard hit. The refugees had to be housed and fed. The government declared that it welcomed the refugees with open arms, but the actual reception had to be organized by the municipalities [20].

Because the railway bridge from Dordrecht to the North was disrupted by a collision for days, a relatively large number of Belgian refugees were stranded unintentionally in Dordrecht from 10 October 1914. On the basis of newspaper reports the number was estimated at around six thousand. As a result, ships with refugees were no longer allowed to berth on the Dordrecht embankment. This did not work out well for the approximately 2400 refugees who were lying on the Oude Maas river in nine Rhine river boats. They were meant for Rotterdam, but they were stopped by a Rotterdam police boat at Dordrecht, because Rotterdam was already full. The Municipality of Dordrecht was now obliged to provide 2400 refugees on the river with food, drinking water, and bedding. Only sick people were allowed to go ashore. This drama lasted for two weeks; then the refugees were able return to Antwerp. In the meantime, 1600 English naval personnel were interned in Zeeland-Flanders and some 32,000 interned Belgian soldiers passed through Dordrecht. After enjoying a lunch here, they took the train to their camps. From the end of 1915, a hundred of these Belgians worked at companies in the Dordrecht region. About eight hundred of the Belgian refugees survived until the end of the war [21].

An overwhelming amount of help came from the people of Dordrecht. The *Dordrechtche Courant* wrote on 8 October:

The hustle and bustle at the station here in the city is currently extraordinarily high [...] Young ladies bring baskets with sandwiches and bottles of milk into the third class waiting room, so that the refugees who arrive by train can immediately be provided with the necessary food and drink [22].

A number of refugees went to Zwijndrecht, where the Bloembergen's lived. Most of the hundreds of thousands of refugees, including many Belgian soldiers, usually returned after a short time, but a large number also stayed in the Netherlands, including a thousand or so in Dordrecht [23].

Although the Netherlands remained neutral, the circumstances had a major impact on public order in Dordrecht. As the food shortage increased and prices rose, criminality also increased considerably [24].

At the outbreak of the war in August 1914, Evert was 48 years old and was not mobilized.

In 1918 he was involved in a major merger that would lead to the Amsterdamsche Superfosfaatfabriek and the Verenigde Chemische Fabrieken (ACF/VCF). IGSW became part of the new organisation and Evert became director. It is no coincidence that this merger took place in 1918. In the Netherlands, all kinds of business projects had come to a standstill due to the outbreak of the First World War.

According to his grandson Auke Reitze, Evert had amassed a fortune for himself [25]. In 1924 or 1925, grandfather Evert had told his grandson Nico: "*Grandfather is a rich and beautiful man.*" [26].

He thus belonged to the group of citizens with the highest incomes, who had benefited from the war. Yet there were shortages that affected everyone, including the rich.

During the war period, food distribution was run by a special minister for food distribution, called Posthuma, because of the threat of serious shortages.

The Bloembergen's grew vegetables, including tubers and potatoes, in the large garden of their house, and those crops grew well with the help of the fertilizer from the factory.

According to the family chronicles by Auke Reitze, it was Evert that brought the fertilizer company to this level of success [27]. In a book commemorating the death of his grandfather in 1925, grandson Evert (Nico's brother) wrote: *In him the ASF/VCF lost a great merchant, who was also well considered by the entire staff for his amiable personality* [28].

Utrecht

After the merger of the Vereenigde Chemische Fabrieken (VCF) and the Amsterdamsche Superfosfaatfabriek ASF to form VCF-ASF, the head office was moved to Utrecht. Evert became director of the merged company in 1918. The family lived in a large, stately house across the street from the Maliebaan office and Evert only had to cross the road to get to work. Strangely enough, given his transgression in Leeuwarden, he became an officer in the Order of Orange Nassau in 1925. Perhaps Evert still had good connections, just like when he entered the fertilizer business in 1897. Perhaps his wealth also helped. A third possibility was his willingness to make a poison gas for the State in 1917 (see Father Auke below).

He had a stroke in 1924 and died of a second stroke in Utrecht on 26 October 1925.

After Evert's death, grandmother Toon moved from the big house in the Maliebaan to a smaller house in the Wilhelminapark in Utrecht. In 2006 grandson Nico, in the light of his grandfather Evert's mistakes, considered grandmother Toon to be a good woman [29]. Toon died of liver cancer in 1933.

According to the family chronicles by Auke Reitze, Evert was a *bon vivant*: cheerful, charming, and sociable, a lover of good food and drink. He spent money easily; now and again, he and his family went to Amsterdam to have dinner and go to the theatre, or Evert would go to the well-known health resort in Wiesbaden, Germany [30].

Father Auke

Auke (born in Leeuwarden in 1892) grew up in Zwijndrecht in a large house with a large garden. Not much is known about these childhood years, nor about his high school days in Dordrecht. It is certain that he was more attracted to science than to languages. Even later on, languages were not his strong point. Much later, the children found an old report, which included a **five on a scale of ten** for geography. Their conclusion was, if you get a five for this subject, you haven't been a good pupil.

After graduating from high school, he went on to study chemistry at the Technical University in Delft. The choice of chemistry may well have been planned much earlier by his father Evert. This was often the case at the time: the eldest son would succeed his father in the family business. Auke became a member of the fraternity, but kept a low profile; he hardly ever attended their social gatherings. He lived on the Leeuwenhoeksingel in Delft, near the station, with two close friends: H. C. P. de Bruijn (later known by the children as uncle Hazepé) and Ton Foest. Auke Bloembergen kept in touch with them all his life. De Bruijn eventually became head of the bridges department at Rijkswaterstaat, while Foest held the same position for the Dutch railway company, so together they were in charge of all the major bridges in the Netherlands. Auke graduated in 1914 then had to be employed until 1916.

As he was born on 9 April 1892, he was thus 22 years old when the mobilization was announced. For his military service, he entered at corporal level and then passed the sergeant's exam. He was finally allowed to leave after his father Evert's intervention:

Finally, in Kralingse Veer, a remarkable company was involved, at the request of the government. In 1917, A. Bloembergen, son of the I.G.S.W. (International Guano and Superphosphate Works) director of the I.G.S.W. (International Guano and Super-phosphate Works) was granted so-called leave from the army to make liquid sulphur dioxide in the factory at Kralingse Veer, which was then judged to be a poison gas for war purposes. The production was stored in barges in Hollandsch Diep, but when the armistice became a fact, fortunately without our country having to use this stock, the government returned the product to the factory, which converted it into sulphuric acid. The story goes that General Snijders[1] visited the installation with the result that one of the workers accidentally opened the wrong tap, and a penetrating flow of SO_2 was set up. According to tradition, the general then established a speed record out of the factory [31].

This confirms that the company produced poison gas at the request of the government!

Auke probably met Sophia Maria (Rie) Quint in 1913 at dance lessons in The Hague. At the time, dance lessons were a way of making contact between marriage-able girls from The Hague and students from Delft.

Rie Quint's father was a mathematician and Rie had a very different background from the Bloembergen family of bankers and manufacturers.

Mother's grandparents: Nicolaas Quint (1862–1916) and Diederika Maria Molenaar (1868–1943)

Nicolaas (Klaasje) Quint, after whom grandson Nico Bloembergen was named, was born in Amsterdam on 25 April 1862 from the marriage of Arend Willem Quint and Sophia Magdalena van Aalderen. Nicolaas studied at the Municipal University of Amsterdam and obtained his Ph.D. at that university on 28 January 1888, becoming a Doctor of Mathematics and Physics, with Diederik Johannes Korteweg (1848–1941). Korteweg himself had obtained a doctorate in 1878, with the thesis *On the Propagation of Waves in Elastic Tubes* [32], studying under Johannes Diederik van der Waals (1837–1923) in Amsterdam. Van der Waals was the first professor of physics at the University of Amsterdam and in 1910 was awarded the Nobel Prize for Physics. Korteweg supervised a total of seventeen Ph.D. students, includ-ing the mathematician L.E.J. (Luitze) Brouwer (1881–1966). Nicolaas Quint was Korteweg's first Ph.D. student (Fig. 1.4).

The title of Nicolaas Quint's thesis was *Eddy motions* and it dealt with turbulent fluid motion, or fluid dynamics [33].

Korteweg was an internationally renowned mathematician and became a member of the Royal Academy of Sciences in 1881. Nicolaas Quint's dissertation was original work.

Whereas the thesis by Combertus Pieter Burger was still written in Latin, Nico-laas Quint's appeared in Dutch. Latin was abolished as a compulsory language of communication at universities by the Higher Education Act of 1876 [34].

---

[1] General Cornelis Jacobus Snijders was Commander-in-Chief of the Dutch Land and Sea Forces.

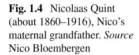

On 25 July 1889, Nicolaas married Diederika Maria Molenaar, born on 5 February 1868 in IJsselmonde. Diederika had not had any education after primary school. Nicolaas became director of an HBS after a 3-year course, and remained there until his death in 1916. He therefore did not choose a career in science.

Nicolaas and Diederika first lived in Dordrecht and later in The Hague. Lien was born on 6 June 1890 in Dordrecht, and Sophia Maria (Rie, Nico's mother) on 2 June 1891, also in Dordrecht, while Arent Willem (Wim) was born on 13 December 1912 in The Hague. Rie grew up in The Hague. They lived in a fairly large house and had one servant.

Nicolaas died suddenly on 24 August 1916, at the age of 54. His Bloembergen grandchildren never knew him. Nico's mother Rie said that she had great admiration for her father, especially because of his interest in culture and science (Fig. 1.5).

In the eyes of the grandchildren, Diederika was a sweet, old-fashioned grandmother. In 1942 Diederika's house in Verhulststraat in The Hague had to make way

Fig. 1.5 Diederika Quint-Molenaar (about 1870–1943), Nico's maternal grandmother. *Source* Nico Bloembergen

for the 'Atlantic Wall' of the German occupiers, and after consultation with the family, Diederika moved to a guesthouse in Bilthoven, where the Bloembergen's also lived [35].

Mother Rie

The following is largely based upon the family chronicles by Auke Reitze [36].

Rie must have been an intelligent, diligent, cheerful, and generally very pleasant girl. She was always surrounded by friends and spent more time with them than at home. Rie's mother did an excellent job of housekeeping, but her intellectual baggage was perhaps too limited for Rie. With Rie's father things had clearly been different. This does not mean that the relationship between mother and daughter was bad. To the end, Rie showed her mother much love and care.

After primary school Rie went to the Meisjesschool (MMS, Middle School for Girls) at the Stadhouderslaan in The Hague. Modern languages were the main dish on the menu at the MMS. Rie's love for these languages and their literature would have come from these years, and it was a love that she would keep for her whole life.

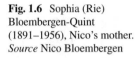

Fig. 1.6 Sophia (Rie)
Bloembergen-Quint
(1891–1956), Nico's mother.
Source Nico Bloembergen

Rie went on to study French because she thought it was the most beautiful language. At that time it was not possible to study modern languages at the universities, especially if one did not have a final exam in grammar school. The only possibility was to take the course for the Secondary Education Certificate (MO). She studied at the Nutsseminarium in the University of Amsterdam (Fig. 1.6).

However, living in rooms was too expensive. She stayed a few days a week with Nicolaas Quint's sister Caro Quint, who lived in Amsterdam at the time. This aunt lived until 1949. She was dear to Rie and her children. Caro Quint also spent her last days in Bilthoven, which illustrates how much the older family members relied on Rie. Rie's studies went smoothly and without problems. She did MO-A and -B and thus ended up with full teaching qualifications.

During her studies she got to know Auke Bloembergen and her life took a completely different turn, in a way that was quite customary for girls at the time. It was not to be a life in education or science, like her father's, but a family with a husband and children.

Father Auke and mother Rie. The Hague, Zwijndrecht, Dordrecht, Utrecht, Vlaardingen [37]

Auke and Rie were engaged in 1914, after Auke made his marriage proposal. Rie asked for a week to think about it and only then did she make the promise. After Auke's death, Rie told her son Auke Reitze that she did not know whether she would have accepted the marriage proposal if Auke had come from a less wealthy family.

Rie came from a family of civil servants and the civil service was not a cash cow in those days. The generous lifestyle of the Bloembergens in Zwijndrecht was new and also attractive to her. There had been some hesitation on the part of Rie's parents: would Rie, who had grown up in a family that moved in the atmosphere of education and science, be able to flourish in an environment of manufacturers? However, the wedding had to wait, because Rie still wanted to finish her MO-B French studies and Auke his studies as a chemical engineer. Auke joined the army, but after just over two years, upon the intervention of his father Evert in 1917, he was dismissed from his military service. Rie followed a six-month marriage course at the Huishoudschool (school of domestic science) and often went to Zwijndrecht in the weekends. Once a week Auke was allowed to eat at the Quint family home in The Hague. On 26 June 1917 Auke and Rie married and went to live in a small house in the Dubbeldamseweg on the outskirts of Dordrecht (Fig. 1.7).

After that, working life began for Auke, who remained in fertilizer from the beginning in 1917 until the end in 1953. Auke started at the factory run by his father

Fig. 1.7 Auke Bloembergen (1892–1955), Nico's father. *Source* Nico Bloembergen

Evert in Zwijndrecht and, after the merging of the factories, worked for several years at the head office in Utrecht.

On 22 April 1918, his son Evert was born in Dordrecht and on 11 March 1920, Nico, also in Dordrecht. Shortly after Nico's birth, the family moved to Parkstraat in Utrecht, a street with late nineteenth and early twentieth century mansions on both sides, close to Auke's parents on Maliebaan.

In 1921 or 1922, Auke became factory director in Pernis (the largest and most important factory) and the family moved to a house on the Schiedamseweg in Vlaardingen. On 22 April 1923 Diederika (Diet) was born in Vlaardingen and on 26 September 1924, Anthonia (To), in Vlaardingen. Shortly after To's birth, Auke returned to the head office in Utrecht. After the death of his father Evert on 26 October 1925, he became director.

Culture, prosperity, genealogy

How did Auke and Rie's family view Evert's misdeeds in Leeuwarden? The family chronicle by Auke Reitze, Nico's youngest brother, makes the following remark:

It must have been a dramatic event for the family, and it must also have had an impact on our upbringing, particularly in the area of sexuality. The event was apparently so dramatic that father and mother (who must have been aware of this) never spoke to us about this family secret during their lives. Not even when the opportunity arose. I once asked my mother: why did grandpa Evert leave Leeuwarden? After a long - and in retrospect meaningful - hesitation, she said that he saw no future in banking. We only heard after her death [in 1956, RH]. *and from bits and pieces about what happened* [38].

Yet there was openness among other family members. Auke Reitze wrote about this too:

The Englishman's family, the family of father's sister Mity, where there were also six children, proved that things could be done differently. There, as niece Iefke told me a few years ago, the children were already told around 1950 without this leading to any problems.

Nico's parents were born in the nineteenth century. According to Auke Reitze, the parents transferred the norms and values of the wealthy citizens of that century to their children [39]. The behaviour of grandfather Evert can also be seen in the light of these norms and values.

Evert's mistake had considerable consequences for his descendants. He had to give up his banking aspirations and it made life impossible for him in Leeuwarden. Evert had to go west and ended up in the fertilizer business. His offspring did not grow up in Leeuwarden, but in the west, and they also married in the west.

Evert and his descendants broke with the family tradition of banker. After Evert, the Bloembergen's remained representatives of the liberal bourgeoisie: Evert, his son Auke, and his grandson Evert became successful factory directors. Voting for a liberal party was always a matter of course, as it was also with Auke's children. We should not think about liberalism as manifested in the political parties, but rather as it is experienced by the voters of those parties, as Aerts had said earlier:

[…] pragmatically minded people who value self-interest, do not want too much interference from the government, have little interest in intellectualism, and are not very principled [40].

What also remained was the wealth, first from the bank and later through Evert's job in fertilizer.

In addition, there were the influences, genetic or cultural, through the female line. There was great-grandfather Combertus Pieter Burger, who had obtained his doctorate in mathematics in 1851. The second scientist was grandfather Nicolaas Quint, who also obtained a doctorate in mathematics, in 1888. The cultural influence of grandfather Quint was channeled through mother Rie and her interest in science and culture, which will be discussed in the next chapter.

References

1. Plaisier L. Bloembergen: een Leeuwarder familie in bankzaken. Fryslân 7 (3) 3–7 (2001)
2. Plaisier L. Bloembergen: een Leeuwarder familie in bankzaken. Fryslân 7 (3) 3–7 (2001)
3. Centraal Bureau voor Genealogie. Nederland's Patriciaat. Genealogie van bekende geslachten. 77(1993)
4. Leeuwarder Courant 21 February 1887
5. www.parlement.com
6. Plaisier L. Bloembergen: een Leeuwarder familie in bankzaken. Fryslân 7 (3) 3–7 (2001)
7. Leeuwarder Courant 20 October 1912
8. Burger CP Specimen academicum inaugurale de solutione problematis Keppleriani. Dissertation. State University of Leiden, Leiden, 1851
9. Burger CP. Wiskundige Aardrijkskunde (Foundations of Mathematical Geography). Engels, Leiden (1860)
10. Burger CP. Iets over onmeetbare getallen en bekorte rekenwijzen bij het werken met decimale breuken. Engels, Leiden (1865)
11. Lennep MJ van. Bloembergen. Jierboekje fan it Genealogysk Wurkforban (1958)
12. Plaisier L. Bloembergen: een Leeuwarder familie in bankzaken. Fryslân 7 (3) 3–7 (2001)
13. Couperus L Eline Vere. Een Haagsche Roman. Van Kampen en zoon, Amsterdam (1889)
14. Anonymus. De Gids 53 (1889), II, 197
15. Mathijsen M. De gemaskerde eeuw. Querido, Amsterdam (2002)
16. Bloembergen AR. Het gezin van Rie en Auke Bloembergen 1917–1956 (The family of Rie and Auke Bloembergen 1917–1956). Private Management, Wassenaar (2003)
17. en.wikipedia.org
18. defensie.nl/nimh/geschiedenis/tijdbalk/1914-1945/mobilisatie_in_nederland_nav_het_uitbreken_van_de_eerste_wereldoorlog_%281_augustus_1914%29
19. mei1940.nl/Voor-de-oorlog/Mobilisatie_1914-1918.htm
20. Roodt E de. Oorlogsgasten. Vluchtelingen en krijgsgevangenen in Nederland tijdens de Eerste Wereldoorlog. Europese Bibliotheek, Zaltbommel, 2000
21. www.eerstewereldoorlog.nu/blog/dordrecht-tijdens-de-eerste-wereldoorlog/
22. Dordrechtsche Courant 8 oktober 1914
23. Boogaard A vd. Belgische vluchtelingen in Dordrecht. On www.wereldoorlog1418.nl
24. Kooij P and Sleebe V (eds). Geschiedenis van Dordrecht. Verloren, Hilversum (2000)
25. Bloembergen AR. Het gezin van Rie en Auke Bloembergen 1917–1956 (The family of Rie and Auke Bloembergen 1917–1956). Private Management, Wassenaar (2003)
26. Interview Rob Herber with N Bloembergen, 15 December 2007

27. Bloembergen AR. Het gezin van Rie en Auke Bloembergen 1917–1956 (The family of Rie and Auke Bloembergen 1917–1956). Private Management, Wassenaar (2003)
28. Knapp WHC. Kunstmeststoffen der ACF-VCF. Utrecht (1933)
29. Interview Rob Herber with N Bloembergen, 7 December 2006
30. Bloembergen AR. Het gezin van Rie en Auke Bloembergen 1917–1956 (The family of Rie and Auke Bloembergen 1917–1956). Private Management, Wassenaar (2003)
31. Bloembergen AR. Het gezin van Rie en Auke Bloembergen 1917–1956 (The family of Rie and Auke Bloembergen 1917–1956). Private Management, Wassenaar (2003)
32. www.genealogy.math.ndsu.nodak.edu/id.php?id=7731
33. Quint N. De Wervelbeweging (Eddy motions). Dissertation Universiteit of Amsterdam, Amsterdam (1888)
34. Roelevink J. Het Babel van de geleerden. Latijn in het Nederlandse universitaire onderwijs van de achttiende en de negentiende eeuw (The Babel of the scholars. Latin in Dutch university education of the eighteenth and nineteenth centuries.) On: dbnl.org/tekst/_jaa003199001_01/_jaa003199001_01_0004.php
35. Bloembergen AR. Het gezin van Rie en Auke Bloembergen 1917–1956 (The family of Rie and Auke Bloembergen 1917–1956). Private Management, Wassenaar (2003)
36. Bloembergen AR. Het gezin van Rie en Auke Bloembergen 1917–1956 (The family of Rie and Auke Bloembergen 1917–1956). Private Management, Wassenaar (2003)
37. Bloembergen AR. Het gezin van Rie en Auke Bloembergen 1917–1956 (The family of Rie and Auke Bloembergen 1917–1956). Private Management, Wassenaar (2003)
38. Bloembergen AR. Het gezin van Rie en Auke Bloembergen 1917–1956 (The family of Rie and Auke Bloembergen 1917–1956). Private Management, Wassenaar (2003)
39. Interview Rob Herber with A R Bloembergen 2 February 2007
40. Aerts R. De VVD en haar ideologische wortels. ru.nl/wetenschapsagenda/huidige_editie/vm/jaargang_1/deskundig_remieg/

Chapter 2
Bilthoven: Playground for the Elite (1925–1938). Primary School and High School

It is a warm day at the end of May 1937. On the hockey field of the Stichtsche Cricket en Hockey Club (SCHC), surrounded by trees and houses, on the Vermeerlaan in Bilthoven, a home match of the men's team is under way. The colors of the club are red and blue. There are not even a hundred spectators around the playing field. The players are being cheered on enthusiastically: "Go on, go on SCHC". Nico Bloembergen is halfback. He plays a little woodenly. He is not a scorer, but supports the attacking players and is part of the first line of defense. That way he always has something to keep him occupied.

2.1 The First Years of Nico's Life

As mentioned earlier, Nico was born in Dordrecht on 11 March 1920. In 1921 or 1922 he moved to Vlaardingen. As a baby he contracted Werlhof's disease, an autoimmune disease in which the blood platelets are broken down.[1] The disease disappeared of its own accord, but returned in 2000. Nico was not considered particularly smart in his early years. He learnt to speak rather late, and a German servant of the family called him "*der dumme Kiek*", the dumb cluck [1]. According to his mother Rie, however, Nico was thinking, and that gave a false impression. But Nico immediately shed that nickname when, at the age of four, he beat an inattentive uncle at chess! And suddenly Nico was coming out with perfect sentences.

---

[1]Immune thrombocytopenia (ITP) is a type of thrombocytopenic purpura defined as an isolated low platelet count (thrombocytopenia) with normal bone marrow and the absence of other causes of thrombocytopenia. It causes a characteristic purpuric rash and an increased tendency to bleed en.wikipedia.nl.

© Springer Nature Switzerland AG 2019
R. Herber, *Nico Bloembergen*, Springer Biographies,
https://doi.org/10.1007/978-3-030-25737-8_2

Auke and Rie's family moved again in 1925. This time not to Utrecht, but to Bilthoven [2]. Auke and Rie would stay there until the end of their lives and their children would grow up there, too. Indeed, Nico spent the rest of his youth in Bilthoven.

2.2   De Bilt and Bilthoven

In 1832, the village of De Bilt was still an agricultural community that provided services to farms and country estates. A small core of craftsmen was established around the inns and restaurants in the Steenstraat. This street was the first paved road in the Netherlands. The first cadastral map of De Bilt from 1832 clearly shows the large estates [3]. The land suitable for agriculture and cattle breeding in the south was divided into small parcels. Bilthoven was built on rough ground, i.e., the sand dunes, heathlands, and forests which still covered large areas north of De Bilt. These poor sandy soils were barely usable for agriculture and cattle breeding and were not inhabited. Doedens and Cladder [4] gave a nice description of De Bilt from around 1900. The restaurants and cafés in the Dorpsstraat (formerly Steenstraat) were still there. In 1889/1890 there was a beer house, five pubs, and four inns (where alcohol was also served), all within a distance of 600 meters! There was a strict hierarchy in the village, with the aristocrats, the owners of the estates, at the top. Among them were the notables, and a few of the more important entrepreneurs, the doctor, the notary, the school head master or mistress, the director of the post office, the town clerk, and the tax receiver. They had the right to vote and often lived in the center of the village, around the town hall built in 1883, where power was concentrated. Among them were the other inhabitants with an official function, for example the town councilors, the teacher, and the postman.

The (lower) middle class was formed by the coachmen, often with higher income than the 23 people with sideline occupations as shopkeepers (hairdressers, bakers, tailors, milkmen, butchers and people involved in other, less clear forms of business), but there were accounts of the arrival of the tram line built in 1879, sign of a new era. And there were independent craftsmen such as cartwrights, painters, masons, a saddler, an interior decorator, and ten carpenters, an indication that the building industry was on the up.

At the lower end of village society we find a few factory workers, but also servants, servants, and more servants. There were many peasant boys on the forty farms, along with laborers for forestry and for the large landowners.

But even lower still came the poor. They had to rely on support from the municipality, but applications for support were usually rejected. The municipality referred the applicants to the diaconate of its own denomination (Fig. 2.1).

In 1860, the Nederlandsche Centraal Spoorwegmaatschappij (NCS, the Dutch Central Railway Company) began construction of the Utrecht—Amersfoort—Hattemerbroek railway line and it was inaugurated in 1863. The station stood in great solitude some 3 km from the centre of De Bilt. Then around 1900, a few villas appeared there, and in 1903, roads were built.

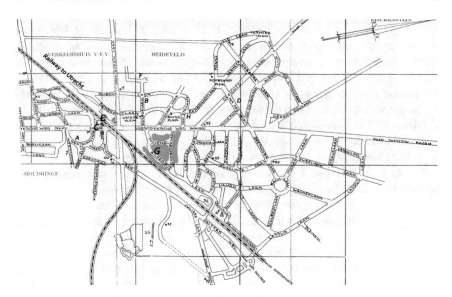

Fig. 2.1 Map of the village of Bilthoven 1930. *Source* Tourist Information De Bilt-Bilthoven, 1930. **A** First house of the family in Bilthoven, Nachtegaallaan 20, 1924–1926. **B** Second house of the family in Bilthoven, Van Ostadelaan 9. Nico lived here from 1927 to 1946. **C** Village center with shops. **D** Field hockey club SCHC. **E** Railway station. **F** German Flak (anti-aircraft). **G** Area occupied by the German 88 Army Corps. Sperrgebiet (restricted area). **H** Primary school

Johan Willem Beijen, banker, minister, and ambassador, wrote a booklet describing the first years of his youth, in which he mentions De Bilt station. Johan Willem's father, Karel Hendrik Beijen, was the secretary of state in charge of the railways and lived in a large villa not far from the railway line. Here are some extracts from the booklet for the period 1902–1910 [5].

I lived in Utrecht, where I was born. In 1902 my parents moved to what is now called Bilthoven (but at that time had no more than ten houses) because an appendicitis that was not operated on, but was cured, made me a 'weak child' that needed fresh air. We lived there for eight years, so I can say that I was an outdoor child.

Bilthoven consisted of woods with some heath and dunes. As a child, you could wander freely there, because there was no danger.

Bilthoven – although that was not yet its name – gradually developed in the eight years we lived there, but it still remained an 'outdoor life'. The village had something friendly about it; there was a personal bond between all levels of society, no matter how much they were separated by social restrictions.

The most interesting thing about Bilthoven was the railway. Bilthoven was also called 'Bilt Station'. Around the turn of the century, the Centraal Spoorwegmaatschappij (Central Railway Company) was established as a subsidiary of the Maatschappij tot Exploitatie van Staats-Spoorwegen. It operated railway lines to Zeist, Baarn, and Amersfoort. The branch line to Zeist was in De Bilt, the branch line to Baarn in Soest. [The locomotive, RH] *blew its whistle when the train was*

travelling through a forest on an unfenced railway track. This served to warn hikers
who wanted to cross the railway line and was not intended to scare off imaginary
bison. The difference in the classes on the railways was socially very important. The
Dutch railways originally had three classes: the first, where people sat on red plush,
the second where they sat on green plush, and the third where they sat on wood.
These classes corresponded to the three levels in which Dutch society was divided at
that time. In the trains of the Central Railway Company there were only two classes,
a second and a third, but the second was covered with red plush, because otherwise
the noble commuters from Baarn and Zeist would not have felt at ease.

I lived in De Bilt until I was twelve years old. In 1910, my father thought it had
become so built-up that he could just as well live in Utrecht again.

The interesting thing about living in a village that had only recently come into
existence and that was still under development is clear from Beijen's description.
After 1918 Bilthoven (as the town council had decided the village would be called)
grew rapidly: houses, shops, small industries, banks, and offices were built everywhere. Bilthoven even outstripped the old village of De Bilt to such an extent that
the municipality decided to move the town hall to the north. The 'personal bond
between all levels of society' only existed in the eyes of the elite. At that time, the
second class, but especially the first class, did fit in with the higher ranks: workers and
middle class were dependent on the higher ranks for their income and condemned to
a certain amount of subservience.

In Bilthoven, less luxurious houses than the big mansions hadn't been built since
the beginning.

Between 1872 and 1876, in what would later become the center of Bilthoven,
sixteen simple one-room houses were built. After that, a number of small houses were
built by the municipality and the *Foundation for the Improvement of Public Housing*.
This foundation came from a Protestant group in Bilthoven, who felt that the actions
of the government were incompatible with the gospel of Jesus Christ. In the group's
Community Gazette of 6 July 1922, reference was made to the 'shameful housing
situation in Bilthoven', in this case the slums near the former Jewish cemetery.

The garden village in Bilthoven South (south of the railway line) was a neighborhood intended for workers, and the first houses were built in 1917 by the *Algemene
Biltse Woningbouwvereniging* (General Housing Association). The district was built
in the same way as the expensive districts in Bilthoven North, along winding lanes
with gardens. The scale of Tuindorp was smaller: the houses were (much) smaller,
the lanes were shorter, and there were fewer plants.

Many organizations and associations were founded between 1910 and 1940. The
Stichtsche Cricket and Hockey Club (SCHC) mentioned at the beginning of the
chapter, was very important for the Bilthoven elite, and indeed it would also prove
to be very important for the Bloembergens.

Some of these associations and societies were based on the Christian faith and
there were many different faith communities. Nico's mother Rie was a Baptist and
had contacts in that community.

The Quakers' Children's Workshop community, set up by Kees Boeke, was also
of some significance to the Bloembergens.

The Quaker community (known officially as the *Religious Society of Friends*) is mainly characterized by the way quakers view life and others, rather than by its members' religious beliefs [6]. Key points were their pacifism and their refusal at the time to pay taxes used to finance the war.

Kees Boeke and his wife Betty Cadbury had initially set up a meeting point for fellow believers and peace activists. In 1924 a Montessori school was built with a subsidy, partly financed by capital put up by Betty's family (famous for Cadbury's chocolate factory). In 1926, the Boekes themselves began to teach their daughters, because they refused to pay school fees. They considered it to be a tax and, as a matter of principle, they would not pay it. Slowly but surely this education was extended to other children as well. The class was called 'The Workshop' and apart from learning to read and write, there was a focus on social and creative education, something that was completely lacking in conventional schools at the time [7]. We shall meet Kees Boeke, the Cadbury's, and the Quakers again in Chap. 5.

In the old village of De Bilt, migration was high: ten percent of the population was actually on the way out. A sample of names beginning with the letter S, comparing data from Bilthoven's address books in 1922 and 1927, shows that 41% of the people or families had already moved out of Bilthoven within five years [8, 9].

Class consciousness remained very strong until around 1960. People lived 'on social standing'. Bilthoven had higher standing than De Bilt. Bilthoven North was more chic than Bilthoven South, and villas were of course more exclusive than workers' cottages, and so on.

The noble estate owners held onto their social duties for a long time yet; mayors also belonged to the patricians or nobility until 1971 and they belonged to the same group as the landowners. It was remarkable that, until around 1950, the nobility still held an important position in the municipality, and there was even then a considerable distance between the nobility and the rest of the population.

Contacts and contradictions

How did personal contacts work in Bilthoven? That depended on the point of view of the person the position they belonged to. Earlier we saw that there were considerable differences between the nobility and the farm workers. But what about the increasing population living in the workers' cottages and the group living in the villas?

In 2003, Auke Reitze Bloembergen wrote in the family chronicle about the typical village character of Bilthoven. Auke Reitze was 76 when he wrote this:

Bilthoven was a real village. And it remained so until after the war. Most of the lanes had no pavements and no cycle paths. Particularly in Bilthoven North, there were undeveloped forest areas everywhere. There were no old people's flats or any other flats. The shops in the village remained rather simple. [....] And at first there was no secondary education either. But there were a lot of nice things for children: a small hockey club, private tennis courts, a natural swimming pool, quiet lanes to play in, and a moor with a lake. The atmosphere corresponded to this: it was friendly and remained on the scale of a village. But it was nevertheless a commuter village. For a large part, the labor force worked elsewhere. As a result, the social cohesion that is

Fig. 2.2 Evert and Nico Bloembergen (right) in 1922. *Source* Nico Bloembergen

often found in old towns and villages was lacking. But all in all, it remained a good village to live in for quite some time [10].

However, the 'pleasant things' that are mentioned were only there for the elite. For example, the hockey club worked with a strict voting system. People were only allowed in if they had been recommended and then passed the vote. Of course, people from workers' houses were never nominated: members would not nominate their staff to play hockey together! Imagine! The residents of the workers' cottages would never have expected to play hockey either: that was for the elite, not for them. Moreover, membership and equipment were far too expensive (Fig. 2.2).

2.3   The Bloembergen Family in Bilthoven—Primary School (1926–1932)

The head office of the fertilizer company VCF-ASF was moved to Utrecht in 1919 and Auke became director there in 1925.

Auke and Riet Bloembergen came to live in Bilthoven in 1925 with their children Evert, Nico, Diet, and To. Bilthoven, as described earlier, had only existed for a few decades and was still in full development in the 1920s. Auke and Riet's choice of Bilthoven was probably prompted by the fact that Auke preferred the countryside, something he had experienced in Zwijndrecht in his youth, to life in the city of Utrecht where his work was. Another possibility is that he preferred not to live close to his parents in Utrecht. Auke would thus cycle three kilometers to De Bilt and take the tram or bus to his office in Utrecht.

The Bloembergens lived a five-minute walk from the station. From Bilthoven it was only ten minutes by train to Utrecht Central Station, so the shops, restaurants, and theatres were closer than for many who actually lived in Utrecht.

The family ended up in Bilthoven South, near the village center. The villa *De Nachtegaal* (The Nightingale) in the Nachtegaallaan was chic for that part of Bilthoven, but it was not as chic as Bilthoven North.

Herbert was born here in 1926. In 1927, Auke became a manager at ASF-VCF. Auke had received part of the inheritance of his father Evert (who had died in 1925). This enabled him to acquire a new house relatively cheaply in 1927. This larger house, called *De Pan* (The Pan) in Van Ostadelaan 9 was in in Bilthoven North, the posh part of Bilthoven. The children, including Nico, would stay in this house until they eventually had to flee. In 1927, the youngest son Auke was born here (he later changed his name to Auke Reitze to distinguish him from his father Auke).

Van Ostadelaan was a quiet lane, leading on to the Soestdijkseweg.

At first there was no cinema and no good bookshop in Bilthoven. Bicycle paths and sidewalks were absent. For children it was also an ideal place to play: near their home there was a garden which contained nothing of value, and on the street there was hardly any traffic. Everywhere were undeveloped areas of forest and heathland, so these were also places where the children could roam freely. The situation was still very similar to the circumstances described by Beijen earlier in the century [11].

The natural swimming pool and the *De Biltsche Duinen* (De Bilt Dunes) with a children's playground, paddling pool, separate pool for women, slide, and canoe pond for the children, was opened in 1933, with a restaurant for the parents, and for many years it would receive up to 9000 visitors a day [12].

Primary School
The Bloembergen children all went to the Van Dijck school, which had opened in 1916, only a ten minute walk from the parental home. What kind of school was it? Maartje Balk, was a former student [13]:

Mrs. Balk lived almost her entire life in Bilthoven. In the pre-war depression years, from 1932 to 1937, she attended the Van Dijck school as a pupil. The objective of the school was to teach the pupils as much knowledge as possible and to prepare them for secondary education, preferably HBS or Lyceum. We never did a nature walk or had a sports day: there was an annual school trip and a great many plays were performed in the gym. The classes were large; on average about 40 students per class. In those years there were also many children of Dutch people from the Dutch East Indies, who were on leave here. [...] Learning was important; the school was performance-oriented. There was actually no personal attention for the child. Those who couldn't keep up were sent to the back of the classroom. No attention at all was paid to things like word blindness or bullying.

According to Herbert Bloembergen, a class at the Van Dijck school was an exclusive club: children who joined in the second class (now group 4) were allowed to take part, but they never really belonged to it [14]. The Van Dijck school was in many ways the opposite of the previously mentioned Werkplaats Kindergemeenschap started by Kees Boeke.

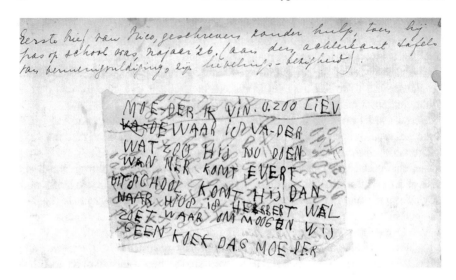

Fig. 2.3 Nico's first letter, 1926. His mother wrote: 'First letter from Nico. Written without help, when he had just started school, autumn 1926' (with tables of multiplication on the back, his favorite activity). *Source* Auke Bloembergen

Nico in primary school, endearing but already very smart

The head of the Van Dijck school was P. Tazelaar, a passionate man of small stature [15]. Despite its performance orientation, the school was far too easy for Nico. More-over, initiative was not welcome there. At the age of six he was already fascinated by numbers and multiplication tables. When had only just finished school, in the autumn of 1926 (he was six, an age at which most children could not yet read or write), he wrote a note to his mother (Fig. 2.3).

MOTHER I LOV YOU SO
WHERE IS FA-THER
WHAT WOULD HE DO NOW
WHEN COMES EVERT
OUT OF SCHOOL WILL HE COME
HOME IS HERBERT
SWEET WHY WE
SHOULD NOT BISCUITS BY MO-THER

On the back of the note, he wrote out the multiplication tables, his favourite activity. Because he was bored in the first class, he learnt the Greek alphabet from front to back and vice versa from an old engineering almanac. He often did his homework twice out of sheer boredom. Nico was obedient or even meek, and felt satisfied with his high marks.

He ended his time at the Van Dijck school in 1932 with the following results [16]:

Calculation	10
Dutch language	9
Geography	9
History	9
Knowledge of French?	Yes

But Nico was not musical. When the first class had singing lessons, he was made to stand in the corridor, something that left a mark on him throughout his life (Fig. 2.4).

In the summer of 1932, when Nico was twelve years old, he got a pain in his foot, probably because he was struck by a hockey stick or a ball, and he immediately caught a 'cold'. His general practitioner made the wrong diagnosis and it turned out to be osteomyelitis.[2] The GP operated, but did so too early and the inflammation

Fig. 2.4 Nico at the primary school in 1928 (second row, second from left). *Source* Nico Bloembergen

---

[2]Osteomyelitis is an inflammation of the bone marrow. The inflammation is usually caused by bacteria. These bacteria can reach the bone through an open bone fracture or through the bloodstream via an inflammation elsewhere in the body. The inflammation is usually in the long bones. The condition is more common in children than in adults. A bone marrow inflammation is a serious condition if the treatment is not started early enough, in which case the infection can spread further in the bone. An abscess (a cavity filled with pus) can then develop, which can turn into a chronic inflammation, and this is much more difficult to treat.

spread from the leg membrane to the bone. Nico tells us: "*I shouted sometimes, cried at night: I was in tremendous pain.*" The general practitioner said: "*Oh, just be a big boy.*" Antibiotics were not available yet. Nico went with his parents to see a surgeon, and an inflammation was observed on the X-ray. The surgeon cut away half the heel bone and it eventually healed. Nico had to ride a tricycle, because he couldn't walk, and many people in Bilthoven thought: "*That poor boy must now spend his life on a tricycle.*" [17]. His father Auke made a boat for him, called the *Ukkie* (Small One).

Nico's Mother Rie [18]

In the recollections of Auke Reitze, Rie was a sweet and caring mother. Herbert did not find her strict either. She was always present when the children came home from school, from Evert's primary school to Auke Reitze's gymnasium, when he was over 20 years old. Rie's attentions sometimes went a long way. In the Hunger Winter of 1944–1945, Herbert was in bed with pleurisy from October to May.[3] Rie came three times within one hour to bring Herbert (18 years old) a glass of warm milk. Nico (then 24), who was ill in the same room, said: "*Goodness mother, you might as well bring three glasses at the same time.*"

As Nico tells us: "*My mother was always busy doing the housekeeping and caring for the smaller children.*" [19] Rie was assisted in his upbringing by a nanny.

Although there were servants present, Rie did not leave home much. Love and devotion characterized her behavior and the children could have a good in-depth or intimate conversation with her. The downside was that it was difficult for her to let go of the children.

Rie did not do any sports except skating, although sports would later play an important role in family life. Hockey was something for noisy rich girls. That was not for "*our kind of people*", she said later. By 'our kind of people' she meant civil servants and teachers, thinking of the period of her life before her marriage. Sometimes she went on canoe trips with her husband Auke, although her heart may not always have been in it.

Rie had a clear set of norms and values: devotion to duty, discipline, and finishing the things you were doing; those were important things for her. She was a total teetotaler all her life. Especially in later years, her care was not limited to her own family, but also to other family members and acquaintances. After the retirement of her husband Auke in 1953, six or seven single women would come to eat every week. For many years, Auke and Rie held one or two weekly gramophone record evenings at home for about twenty people. The Bloembergens were firmly anchored in the Bilthoven elite.

The foundation of Rie's life was her faith. She regularly read the house bible, as evidenced by the many penciled annotations in it. Her mother belong to the Dutch

---

[3]Pleurisy, also known as pleuritis, is an inflammation of the membranes that surround the lungs and line the chest cavity (pleurae). This can result in a sharp chest pain when breathing. This pain may be a constant dull ache. Other symptoms may include shortness of breath, cough, fever, or weight loss, depending on the underlying cause. en.wikipedia.org.

Reformed Church, but during her secondary school years she had become fascinated by the Mennonite preacher Sicco Lulofs (1849–1927) in The Hague [20]. She remained loyal to the Baptist faith throughout her life.

The *Doopsgezinden* (Mennonites, officially the General Mennonite Society) are the Dutch branch of what is internationally called the Mennonites [21, 22]. They are the followers and kindred spirits of Menno Simons (1496–1561), who held specific views on baptism. Instead of having children baptized, as is still common in most other churches, they only recognize (adult) baptism on a voluntary basis. In the Baptist Church, anyone who wants to can be baptized and thus become a full member of the denomination, writing her or his own confession. From the eighteenth century onwards, many Dutch Baptists shifted more and more to the liberal end of the spectrum as far as orthodoxy was concerned. In the second half of the nineteenth century, many orthodox Baptists switched to the Reformed Church, which was founded in 1892, giving the Baptist Church an even more liberal character. The Anabaptists constitute a Christian religious community in the Netherlands - better known as the Baptist Brotherhood - a contemporary and open grouping, without hierarchical structures, where men and women ('brothers' and 'sisters') are considered equal. During the pillarization in the first half of the twentieth century, they belonged to the pillar which also included the *VPRO* (formerly, *Liberal Protestant Radio Broadcasting*).

Rie was a member and for many years chairperson of the Baptist Sister Society in Bilthoven and also a board member of the Baptist brotherhood house in Bilthoven. The children were sent to children's clubs and catechism. All six children were baptized around their eighteenth birthday, the usual age for Baptists. The author of the family chronicle, Auke Reitze, thinks that Rie got the children so far because of her personality and attitude to life, and because of the inspiring and non-intrusive example of her own faith. There is a collection of Mennonite songs with Rie's handwriting in the front: *N. Bloembergen, 11 March 1947*. Apparently, a birthday present for Nico's 27th birthday. Her husband Auke was eventually influenced by Rie's faith. Auke was baptized when he was about sixty, and subsequently became chairman of the Bilthoven Baptist Circle.

Rie read a lot; reading was an essential part of her existence. Of course, she read French (she had obtained the MO French certificate), and in particular the nineteenth-century writers, but later also Jean-Paul Sartre, Simone de Beauvoir, Albert Camus, and Marguerite Yourcenar. She also read other literature from around the world, such as Johann Wolfgang von Goethe, Friedrich von Schiller, William Shakespeare, the Lake Poets William Wordsworth, Samuel Taylor Coleridge, and Robert Southey, Leo Tolstoy, Dante Alighieri, Francesco Petrarca, and many more. During her studies Rie had done a minor in Italian, and in the 1930s and 1940s she read Dante's Divina Commedia with a group of four ladies in Italian. She was particularly fascinated by poetry in Dutch literature: Jacques (J.C.) Bloem, Martinus Nijhof, Maria Vasalis, and Ida Gerhardt. This attention to poetry was in keeping with her romantic, dreamy, and sometimes a slightly woolly personality.

Rie did not do any scientific work after her studies. But, probably through the influence of her father, she did have a great admiration for science.

According to Auke Reitze, at the children's primary school (the Van Dijck school), she asked if there were any tests at school and she was a member of the parents' association. Nico reported that his mother encouraged reading, literature, and French.

Fig. 2.5 Nico's mother Rie and father Auke. *Source* Nico Bloembergen

According to Nico, Rie raised her children with the Calvinist work ethic, formulated by Max Weber. The latter declared that devotion to work and rational thinking were the fundamental basic elements of modern capitalism, and this was characteristic of the Puritans who made work a spiritual calling [23]. Put simply, the work ethic just declares that hard work is a vocation.

It is no coincidence, therefore, that all six children studied, that five of the six completed their studies, and that three of them also obtained their Ph.Ds (Fig. 2.5).

Nico's Father Auke [24]

Auke worked at the Utrecht office during the day and was usually home after dinner. Opinions are divided about his work. Nico believes that his father found the work most interesting as factory director of a chemical company, the superphosphate works at Pernis, but that he was not very happy with his job at the head office. Evert disagreed; he thought his father enjoyed working there. Auke certainly did much for the social interests of the staff. In the early 1950s, he made it possible for workers and their families to spend a week in a bungalow park. The plan was first received with suspicion by the employees, so their wives were taken by bus to the park to show them what it looked like. At the same time, he organized a winter sports trip for staff from the head office to Sankt Moritz in Switzerland.

During a dinner on the occasion of "75 Years of Superphosphate" on 4 September 1953, R. Rook, owner of the factory in Kralingse Veer, said:

Mr. Bloembergen,

I just said a moment ago that it was nice to work under your guidance. That is absolutely true: it was and is pleasant. But when you first came to the factory in

Kralingse Veer 37 years ago, we held our breath and said: "What does this greyhound know about this plant?", Because now you are here and before you were there. We thought, "It'll give us grey hair before our time." But it wasn't that bad in the end. Auke collected art from living painters such as Carel Willink, Dik Ket, Pijke Koch, Charley Toorop, Raoul Hijnckes, Willem Bastiaan Tholen, and Wim Oepts. Auke and Rie loved the contact with artists: they were friends with the sculptor couple Bertus Sondaar (1904–1984) and Ton Sondaar-Dubbleman (1907–2000), who lived in Loenen aan de Vecht. Bertus was also a singer and interpreter of French songs, which provided an extra bond.

Of course, as a lover of technical matters and a wealthy businessman, Auke was one of the first to own a car. Between the wars, the car became a new possibility for increasing one's mobility.

> The different forms of transport created a culture of mobility, which means nothing more than that there was general appreciation and interest in mobility. This became of particular importance for the spread of the car. Initially, it was a replacement for the coach, bicycle, or motorcycle, but from 1923, it became a replacement for rail, truck, and horse and carriage. In the early stages the car was associated with sportsmanship, health, exclusivity, speed, masculinity, and elegance [25].

A number of associations evoked by cars, such as sportsmanship, speed, and masculinity, certainly applied to Auke. In 1930, he purchased a Ford two-seater, then considered elegant. It was not until 1938 that elegance was exchanged for the solidity of a family car, a Chevrolet.

Auke could draw nicely. Nico once had to draw a regular dodecahedron for his studies and his father's contribution to the final result greatly improved it.

What Auke probably liked the most was going on outings, both on water and on land. He did a lot of canoeing, skating, and sailing trips on the water, and walking and cycling trips with Rie and the children. Canoeing was especially important to him in the period 1930–1940. In 1938, after a canoe trip in Sweden, Auke wrote in the Dutch magazine *De Waterkampioen* (The Water Champion):

The best form of tourism is the camping trip; the best form of camping trip is a trip on the water; and the best form of camping trip on the water is a canoe trip. I can prove these three statements, but I won't, because the editors don't want to have articles of a controversial nature in the W.K.. I would like to add another statement: the most beautiful country for a canoe trip is the Netherlands.

He enjoyed exploring something new every day, spending the night in a different place, and organizing and experiencing adventures. Auke liked to take the children with him, but actually only Nico was almost always there between 1930 and 1940. Canoe trips were planned months in advance and a certain minimum distance had to be covered every day. So it was not just relaxation, but also performance; performance which was encouraged and sometimes even imposed. In 1930 Auke was awarded the prize of the Touring Commission 1930, received for canoeing the greatest distance 'without sailing or towing'. This prize was awarded by the *Amsterdam Kanovereniging* (Canoeing Association).

In 1938, after his final exam, Nico travelled with two classmates and friends on a banana boat from Rotterdam to Gothenburg in Sweden. His father Auke followed with the luggage. The company travelled by train from Gothenburg to Kil where they began a canoe trip of 80 km.

Auke wrote in *De Waterkampioen*:

The trip along these three lakes took us through the region of Gösta Berling, rich in memories of the many legends Selma Lagerlöf[4] *incorporated into her books. She herself still lives there ... [...]*

In the regions we visited, strangers were a rarity. The natural result is the unspoilt nature of the land and people. That may be different in the more popular tourist centers. After all, it should not be forgotten that, by canoe, one travels in the interior of the country and well off the beaten track, and the country is very sparsely populated. The woodlands stretch out for miles and miles without the presence of another mortal, but along the waterways there is always more life and movement.

After an exhausting trip, during which the canoes were dragged over seven waterfalls and long sections of wild water, Auke wrote:

And yet we have not regretted a moment of it. It has been such a great experience that we wouldn't have missed these adventures for anything, and we were happy to have three days to do the trip, using them up gradually as we plodded from one waterfall to the next. [...]

A pleasant day in Gothenburg was a worthy end to this holiday trip, of which we will keep such pleasant memories. Certainly, there will be some readers who wouldn't feel like spending their holidays as 'luggage carriers', and who may wonder what kind of mentality could actually have inspired us. This should not prevent them from including Sweden on their wish list for canoe trips. There are plenty of tours, where all that toil can be avoided, and where one can get as much pleasure as we have experienced on our trip.

When asked about it in 2011, Nico said he had found it a successful holiday.

Auke was a game fanatic: he played games like chess, checkers, and other board games, as well as bridge, but he always wanted to win. He stopped playing bridge because his wife Rie was better at it. He was president of a bowling club for over ten years. Every Tuesday evening, he went to the bowling club in a room at the Hotel Brands on the Emmaplein in Bilthoven.

Unlike Rie, Auke was not someone with a great social life. He had few friends and hated birthdays and polite visits. He didn't want to join the Rotary Club and had no use for an old boys network!

Was there still time left, after his work and his many hobbies, for Auke to raise the children? At that time (before the 1950s), raising children was the prerogative of the mother, possibly with the help of nannies. The Bloembergen family was no exception. Auke only interfered 'with the big picture'. He never dressed the children when they were small, put them to bed, or read stories to them, and he was often absent from dinner. When the children were sad, their mother Rie was there to comfort them. Auke was not interested in how things went at school. As we have seen, in-depth conversations with the children were conducted by Rie, not Auke. Herbert recounts

---

[4]Selma Ottilia Lovisa Lagerlöf (20 November 1858–16 March 1940) was a Swedish author and teacher. Shew was the first female writer to wind the Nobel Prize for Literature. en.wikipedia.org.

how his father Auke had been on a boat trip to South Africa for several months and was waiting for him on a bench at the primary school. Herbert didn't recognize the tanned man and Auke found this most amusing. But when Auke was at home, he made his presence felt with stories, jokes, teasing, and bickering.

He also provided many toys: an electric train, construction games, a billiard table, and a table tennis table, in short typical boys' toys. The biggest toy for him and the children was of a completely different nature: the house he bought in Loosdrecht in 1936.

Although Auke had money for all his hobbies, he and Rie were unnecessarily frugal, according to Nico. In this way they saved on food and outings [26].

Auke could be difficult and moody. He sometimes commented on the food and then left the table angrily. He was not always kind to Rie either and that's why Rie really wanted to profit from her French education and go on a holiday to France. Her son Auke Reitze remarks: *"But father damned it."* In his spare time he went his own way: he went to winter sports with his family and left Rie with the little children. He made most of his trips with the children, but without Rie. When the family was in Loosdrecht in the summer, he travelled up and down to work every day, because he didn't want to sacrifice his holidays for a cozy family life. He was slightly fed up with all those children, that is, the little children.

Moreover, Auke was a bully.

He made fun of his daughter Diet for what he called her raised stork's beak and his son Herbert was compared to a hippopotamus. Bullying girls (friends of the boys) was the order of the day. And guests had to go along with it. According to Auke Reitze, the bullying was often not fun and sometimes really went too far. Sometimes one of the children would be rolled up in the mat from the corridor with his arms pinned down his sides and then held against a wall. The unfortunate child just had to try to get free again. This was not done as a punishment, but gratuitously. Auke Reitze says: *I still shudder when I think about it.* Piet Hondius, a friend of Herbert's, tells us: *I was deeply impressed that Auke was not allowed to play hockey for a while because he had had a five on his report.*

Auke often set his children up. According to Herbert, the children were trained for life through the efforts of their father Auke.

According to Piet Hondius [27]:

A particularly special occasion was dining at the Bloembergen family home on Herbert's birthday. Herbert and Auke liked to use this opportunity to put pressure on the friends they had invited. Following certain remarks, they gave father Bloembergen the opportunity to go into things in more detail. He would do this with great pleasure - the famous Bloembergen atmosphere. The best counter-attack for the friends was to let go of something about Herbert and Auke, e.g., about the ladies they had been to a party with. Father Bloembergen would immediately jump in.

If you were well in with the family you would also come into contact with their sailing center 'Mas Afuera' between the Drecht and the Tweede Plas. Performance was also an important element there. Firmly animated by father Bloembergen. Staying in Loosdrecht was always a great privilege, also for Sunday sailors. There was

a card game, played against the clock, and the Bloembergens were so fast that they could watch their own cards and those of their opponents, who had to 'give up'.
Yet father Auke could also be generous, especially for the big (sometimes already adult) children. After his final exam, Evert was allowed to go by boat to South Africa for three months; Nico was allowed to study in the United States after the war; Herbert was allowed to re-sit his exams; Auke Reitze was allowed to sit on the committee of the Utrecht Student Society PHRM for a year, and three children were allowed to go on to obtain a doctorate.

Nico's brothers and sisters [28]

This description is largely based on the family chronicle by Auke Reitze. However, he not only wrote down his own memories, but also those of all the brothers, while unfortunately, both sisters were already in nursing homes at the time of the recording and could no longer express their recollections.

After primary school, a secondary school had to be chosen for the children. It was only in 1935 that Bilthoven set up a secondary school, Het Nieuwe Lyceum (a grammar school with an HBS department, i.e., not intended to prepare for university).

The children's father Auke did not see much point in learning the dead languages, Latin and Greek, but their mother Rie wanted them to go to the "gymnasium" (a preparatory school for university, like a grammar school) and her opinion was decisive. All the children would go to the gymnasium. Evert went to the Baarns Lyceum in 1930, because the whole of his class from the primary school, the Van Dijck school, went there. Nico had to go to the Stedelijk (Municipal) Gymnasium (USG) in Utrecht to avoid rivalry (for example, if Evert had gone to the same school, Evert and Nico would have been in the same class). Diet and To also went to the USG, Herbert to the Baarns Lyceum (the HBS department), and Auke Reitze to the USG. According to Evert, his father Auke later called the Baarns Lyceum a 'luxury school'. Indeed, Evert and Herbert, who were at the Baarns Lyceum, found the USG to be a rather poor school. Although according to Evert, the Baarns Lyceum could better have been called "posh". They all went to school by bike, which took half an hour. During dinner, they would bicker over whether the south-west wind was stronger in the morning or in the evening.

Evert was and remained the oldest. As a boy, he asked his mother Rie: "*Will I always be the oldest?*"and Rie replied: "*Yes, boy, you will always remain the oldest*". A woman once asked him how old he was. Evert replied: "*I am twelve, but I have the development of fourteen.*" He was energetic and strong in body and mind. He was also cheerful and outgoing. Evert often took the lead and was ambitious; he decided to make a career early on. He already knew what was best and was like a second father to the family. According to Evert himself there was a division in the family: on the one hand the big boys, Evert and Nico, and on the other, the little ones Herbert and Auke Reitze. And the girls? Well, they were just girls. It was convenient for Rie that Evert supervised Nico, because she had her hands full with the three or four little ones. And Evert did indeed take charge: once he grabbed Nico by an earlobe and locked him up in a dark room. Nico was bullied by Evert and, according to Evert, Nico never forgot that. Evert could also be nice with the little ones and treated them to trips or food. In the house and in Loosdrecht, he was the pacesetter. During his years of study, he brought club friends home with him and they eventually became study friends. Evert did not have a good relationship with Nico: he found Nico "selfish" [29].

Evert remained the older brother all his life, from whom the others could seek advice. He would also advise unsolicited whether or not something was worth doing.

According to the family chronicle, Nico was very different from Evert. Just like Evert, Nico had a considerable drive to perform as a child, but also later. Nico wanted to get the highest possible marks at school, and would get excited when he discovered a mistake in a test afterwards. He did indeed achieve those high marks. Even when they were in primary school, his father Auke had introduced a reward for high marks. But he had had to withdraw the scheme when it became too expensive due to Nico's amazing marks.

Nico had no need to lead, to stimulate family activities, to be fun with the little ones, or to act as a deputy father. He also lacked Evert's cheerfulness; he tended to be gloomy and could be very moody, which sometimes made the family laugh. Once he shouted from upstairs, lisping somewhat: *"Goodness, mother, all my suits are moldy"*. He was also less sociable than Evert, although he certainly did have friends. Nico tended to be withdrawn. His father and mother understood that he should not always be with his books and ensured that he also took part in family activities. They encouraged him to do plenty of sports. Nico's brothers and sister were instructed to take him with them to play games and do sports when he was spending too much time with his books. Fortunately, this was a success in such a sporty family. Nico, for example, was only allowed to do his homework if he had played hockey first. And through all those canoe trips with his father Auke, Nico was familiar with the Dutch rivers, lakes, canals, and locks. He was well acquainted with the geography of the waterways.

So how did Nico stand in the family? He was certainly not a misfit (which would have been easy enough with his talents). He took part in the games, discussions, battles, and bullying in the house. His siblings could still beat him in these games. He participated in the sports activities and he had his own friends, some of whom were family friends. Although his relationship with Evert was not cordial, according to Evert, Nico did pick up certain things from him, such as sports and other activities. At a young age, they each went their own way. The age difference with his two youngest brothers was too great to allow real warmth and camaraderie, and he also had no special ties with his two sisters.

Diet was a doer. She excelled at school, but just like her father, she had trouble sitting still. As a five-year-old, she had already made breakfast for eight people. She was very performance-oriented and was the best of her year at the gymnasium. She was the only one of the children to have piano lessons. She liked to talk and was indeed very talkative, cheerful, and sociable. Herbert found her somewhat fierce and mature for her age. Diet was a competitor for the boys, but she did have a special bond with Evert.

Nico's other sister To had problems as a small child. Her fontanel would not close and an operation was needed to remedy the problem. Her mother Rie concluded that

To had poor health, so she received special attention and extra nutrition. She was pampered and did not do 'rough' sports such as hockey and canoeing. She could play tennis, but she didn't have to win. At the "gymnasium", she got on well and, just like mother, was a language person. However, she did not get along well with her older sister Diet, and preferred to be with Herbert and Auke Reitze. According to Herbert, that was because To was one of the few in the family who could face losing. She was quiet, wiser than the others.

Herbert, known as Beppo when he was a boy, was the nicest. The successive nanny's and maidservants loved him. He was the most socially oriented of the children. In primary school he was the center of a circle of friends, and those friends may well have appealed to him more than the Bilthoven family. He eventually made it to the high school with reasonable marks. Later he organized and directed a lot. Herbert had the same active, restless nature as his father. Herbert was closest to his younger brother Auke Reitze, but was proud that he could make him cry at any time. Auke had to do what Herbert said.

Auke Reitze, the youngest, was nicknamed Kloer and was not only the youngest, but also the smallest. He did not have much physical strength and had to make up for it with his mouth. He could do that well; indeed, when he was ten he was sometimes called 'Big Mouth' The family said that he should later become a lawyer. Outside the house he was a braggart. He could be moody and bad-tempered. In sport he was a fierce hit, but apart from that he was no doer. At school he got on well, but he was no star.

The Bloembergen family was a close community and according to Auke Reitze even a kind of clan. The children had very sheltered lives. The depression in the early 1930 s went unnoticed. Within the family there was fierce competition, initiated by the father Auke, who himself also competed with his sons and his daughter Diet, while Rie and To were left to their own devices.

Friends of the children had to pass the competition tests before they were accepted. Some quit.

The hockey club SCHC—joining the crème de la crème

Evert became the first of the family to join the local hockey club, the Stichtsche Cricket and Hockey Club or SCHC, when he was twelve years old. The club was founded in 1906 by (members of) the Utrecht Student Corps (USC) and officers of the Kromhout Barracks in Utrecht. Many members of the USC also became members of the Utrecht Student Sports Association Sphaerinda and/or SCHC. The good hockey players came first, the others played in Sphaerinda, in the fourth or seventh team. Evert was important as an organizer and director, and among other things as editor of *De Stichtenaar* (The Founder), the club magazine.

The club had more than two hundred members in 1937. Nico, Diet, Herbert, and Auke Reitze also played in SCHC, all in the first team. Nico was a very enthusiastic hockey player and still thought it important in 2006, as was mentioned in the SCHC jubilee book. According to Auke Reitze, Nico played rather spasmodically. When Nico studied in Utrecht (1938-1943), he came to Bilthoven every weekend to play hockey. Later, when he came back to the Netherlands for his Ph.D. (1947–1948), he started playing hockey at the SCHC again.

The Bloembergens played an important role in the hockey club for years. Hockey dominated family life. At the weekend, they were always coming and going from the hockey field five minutes away. Dinner conversation was often about hockey. Auke Reitze tells us: "*It was just a hockey family. But unlike now, where parents are actively coaching and queues of parents are making their appearance at the competition, parents used to go and see; that's what it was like.*" [30]

According to Piet Hondius:

The house on the Van Ostadelaan was on its way to the football field and was a 'mooring' for football and other activities in the park on the heath. And it turned out that the Bloembergen family was a family with four sportive gentlemen (including Evert and Nico) and two sisters, of which the eldest, Diet, like her brothers, was quite a hockey star. The youngest sister, To, was not such a sportswoman, but she was the sweetest of the family. You could see from the garden that their sporty sides and their dog Tommy had left their mark. All the brothers and sisters sprang into action, often in a clearly audible way, on the fields of the SCHC. They were quick-witted and witty in tongue, too, and their intelligence showed through their tremendous speed of reaction. Herbert and Auke, trained by their brothers, helped to create the typical SCHC atmosphere among their contemporaries, distinguishing the club from other hockey clubs. Corps members from Utrecht, who were well represented in the different teams, contributed a lot to this [31].

The above shows that, in the period described here, the hockey club had an important function in the elite community of Bilthoven (and Utrecht). The Bloembergens belonged to the core of the club. There were close ties between the SCHC and the Utrecht Student Corps. For the Bloembergen children, the SCHC was the center of their social life. The Bloembergen children were so famous for their performance at the club that their mother Rie was greeted on the street by boys who were quite unknown to her, just because they knew that she was the mother of the brothers in the SCHC first team.

The world outside Bilthoven

At the end of the 1920s there was a global economic crisis which lasted for ten years [32, 33].

There is nothing in the family chronicles about the Great Depression. Nico comments: "*As children, we were not directly affected by the economic crisis. We heard about it in conversations with the adults. Our family never lived in luxury, but we had a house at the Loosdrecht Lakes with canoes and sailing boats. In the winter, there was always hockey at the SCHC. So no poverty.*"

And Auke Reitze in his family biography:

And the great evil of the world remained at a distance. Take, for example, the economic crisis of 1929 and the years that followed, which led to high unemployment and poverty in the Netherlands and which also deeply affected father's business. As children in the relatively prosperous Bilthoven we didn't notice any of this.

2.4 Bloembergen at the Utrecht Municipal Gymnasium (1932–1938)

The best in everything

In 1932, after the primary school in Bilthoven, Nico started going to the Utrecht Municipal Gymnasium. Every day he would ride about 8 km on the bike.

In 1876 the Hieronymus School, a so-called Latin school, became a grammar school due to the new law on higher education [34]. Until then, it had been a preparatory school for the University of Utrecht.

When Nico first went to the USG on 8 September 1932, it was located in an old building at the Janskerkhof. In October 1932, a new building was opened and all the members of the board of governors, teachers, and students (including Nico) walked in procession to the new premises. In 2012, the school is still located here. In his speech as president and member of the board of governors, the mayor mentioned that those present approached the sun via the Zonstraat (Sun Street) [35]. The opening was accompanied by an exuberant party that lasted three days. The highlight was a performance of Sophocles'*Aias* by the pupils. The new building was designed by the architect Planjer (1891–1966) and the streets where it is located were given the appropriate names Homeruslaan and Minervaplein. The building was bordered at the rear by Nicolaasweg (!). The new building had a wing with laboratories for physics, chemistry, and biology and a room for geography. The neighbourhood formed the boundary between the districts of Oudwijk and the chic Wilhelminapark. In the latter neighborhood was the Maliebaan, where Nico's grandparents Bloembergen lived. At noon he sometimes went to see them.

Nico obtained the following marks in the USG entrance examination:

Calculation	10
Dutch language	8
Geography	8
History	10

In the first year, the pupils took seven subjects; for the sixth grade science which Nico followed, there were eleven. It is noteworthy that the beta study program was designed for *those who wished to obtain a certificate exclusively for the faculties of medicine, mathematics, physics, or law*, while the alpha orientation was for the *faculties of theology, law, or the arts and philosophy*. So for law, one could choose between alpha and beta! In the first few years, classical languages were dominant. In the first year the pupils got eight hours of Latin, compared with three hours today. In the second year, five hours of Latin and six hours of Greek, compared with three hours of each today. In the sixth class, which was the exam year, a further three hours of Latin and three hours of Greek were given. Today it is four hours of Latin

Fig. 2.6 Former town secondary school (gymnasium) in Utrecht, where Nico studied from 1932 to 1938. *Source* Antoine, Wikipedia NL

and/or Greek. However, Greek is out of favour as an examination subject. By way of comparison, there are now about twenty subjects in the sixth grade, some of which are optional. The gymnasium and the HBS were real schools in the 1930 s. In addition, the HBS and especially the gymnasium were also schools for the elite (Fig. 2.6).

At the time Nico thought himself at school a broadly developed person because of his various activities, but later he realized that he actually knew nothing about social relations, psychology, governance, and politics. He also knew nothing about music; he only came into contact with it later. And as far as history was concerned, he had not come any further than the Netherlands in the eighteenth century. Looking back, he considered this lack to be a considerable shortcoming in his schooling [36].

On the Board of Trustees of the school from 1932 to 1938 we find well-known names such as Utrecht's mayor Mr. Dr. G. A. W. ter Pelkwijk and Prof. Dr. H. Bolkestein, professor of Ancient History and Rector Magnificus of the University from 1934 to 1935. In the school year 1937–1938 the school had seventeen teachers for academic subjects and three for drawing and physical exercise (gymnastics). Eleven of the seventeen teachers for academic subjects had doctorates! The level of substantive expertise was much higher than it is today. On the other hand, the educational skills of the day were hardly demanding.

According to Jan Visser, former teacher of classical languages at the USG, the following considerations came into play [37]:

In those days, education had to be rock solid. The USG was no exception: the proportions were fixed. Aims were high. The grammar school was predominantly liberal, with straightforward Protestant elements. Catholics were also present. And on the left side of the social spectrum there were also a reasonable number of social democrats. Humanist, but also Christian, that is how the school could be characterized, and tolerance was a high virtue. A liberal, general school in a time of rather strict pillarization.

The most important teachers for Nico were those in physics, mathematics, and chemistry. He learnt a lot from his physics teacher Dr. Abraham Davis Nathans (1888–1948). Nathans obtained his Ph.D. in 1923 with Prof. Dr. Leonard Ornstein (see Chap. 3) at the University of Utrecht, with a thesis entitled *Overview of the application of complex functions to physical issues.* Nathans had a solid scientific background: Ornstein was one of the leading physicists in the Netherlands, but Nathans also had a didactic background; he wrote three textbooks about teaching physics. His books *Nathans and Lindeman. Physics Part I and II* and *Nathans Interference and polarization* were mandatory at the USG. According to Jan Visser, former pupils did not find Nathans an inspiring teacher [38].

The Hertz experiment (Heinrich Hertz, 1887) was demonstrated as part of the physics practical course (a novelty). A high AC voltage was generated with a battery, a capacitor (a so-called Leiden jar), a coil, and a switch. This type of coil was introduced by Heinrich Ruhmkorff in 1851 [39] and still lives on in the ignition coil of a car, where it serves to generate the spark between the electrodes of the spark plugs [40]. The voltage can discharge via a spark gap, which can be as wide as 30 cm. This is the transmitter. The Hertz experiment consists in placing an open coil with two knobs at the ends, called the receiver, at a distance of one meter or less. In this coil, electromagnetic transmission causes a voltage build-up with the same frequency as that of the arc. The voltage can discharge again between the buttons of the coil [41]. Electromagnetic waves can thus propagate from the Ruhmkorff to the coil. Today, all radio stations, TV stations, and mobile telephony work on this principle. There is a transmission mast where electromagnetic radiation is emitted and a receiver, the radio, TV, or mobile phone, where the radiation is received.

Dr. D. E. J. J. Schrek, the mathematics teacher, was good enough and Nico learnt a lot from him [42]. Since 1942, Schrek has written a number of articles about the history of mathematics, especially in the magazine Euclid. Commenting on Schrek, Jan Visser tells us:

He was a real gymnasium teacher. He not only dealt with the curriculum, but also said something about the history of mathematics, for example, the work of Euclid or Pythagoras. What's more, he also had the right image: reliable and hard-working.

Dr. H. R. Bruins for chemistry was a very good teacher, according to Nico [43]. Nico had to analyse distilled water at his laboratory and was surprised to find that, after evaporation, there remained a residue. However, these were the remains of salts which Nico himself had added to test certain reactions. According to Jan Visser, this teacher was considered inspiring and even friendly by former pupils. Bruins was musical and was an above-average violin player. He had a balanced character and a broad general knowledge.

Dr. J. C van der Steen, who taught natural history (biology), was something of a phenomenon according to Jan Visser [44]. A good teacher, he also directed the plays at the USG, despite his somewhat uptight appearance. He also loved excursions, but especially outdoors, for example by bike to Driebergen. There was then a kind of battle played (his best known book is: *The plant and animal world, Panorama of Biology, Universiteit voor Zelfstudie*, 1908). He loved games and played 'winking games' with the pupils in evening classes. That was somewhat risky at the time and the students found it rather exciting.

The Rector, Dr. A. H. Kan, was teacher of ancient languages and Nico's teacher in the sixth grade. (Kan translated the work of Thomas More and was involved in other translations, such as translations of Erasmus). Kan's wife was friends with Rie Bloembergen, Nico's mother. J. S. E. Zwart, the geography teacher, could not explain a lunar eclipse, so Nico stepped in and did it in his place [45].

While Nico was in the lower years, the grading system used for the report ran from 1 to 5, where 5 was the highest grade. The lowest grade Nico got on his report was 4 (hence, 8 in the current system) for geography. For mathematics, he had a 5 on all reports. From class III onwards, the current grading system was used.

Meanwhile, evil continued in the world. Germany had lost the First World War and was burdened with exorbitant war damages by the victors France and the United Kingdom. When the global economic crisis broke out in 1929, the country became even more impoverished and fell prey to political struggles and a street war between the communists and Nazis. In January 1933, Hitler seized power and his political opponents, the communists and social democrats were hastily eliminated by a strategy of terror. Then it was the turn of all the other organizations that did not go along with the Nazis. There was no longer any freedom of expression, and with the abolition of all other fundamental rights, dictatorship became a fact.

Auke Reitze writes in the family biography [46]:

In 1934 [...] Nico went as a fourteen-year-old boy with a youth [hockey, RH] team to Magdeburg in Germany to play hockey against a German team. He was impressed by the motorways, at the time still virtually unknown in the Netherlands, but 'pretty shocked by what was already going on there [47]. Some of their German opponents were enthusiastic about Hitler and said to the surprised Dutchmen: you have colonies and you don't have to fight for 'Lebensraum'.

After his first two years, in 1934, Nico was able to choose his own future for the first time, and he chose the beta program, which focused on science.

In classes III through VI, he also took physics and on all final reports for physics he got 10! The same went for mathematics. In class IV, he also took chemistry. In class VI, the year of the final exams, he got 9 twice and 10 four times for these subjects. This does not mean, however, that he got low marks for the other subjects: from class IV onwards, the lowest marks he got were a number of sevens for Dutch, French, German, and Dutch history.

Nico was invited to become a member of a 'secret' society called *Mentor* whose members were students of the highest classes. *Mentor* was founded in 1878. They played at being students there, imitating rituals from the student corps. Snobbery was the dominant theme. Nico was "ab-actis" (secretary) in 1936 and 1937. Members received a medal and met at other members' homes. Former members, who had completed their final exams and were therefore no longer at the USG, were called honoraries and would still visit regularly. They had discussions about Leibniz, Kant, and Einstein and read poems by Homeros and Rilke.

Jan Caro also spent six years at the gymnasium, initially in the same class, and would remain a lifelong friend of Nico's.

He recounts as follows: *"We were bosom friends. Nico had a big mouth in the classroom. He was already brilliant. Nico and I were paired by his parents, because he always had his nose in his books and they hoped I could make sure he did something else. Mentor was more of a debating society. Literature did not come first. Who else was in it: Wils Bruning and a few other strange characters."* [48]

Wils Bruning was also a friend of Nico's.

Nico remained an honorary member of Mentor until he obtained his doctorate in 1948.

Suusje Kramers, a daughter of the professor of physics, Hans Kramers, at Utrecht University was also in the first class. Kramers was professor there from 1926 to 1934. Nico in 2006:

He [Kramers, RH] *then lived in Bilthoven and he offered me a ride home once, after Saturday morning classes. I don't know why I didn't have my bike, but at least I got to do the ride back to Bilthoven with Suusje Kramers. That was my first meeting with Kramers. Later I saw him regularly in his house in Leiden* [49].

We shall meet Kramers again in later chapters.

Suusje was a school friend, but Nico had no other friends at school.

Nico was interested in theatre. In 1937–1938 he was Praeses (chairman) of the theatre association *Lampome*. He played a role in the performance of Oscar Wilde's *The Importance of Being Ernest*.

Mother Rie showed her interest in Nico's school (which would later be his sister Diet and brother Auke Reitze's school) by joining the parents' association of class IIa in the school year 1934–1935.

Nico had his own chemical laboratory, which he got from his father Auke. It was 'located' in a shed at his friend Wils Brunings' grandmother's. There the two friends did some experiments with copper shavings and sodium, which had to be stored in kerosene. So he already knew a lot about chemistry, such as the difference between bases and acids, before he even started chemistry at school.

On 11 March 1938 Nico turned 18.

At 8.45 am on this day, Hitler gave the order to enter Austria the next day. There was little opposition in Austria, and indeed there were cheering crowds that gave the Nazi salute. Arthur Seyss-Inquart became governor of the province of Austria (in Nazi terminology *Reichsstatthalter Ostmark*).

In the final exams, Nico was top of his class. His final exam marks were:

Greek	9
Latin	10
Dutch	8
German	8
French	8
English	8
Algebra	10
Geometry	9
Trigonometry	9
Physics	9
Chemistry	10
Biology	9

His physics teacher Dr. Nathans wanted to give him 10, but the dedicated Prof. F. Zernike (who was to receive the 1953 Nobel Prize for Physics), said "*we not going to start with a 10*". Nico was the first person to take the oral exam in physics and apparently Zernike did not want to give a 10 to the first person that came along. Later Zernike visited Harvard and Nico spoke to him about what was in his opinion too low an appreciation. But Zernike would say nothing more about it [50].

According to the school tradition, he was allowed to give a speech to the whole school in July 1938. This was called a 'valedictory' and it was held in the Pieterskerk, the oldest church in Utrecht, in Romanesque style, with murals from the eleventh century and reliefs from the twelfth. Nico gave his speech according to custom in dress suit with a white bow tie and top hat. His speech referred to several quotations from the Flemish writer Felix Timmermans. In the novel *Pallieter* (1916), we find the following:

And over the Nethe, the great, tomato-red sun had blossomed like a joyful surprise out of all that whiteness. […]

It was too clean here to go down. But the white light gilded, and the sun grew bigger and bigger, and red. The red adorned the clouds cheerfully and rolled over the world. (Fig. 2.7)

Nico had elocution lessons, because he expressed himself in the Zwijndrecht dialect, where the 'o' was pronounced more like 'au', and of course that did not fit in with the elitist Bilthoven of the day. The family later pestered him with his overly neat statement of the 'o's in these quotations.

Nico comments: *This literary passage is interesting from a physical point of view. Reddening is a physical phenomenon. The blue and green components of the white sunlight are removed from the sunbeam by diffraction due to small drops of water in the atmosphere. The fact that the sun looks bigger is a physiological effect. The solar disk does not change in size. The same optical illusion occurs with the almost full moon, which appears larger when it emerges next to trees on the horizon than when it is high in the sky* [51].

Fig. 2.7 Bloembergen in
1938 after his valedictory in
Saint Peter's church in
Utrecht. *Source* Nico
Bloembergen

The Rector Dr. A. H. Kan, addressed Nico:

Bloembergens tend to aim high, but often hit the target.

In the foreword to his 1948 dissertation, Nico wrote:

I am still grateful that I began my studies at the Utrecht Municipal Gymnasium [52]

In retrospect, Nico found that the USG offered an excellent program in the natural sciences, especially in the last three years. According to Nico, most pupils found the curriculum traumatic, but he enjoyed it. He liked Latin as much as he liked mathematics, but his favorite subjects were physics and chemistry. He liked Dutch literature the least. It was certainly not the case that he was 'predestined' to continue in natural science; at some point he was considering studying languages. This was because, despite its excellent reputation in the natural sciences, the USG did not encourage careers in this area. However, the teaching of natural sciences in the higher classes tipped things that way and in the end he chose to go on with that.

Quantum mechanics and nuclear physics were not yet taught at school; it was just mechanics, electricity, and a little atomic physics.

2.5 Intermezzo: Brilliance, Predisposition or Education?

In primary school, as mentioned before, Nico obtained the highest grades. The gymnasium also had no secrets for him: his average for the twelve final examination subjects was 8.9, and in physics, mathematics, and chemistry his average was actually 9.4.

There is no doubt that he already showed great cognitive skills in primary school. These are skills in the field of perception, thinking, learning, language, and motor skills. He had no trouble thinking, learning, and using language. He trained himself in a number of sports: canoeing, sailing, hockey, and ice skating, and he always wanted to be the best. In canoeing, his skills focused on long distances; in hockey he ended up in the first team. I won't yet comment on his skills in the field of perception during this period. In Chap. 14, I shall return to this under the heading *Scientific Intuition*. It is obvious that Bloembergen had a high intelligence. A word often used today is giftedness [53]. In childhood, high ability manifests itself as a developmental advantage, such as Nico learning the Greek alphabet at the age of six. If the cognitive developmental level of young children is determined using a validated test, the result can be seen as an indication of the child's abilities. This 'potential' can then lead in due course to high performance in certain new areas; for the few in more than one area. It should be emphasized, however, that giftedness manifests itself differently in every period of life. A primary school child has different interests than secondary school pupils, and so on. High potential is no guarantee of extraordinary performance as an adult.

Highly gifted young people achieve high marks in almost all subjects during the school period[5] provided that they 'fit' into the context of a classroom education [54].

Bloembergen did indeed achieve high marks in all subjects.

In addition, talented young people are characterized by their participation in many extracurricular activities. Nico played hockey, was a member of the Mentor Society, experimented in his own chemical laboratory, canoed, sailed, and skated.

Giftedness can manifest itself in certain behavior, of which a number of characteristics are described above. Nelissen and Span give more examples of positive characteristics, such as understanding and remembering difficult information when interested [55]. According to Nico's youngest brother Auke, Nico was 'monomaniac' in his interest in physics and only took part in the conversation if there was a connection with physics. Another characteristic is the achievement of high performance in a certain area.

---

[5]However, many gifted children fail at school because they do not feel at home in a class. For example, homework is badly done or not at all, and the child hates school instinctively or avoids new learning activities for fear of failure.

A stimulating environment like the one at the Bloembergen home and at the gymnasium is necessary but not sufficient. Highly gifted children need to develop their abilities further [56]. The child can initiate this development by showing commitment, involvement, interest, and curiosity. Giftedness and performance do not come naturally; hard work is required. A statement by Bloembergen in an interview in 2004 reads:

[As a child, RH], *I always enjoyed a challenge* [57].

"Practice" is the key word here, indeed the secret is hours of practice. Ericson et al. [58] showed that there is a connection between achieving a certain level as a musician, for example, and the number of hours of practice. Through prolonged practice, the difference with others increases over the years.

Over the years, the influence of family circumstances decreases: over twenty years, the influence on the IQ decreases from 40 to 0%. On the other hand, the influence on the IQ of origin actually increases [59].

There is no doubt that Bloembergen was gifted. His high grades in primary school and at the gymnasium, his focus on performance, the fact that he always had a book with him, his focus on one subject, his pleasure in facing challenges as a child: everything points to giftedness. Was that innate? Learning to speak, being able to beat an uncle at chess so early on, at the age of four, each of these might indicate an innate giftedness. On the other hand, the giftedness of Nico Bloembergen developed further through the stimulation he received in his family and at the gymnasium, and not least through his habit of seeking out challenges.

References

1. Interview Rob Herber with E Bloembergen, 20 November 2006
2. Bloembergen AR. Het gezin van Rie en Auke Bloembergen 1917–1956 (The family of Rie and Auke Bloembergen 1917—1956). Eigen Beheer, Wassenaar (2003)
3. Werkgroep kadastrale atlas provincie Utrecht/Historische Kring D'Oude School. De Bilt in 1832 (1999)
4. Doedens A, Cladder J. Een dorp betrapt: fraude in De Bilt. 14: 60–64 (2005). De Biltse Grift
5. Beijen JW. De zin van het nutteloze (The meaning of the useless). Rarekiek van de 19e eeuwse jaren der 20ste eeuw. Donker, Rotterdam (1970)
6. www.quakers.nu/
7. Heijden J vd, Drees E. Kees Boeke en de Werkplaats: "Wij gaan een schooltje beginnen en ik heb al een schrift en penhouder!" De Biltse Grift nr 15 3–5 (1996)
8. Adresboek Bilthoven (1922)
9. Bilthovensch Adresboek (1927)
10. Bloembergen AR. Het gezin van Rie en Auke Bloembergen 1917–1956 (The family of Rie and Auke Bloembergen 1917—1956). Eigen Beheer, Wassenaar (2003)
11. Beijen JW. De zin van het nutteloze (The meaning of the useless). Rarekiek van de 19e eeuwse jaren der 20ste eeuw. Donker, Rotterdam (1970)
12. www.historischekringdebilt.nl
13. Koopman A. In: Pieterse E (red) Van Dijckschool 80 jaar. Een reis rond de wereld. Utrecht (1997)
14. Interview Rob Herber with H Bloembergen, 6 November 2006
15. Interview Rob Herber with E Bloembergen, 20 November 2006

16. Archief Utrechts Stedelijk Gymnasium
17. Interview Rob Herber with N Bloembergen, 7 December 2006
18. Bloembergen AR. Het gezin van Rie en Auke Bloembergen 1917–1956 (The family of Rie and Auke Bloembergen 1917—1956). Eigen Beheer, Wassenaar (2003)
19. Nicolaas Bloembergen. In: Mattson J, Simon M. The Pioneers of NMR and Magnetic Resonance in Medicine, Dean Books, Jericho, V.S (1996)
20. www.genealogieonline.nl/database-bergman/I5828.php
21. nl.wikipedia.org/wiki/Doopsgezind
22. www.doopsgezind.nl
23. Weber M. Die protestantische Ethik und der 'Geist' des Kapitalismus. Tübingen, Mohr (1934)
24. Bloembergen AR. Het gezin van Rie en Auke Bloembergen 1917–1956 (The family of Rie and Auke Bloembergen 1917—1956). Eigen Beheer, Wassenaar (2003)
25. Staal PE, Automobilisme in Nederland. Een geschiedenis van gebruik, misbruik en nut (Motoring in the Netherlands. A history of use, abuse, and usefulness). Walburg Pers, Zutphen (2003)
26. Interview Rob Herber with N Bloembergen, 6 December 2006
27. Hondius P. De Bloembergens van de Van Ostadelaan. Hoog Soeren, 23 January 2003
28. Bloembergen AR. Het gezin van Rie en Auke Bloembergen 1917–1956 (The family of Rie and Auke Bloembergen 1917—1956). Eigen Beheer, Wassenaar (2003)
29. Interview Rob Herber with E Bloembergen, 20 November 2006
30. Interview Rob Herber with A R Bloembergen 2 February 2007
31. Hondius P. De Bloembergens van de Van Ostadelaan. Hoog Soeren, 23 January 2003
32. www.geschiedenis.nl
33. nl.wikipedia.org/wiki/Beurskrach_van_1929
34. www.usg.nl
35. Jille E. Nobelprijswinnaar Nico Bloembergen. Vox Hieronymi 34:10–11 (2006)
36. Interview Rob Herber with N Bloembergen, 6 December 2006
37. Interview Rob Herber with J Visser 30 March 2007
38. Interview Rob Herber with J Visser 30 March 2007
39. people.clarkson.edu/~ekatz/scientists/ruhmkorff.htm
40. nl.wikipedia.org/wiki/Bobine
41. people.seas.harvard.edu/~jones/cscie129/nu_lectures/lecture6/hertz/Hertz_exp.html
42. Interview Rob Herber with N Bloembergen, 6 December 2006
43. Interview Rob Herber with N Bloembergen, 6 December 2006
44. Interview Rob Herber with J Visser 30 March 2007
45. Interview Rob Herber with N Bloembergen, 6 December 2006
46. Bloembergen AR. Het gezin van Rie en Auke Bloembergen 1917–1956 (The family of Rie and Auke Bloembergen 1917–1956). Eigen Beheer, Wassenaar (2003)
47. Nicolaas Bloembergen. In: Mattson J en Simon M. The Pioneers of NMR and Magnetic Resonance in Medicine, Dean Books, Jericho, V.S (1996)
48. Interview Rob Herber with J Caro, 22 November 2006
49. Interview Rob Herber with N Bloembergen, 6 December 2006
50. Interview Rob Herber with N Bloembergen, 6 December 2006
51. Vox Hieronymi 34 10–12 (2006)
52. Bloembergen N. Nuclear Magnetic Relaxation. Rijksuniversiteit Leiden, Leiden (1948). Dissertation
53. Span P. Korte historie van het onderzoek naar begaafdheid in Nederland. In: Span P, De Bruin-De Boer AL, Wijnekus MC (red) Het testen van begaafde kinderen (Testing gifted children). Samson Onderwijsbegeleiding, Alphen aan de Rijn (2001)
54. Nelissen J en Span P (red) Begaafde kinderen op de basisschool: suggesties voor didactisch handelen (Gifted children in primary school: suggestions for didactic action). Bekadidact, Baarn (1999)
55. Nelissen J and Span P (red) Begaafde kinderen op de basisschool: suggesties voor didactisch handelen (Gifted children in primary school: suggestions for didactic action). Bekadidact, Baarn (1999)

56. Nelissen J. De derde factor. Naar een constructivistische opvatting van begaafdheid (The third factor. Towards a constructivist conception of giftedness). Vernieuwing 59(4):7–9 (2000)
57. Griehsel M. Interview with Nicolaas Bloembergen, Nobel Web AG (2004)
58. Ericsson KA, Roring RW, Nandagopal K. Giftedness and evidence for reproducibly superior performance: an account based on the expert performance framework. High Ability Studies.18:3–56 (2007)
59. Ridley M. Nature via Nurture. Harper Collins, New York (2003)

Part II
Studies

Chapter 3
In the Footsteps of Lorentz, Kramers, and Ornstein

Location: the Physical Laboratory of the University of Utrecht, Bijlhouwersstraat, 1939. Nico is following the physics practical for prospective students. The laboratory has received new, so-called rectifier tubes from Philips. Nico's task is to connect the tubes. There are four contacts, two to which a low glow voltage of 4 V should be applied and two to which an AC voltage of 220 V should be connected. The glow voltage comes from the cellar, where there are numerous batteries. But at this point they do something wrong and the tubes go up in smoke … Nico looks on bewildered and practice supervisor Herman Burger is very disappointed.

Nico Bloembergen, from now on referred to by his surname, continued his scientific education at the University of Utrecht. To get an impression of the area where Bloembergen ended up, we shall give a brief description of the university, but first, let us say something about the father of Dutch physics, Hendrik Lorentz.

3.1 Science in the Netherlands

Science has a long history in the Netherlands. The Dutch Golden Age of the 17th century is one of the most fertile cultural periods in the history of Europe. The economic prosperity brought about for a large part by domestic and foreign maritime trade fostered a thriving period of artistic and scientific activity that far surpassed what was happening anywhere else in Europe at the time [1]. Examples from science are: Christiaan Huygens (1650–1700) who developed the theory that light consists of waves, developed the pendulum clock, discovered the moon Titan of the planet Saturn, and constructed telescopes; Anthonie van Leeuwenhoek (1725–1750) who constructed an improved microscope; Willebrord Snel van Royen (Snellius, 1580–1626), a surveyor who calculated the circumference of the Earth using his

© Springer Nature Switzerland AG 2019
R. Herber, *Nico Bloembergen*, Springer Biographies,
https://doi.org/10.1007/978-3-030-25737-8_3

triangle method and rediscovered the law of Snellius, as did René Descartes (French, but living in the Netherlands, 1596–1650) independently of him; Simon Stevin (1548–1620) mathematician and hydraulic engineer; Jan Leeghwater (1575–1650) hydraulic engineer; Zacharias Jansen (1588–1631) inventor of the microscope; Jan Swammerdam (1637–1680), optician and biologist. These are just a few samples among many others. Culturally speaking, philosophy was also an important issue. Although Desiderius Erasmus (about 1467–1536) lived in the 15th and 16th centuries, his humanism had been clearly implanted in Dutch culture since then. René Descartes, mathematician and philosopher, lived for 20 years in the Netherlands, where he wrote his most important publications. His research vision spread across the whole of Europe. Not forgetting the most important philosopher in the Netherlands, Baruch de Spinoza (1632–1677), who propagated the work of Descartes, but also criticized it. Spinoza has been praised by many philosophers in the 19th, 20th, and 21st centuries, including Albert Einstein.

A second golden age of science in the Netherlands began in the 20th century [2]. In 1863 the Minister of Internal Affairs, Johan Rudolf Thorbecke, introduced a new kind of secondary school, the 'Citizen's High School'. Unlike the existing "gymnasium", no Latin or Greek was taught. After an admission exam in Latin and Greek, one could then go on to study at university. In 1917 this admission requirement was no longer needed. Due to the greater emphasis on the study of mathematics, physics, chemistry, and biology, a better preparation was obtained at the Citizen's High School than at the gymnasium [3]. The very first Nobel Prize for Chemistry (1901) was awarded to the Dutch chemist Jacobus van't Hoff (1852–1911), and the following year, the Nobel Prize for Physics went to Hendrik Lorentz (see below) and Pieter Zeeman (1985–1943). The list continues in 1910 with Johannes Diderik van der Waals (1837–1923), in 1913 Heike Kamerlingh Onnes (1853–1926), both for Physics, in 1924 Willem Einthoven (1860–1927) for Physiology or Medicine, in 1929 Christiaan Eijkman (1858–1930) for Physiology or Medicine, and in 1936 Peter Debye (1884–1966) for Chemistry. Many of these men studied at the Citizen's High School before going on to university.

3.2   City of Utrecht

At the time of the Romans, the southern part of what is now the Netherlands belonged to the empire. The Romans built an extensive infrastructure along their border, called the 'limes'. These were mainly built along rivers which were used for transport, together with the land routes or 'via's. One such combination was constructed in year 50 along the river Rhine, with an army camp in the center of the country called Traiectum. On the remains of the Roman camp in the Middle Ages around 500 AD, the Franks built a borough and ecclesiastical center. This developed in the 10th century to an important trading post. In 1122 the city obtained city rights and, until halfway through the 16th century, was the most important city in the Northern Netherlands. In this period of major growth, the church with its striking Dom tower was constructed from the year 1254, and at a height of 112 meters, remained the highest building in the Netherlands until the 20th century. The city still has beautiful canals and houses from this period. Urban expansion continued in the 19th century and in the period after World War 2.

3.3 Hendrik Antoon Lorentz (1853–1928), Father of Dutch Twentieth-Century Physics [4–6]

Lorentz was very well known through his chairmanship of the State Commission in preparation for the closure of the Zuiderzee. When Lorentz died in 1928, there was national mourning. The funeral was attended by Albert Einstein, Ernest Rutherford, and members of the cabinet and the Royal Family, among others. Thousands of people stood along the route of the funeral procession in Haarlem. Flags hung at half-mast and lanterns were wrapped in black cloth.

Lorentz obtained his Ph.D. cum laude at Leiden University on 11 December 1875, with a dissertation entitled *On the reflection and refraction of light*. In 1877 the then 24-year-old Lorentz became professor of theoretical physics at Leiden University. His inaugural speech was entitled *Molecular theories in physics*.

In 1902 Lorentz was awarded the Nobel Prize in physics together with Pieter Zeeman: *in recognition of the extraordinary service they rendered by their researches into the influence of magnetism upon radiation phenomena*.

In 1912 Lorentz was appointed curator of the 'Physics Cabinet' of the Foundation for Management and Conservation of Teylers Museum in Haarlem. He remained professor in Leiden until his successor Paul Ehrenfest (1880–1933) was appointed there in 1912. Lorentz then became extraordinary professor in Leiden.

In 1911, Lorentz became president of the Solvay congress, a meeting where a small number of leading physicists could discuss the latest developments in their field. Lorentz presided five of these congresses until 1927.

The Lorentz Medal is awarded once every four years by the Royal Academy of Sciences (KNAW) as a gold medal to a researcher in recognition of pioneering contributions to the development of theoretical physics. The prize was imposed in 1926 *on the occasion of the fiftieth anniversary of the doctoral thesis of the member of the Academy H. A. Lorentz* [7].

Lorentz is probably the most important Dutch physicist and one of the most important physicists even at the international level, comparable to Newton, Einstein, and Maxwell. Lorentz' work bridged the gap between the classical physics of Newton and Maxwell and modern, relativistic physics. Although Lorentz also contributed to the development of this new physics, his achievements lie mainly in the field of classical physics. Lorentz was also involved in quantum mechanics, the new physics of the very small. He invited Max Planck to the university and published a paper with Einstein on the Planck constant [8]. Lorentz was an exceptionally amiable person. Anyone who asked for his help could count on him. Einstein came, whenever he could, to visit Lorentz in Leiden. Einstein noted in 1953 that many physicists did not realize how important was the role Lorentz played in the development of physics, because his contributions are so thoroughly incorporated into this science that it was difficult to see just how groundbreaking they actually were.

Physicists of the younger generation are for the most part no longer fully aware of the decisive role which H. A. Lorentz played in the construction of the fundamental ideas in theoretical physics. This strange situation has arisen because Lorentz's fundamental ideas have become so much part of their flesh and blood that they are barely able to fully appreciate the boldness of these ideas and the simplification of the physical foundations that he has achieved [9].

3.4   Utrecht University, Sol Iustitiae Illustra Nos

Before the war, Utrecht University, like the other institutes for higher education in the Netherlands, was a closed community within Dutch society. It had its own rules of conduct, within which everyone knew his place. There was a lazy 'professorial culture', while they hardly interacted with the cold outside world. With the exception of the complexes at the end of Biltstraat and the buildings at Croeselaan and Catharijnesingel, everything was still within the canals. Most of the students lived there or near the centre, and most of the professors also lived in the city, around Wilhelminapark. There was peace, order, and a good measure of complacency. Pre-war reports on the future of the university stressed that it should serve the 'non-political' truth. Two basic principles appeared on the banner: humanism and idealism. As intellectuals, we must ensure that the creative work of the mind retains its continuity in the realm of culture [10].

3.5   Physics in Utrecht, the Period Until 1914. From Education to Research

Willem Henri Julius (1860–1925) [11]

In 1888 Julius obtained his doctorate with Buys Ballot[1] on a dissertation with the title *The heat spectrum and vibration periods of molecules of some gases.* In his Ph.D. research, Julius found that the maximum radiation intensity shifted to shorter wavelengths at rising temperatures. This can be observed by watching a piece of glowing iron in the blacksmith's workshop. When the iron is heated, we don't see

---

[1]Christophorus Buys Ballot (1817–1890) is best known for his law: *Standing with your back to the wind, the air pressure on the left is lower than on the right, in the northern hemisphere, while it is the other way round in the southern hemisphere.* In the Netherlands, he was also well known as founder of the Royal Netherlands Meteorological Institute (KNMI).

anything at first, but the iron does heat up. When the iron gets hotter, the colour changes from dark red to red to orange to yellow. Yellow has a shorter wavelength than red and red has a shorter wavelength than dark red. In 1896, Wilhelm Wien (1864–1924) derived his shift law to explain this [12].

Following the example of the astronomer Samuel Pierpont Langley (1834–1906), [13] Julius developed a very sensitive instrument, the bolometer, to measure infrared (heat) radiation. He built a spectroscope, an instrument to study the light spectrum so that it could measure heat radiation. For this purpose he made prisms and lenses out of rock salt. In 1890 he became extraordinary professor of physics at the University of Amsterdam. The research he did during his Ph.D. was the basis for the work that followed.

In Amsterdam he produced a radiometer based on the example of Charles Vernon Boys (1855–1944) to make accurate measurements of radiation—analogous to the micrometer used to measure lengths [14]. Julius manufactured a special suspension system for this instrument, which went into production by a Dutch company.

His research was in the field of solar phenomena, such as refraction in the solar atmosphere and the uneven shift of the spectral lines of the sun. For this purpose the Heliophysical Institute was set up, based in Sonnenborgh.

The laboratory focused mainly on spectroscopic research for astrophysics. The research formed the basis for the work done under Ornstein (1920–1940), which was subsequently applied after the Second World War.

In 1920, due to illness, Julius had to leave management of the laboratory temporarily to his colleague in theoretical physics, Leonard Ornstein. After six months he had recovered, but Ornstein remained in charge.

Julius died in 1925. Einstein wrote: *"With the passing away of W. H. Julius, one of the most original exponents of solar physics has left us."* And he expressed his desire to *"bring anew to the attention of the profession the work of this clear-sighted, artistically fine-spirited man* [15].

3.6 Physics in Utrecht, the Period 1914–1940

Expansion Under Ornstein

Leonard Salomon Ornstein (1880–1941) [16]

In 1898 Ornstein took up the study of mathematics and physics in Leiden. Jan Cornelius Kluijver (1860–1932) was his mathematics teacher. Mathematics was more than a tool for Ornstein; he was especially interested in probability theory and stochastic processes. These are models for processes, i.e., phenomena, that take place in time or space and involve physical quantities that depend on chance. His most important teacher, however, was Lorentz, who had a great influence on his work and his views on science. Ornstein called Lorentz *"the greatest physicist [...] in the Netherlands since Christiaan Huygens."*

Fig. 3.1 Leonard Ornstein. *Source* University Museum, Utrecht University

In 1905 Ornstein took his doctoral exam, after which he immediately started working on his thesis. In 1908 he obtained his doctorate under Lorentz with the thesis *Application of the Statistical Mechanics of Gibbs to molecular-theoretical questions* [17]. Statistical mechanics or statistical thermodynamics was originally an idea of Ludwig Boltzmann (1844–1906), whereby the physical properties of substances are calculated as a weighted average [18]. In 1902 Gibbs (1839–1903) derived theorems which displayed many similarities with thermodynamics [19]. Ornstein showed that the Gibbs method combined the advantages of thermodynamics and kinetic gas theory. Lorentz was a prominent representative of Dutch thermodynamics and Ornstein followed in the footsteps of Lorentz (Fig. 3.1).

Pollen (pollen grains) in a drop of water show irregular motions under the microscope. This was discovered in 1827 by the botanist Robert Brown [20]. The explanation is that very small particles start to move due to collisions with molecules (in this case, water molecules). Einstein was the first to state in 1905 that the average distance travelled was a statistical quantity [21]. Einstein conceived of the motion as a stochastic process. Ornstein published a paper about this *Brownian motion* in 1917. In 1930 he published a major improvement to the theory with George Uhlenbeck [22].

In 1909 Ornstein became a lecturer in mathematical physics, hydrodynamics, and mathematical chemistry in Groningen. He worked with Frits Zernike, who was appointed his successor in Groningen after Ornstein left for Utrecht. In 1914 Ornstein became professor of theoretical physics in Utrecht. In 1915 his lecture *Problems in the kinetic theory of matter* still reflected his earlier work on thermodynamics and stochastic processes, but at the end he briefly discussed quantum mechanics, which was in full development at that time. However, he remarked that there was still no appropriate theory of motion in this theory and that it was not therefore without its risks [23].

In Utrecht, Ornstein began to devote himself increasingly to experimental work. His interest was aroused by the instruments developed by Julius and especially by the objective registration method of Willem Moll (1876–1947). Moll's machine made it possible to accurately measure details in the blackening of photographic plates, and Ornstein saw this as an opportunity to expand Utrecht's expertise in the field of photometry. This blackening of the photometric plates was caused by the spectrograph, an instrument that split light into different colours, just as a prism does with solar radiation.

As mentioned earlier, Ornstein took over the running of the laboratory when Julius fell ill in 1920. In 1925 he changed chairs and became professor of experimental physics. He also became director of the laboratory. In 1926, after a major renovation, the new laboratory was opened on the Bijlhouwersstraat.

While Ornstein had initially shown an interest in Lorentz's theoretical work, he later took an interest in experimentation: *"[I] learned mathematics from the outstanding mathematician J. C. Kluijver in Leiden, and physics from the great Moll in Utrecht."* Ornstein's collaboration with Moll, and later with Burger, was certainly the main reason for the success of the laboratory. Internationally there was great appreciation for the work of Ornstein, Moll, Burger, and Cittert, as evidenced by a letter from Arnold Sommerfeld to Hermann Jordan (dean) in 1925: *"The work of the Utrecht University Laboratory, which Mr. Ornstein has organised so generously with the help of Mr. Moll, Mr. Cittert, etc., has proved itself to be fundamental and will be carried out at the highest level wherever there is an interest in spectroscopy."* [24]

In 1932, in a joint publication in book form by Ornstein, Moll, and Burger, with an emphasis on the measurement methods and equipment developed in Utrecht, various such methods for light and heat radiation were discussed [25].

Ornstein was an exceptionally good organiser within the scientific community. Not only did he take care of the radical refurbishment, but he also successfully lobbied the cabinet and created a kind of scientific production line where theses and publications were produced on an continuous basis. One of those Ph.D. students was, as we saw in Chap. 2, Bloembergen's physics teacher at the gymnasium, Abraham Davis Nathans.

The applied research, for example developing standard methods for light measurement and calibration of standard lamps, became a well-known export item from the laboratory.

Ornstein's own work was more conservative than innovative. Although he gradually became more involved in quantum mechanics, Hans Kramers' biographer Dresden called Ornstein *"a highly opinionated person who had utterly no use for quantum mechanics"* [26].

Ornstein was certainly a strong personality, who managed to argue with everyone outside the laboratory except Zernike. And likewise, with the mathematicians in the faculty, it was no cake walk. Ornstein was strict with his students, but was also highly appreciated for his interest in each as an individual and the success of their research.

According to Bloembergen, everywhere else the study of spectroscopy meant determining the wavelengths of different elements, while in Utrecht quantitative spectral analysis was performed [27]. This means that the intensities were determined, and these in turn were directly related to transition probabilities. Moreover, the Ornstein group determined the transition probabilities and oscillator strengths of all elements in the periodic table. This research accounted for 20 dissertations.

Ornstein also pointed out to students the possibilities for doing applied research. Examples were lighting technology and electrical engineering. Ornstein then brought in special teachers from Philips and from the Kema Laboratories[2] in Arnhem. This shows that Ornstein also had good contacts with industry. Many of Ornstein's students found work at Philips through Ornstein's connections [28].

In October 1939, after the departure of Uhlenbeck, Ornstein gave temporary lectures in theoretical physics. The lectures on experimental physics were dispensed by Milatz for a fee of 100 Dutch florins a month. Maarten Bouman, who would later do his doctoral studies with Bloembergen, was an assistant lecturer to Ornstein.

Hendrik Anthony (Hans) Kramers (1894–1952) [29, 30]

Kramers studied mathematics and physics in Leiden. In 1916 he graduated with Paul Ehrenfest (1880–1933), succeeding Lorentz in Leiden. In the same year, Kramers moved to Copenhagen, where he started working on his thesis with Niels Bohr. In 1919 he obtained his Ph.D. under Ehrenfest with a thesis on the intensities of energy transitions in atoms: *Intensities or spectral lines: On the application of the quantum theory to the problem of the relative intensities of the components of the fine structure and of the Stark effect of the lines of the hydrogen spectrum.*

In 1920 Kramers became Bohr's first assistant at the Institute of Theoretical Physics in Copenhagen. Here Kramers developed his theory of optical dispersion. This laid the foundations for the matrix mechanics of Heisenberg, who would receive the Nobel Prize for this contribution in 1932. In these years, through his work with Bohr, Kramers became an authority in the field of quantum mechanics. In 1926 he was appointed professor of theoretical physics in Utrecht. Ornstein would have preferred Burgers. Moreover, Kramers was a member of Bohr's quantum mechanics school, for which Ornstein had little interest. Kramers received little cooperation from Ornstein in the laboratory [31]. He only published work on statistical mechanics in Utrecht, work he had done in Copenhagen. In Chap. 2 we met Kramers as the father of Bloembergen's classmate, Suusje. In 1931 Kramers also became professor at the Technical University in Delft. In 1933 he published a manual: *The Foundations of Quantum Theory,* [32] and in 1934 he was appointed successor to his former teacher Ehrenfest in Leiden (Fig. 3.2).

Hendrik Casimir wrote about Kramers: "*He tackled problems because he saw them as a challenge and not primarily because they offered a chance of success. As a result, there are few spectacular results in his work that can easily be explained to a layman, but among colleagues he was generally recognized as one of the grandmasters.*" [33]

---

[2]Kema Laboratories perform accredited testing, inspections, and certification of components for transmission and distribution of electricity.

Fig. 3.2 From left to right, physicists Jan de Boer, Hans Kramers, and Eliza Wiersma in 1944.
Source Kamerlingh Onnes Laboratory, Leiden University

In 1947 Kramers was awarded the Lorentz Medal.

Personalities
The spectroscopic research in Utrecht was made possible by the development of their
own measuring equipment. In this respect, they proved themselves a match for what
was done in Leiden. With this development of spectroscopic measuring equipment,
Julius, Moll and Burger established a tradition in Utrecht. Ornstein was above all a
scientific manager and enjoyed supervising students and Ph.D. students more than
doing his own research. Ornstein's guidance was very personal: as far as possible he
would meet each student or Ph.D. student every day to see how things were going,
make suggestions, and … then he was gone again. He replaced some of the older
standard tests with modern experiments, but the elementary first year practical work
showed many similarities with the old collective exercises. Pupils who seemed to
have a special talent for experimental work were picked out and assigned as 'slaves'
to a senior student. The latter, in the doctoral phase, would have an average of two
such slaves at his disposal. This was seen by the younger students as an honor for
them, something like a first assignment in the world of research as it was really
done. They were involved in serious scientific research at an early stage and did not
have to repeat experiments that had already been carried out hundreds of times. Both
scientific research and education were thus led in new directions by Ornstein [34].

Ornstein had great admiration for Moll and could therefore work well with him. Ornstein's collaboration with Burger and Cittert was also good, but cooperation with the theoretical physicists Kramers and Uhlenbeck was cool to neutral. In the faculty, mathematics was outside Ornstein's territory and his relationship with mathematicians was sometimes very poor.

A shift in research

After 1935, research in the laboratory gradually shifted from intensity measurements in atomic and molecular spectra, i.e., atomic physics, to similar measurements for atomic nuclei, i.e., nuclear physics. This required equipment for measuring radioactive radiation, such as Geiger-Müller counters, a Wilson chamber, and beta-spectrometers. A Geiger-Müller counter, often just called a Geiger counter, works with a tube filled with gas across which a high voltage is applied. If the tube is affected by radioactive radiation, namely gamma-radiation or alpha-particles, the gas is ionised and briefly becomes conductive, so that a current can start to flow. This current can be read from a meter or made audible through a speaker which produces a clicking sound. A Wilson chamber is a sealed vessel containing super-cooled, supersaturated vapour. If radiation or particles pass through it, a condensation trail occurs, similar to the condensation trails left by jet engines in the troposphere. These condensation traces can be photographed and thus recorded. With a beta-spectrometer, the spectrum of beta-particles (electrons) can be determined. The equipment was designed by H. Brinkman (1909–1994), J. M. W. Milatz, and D. Th. ter Horst (1919–1976). The ionization chambers required special high voltage transformers, amplifiers, and automatic counters. The contacts with Philips and the KEMA short-circuit laboratory were crucial here. The most important nuclear physics research in Utrecht concerned the nature and energy of the forces that held the neutrons and protons together in the nucleus.

Rutgers used ionization chambers to study the ionizing power and the variation in the range of alpha-particles in air, hydrogen, and helium, and in various metal foils and mica. We shall return to Rutgers later on.

Ornstein did not concern himself much with nuclear physics, but left that research to Milatz [35].

3.7 Bloembergen at the University of Utrecht, Summer 1938–10 May 1940

Classical physics

In the final years at the grammar school, Bloembergen was free for the first time to chose courses from those on offer and opted for the beta program, which was the science stream. Like everyone else, after the final exam he had to make more choices: so how to proceed? His parents set the example, because his mother Rie had taken French and had always encouraged the children to study, while his father Auke was a chemical engineer. Even more important here was the example of his eldest brother Evert: he was studying law at the university. Evert had chosen to study law and not physics, chemistry, or mathematics because he had realized that he could never compete with his brother Nico. So, the choice for Bloembergen's studies was actually a rather easy one. His father Auke thought that Nico should do the hardest thing there was. Nico thought so too, and the hardest thing was physics. That too was a rather easy choice: it was a small step from the ten's for physics at the gymnasium to physics at the university. Bloembergen then had to choose the university. Evert was already studying in Utrecht and Nico also chose the university closest to Bilthoven. Another very small step. He would live in rooms in Utrecht, but go home to Bilthoven at weekends. He would also continue to play hockey in Bilthoven.

So, Bloembergen took a series of small steps towards independence at the beginning of his studies, but remained bound to the security of his family and the village of Bilthoven. So, what did this transition from grammar school to university actually involve?

At the time, university studies consisted of two parts: the candidate examination and the doctoral examination, when one became a Ph.D. student. Studies for the candidate examination usually lasted two years in physics. During this time, lectures and practicals could be attended, but the student had no obligation to do so. The order and time of taking the (oral) examination was also up to the student. The only requirement was that the examinations be taken and that the practicals be followed. After the candidate examination, two years of lectures were taken, while experimental physics usually required three years of experimental work. This experimental work was of a very different nature to the practical part of the candidate examination. The candidate's practical training usually consisted of routine tests, while for the doctoral examination, truly independent new research (as a 'slave') was carried out for a Ph.D. candidate.

The transition from the grammar school to the freedom of the university was a major step. As we shall see, there were many dropouts in the first year, and there were also a number of 'eternal students'.

On 3 June 1938 we find Bloembergen on the list of Novitii (or freshers) of the Utrechtsch Studenten Corps (USC). Following the example of his brother Evert, Bloembergen became a member of the USC. Other new members were Jan Caro, whom we already met in Chap. 2, and J. C. (Jan) Kluyver and M. P. (Bob) Lansberg, with whom Bloembergen would become friends.

Jan Caro lived at Schoolplein 3, Bloembergen at Nieuwe Gracht 151 [36].

Whereas the university was an elite institution with an estimated 2500 students (nowadays there are more than 25,000), within this elite the USC was once again the *crème de la crème*, or at least, so they viewed themselves. In 1816 the board, the Senatus Veteranorum, had founded its own society, PHRM (Placet hic Requiescere Musis, Here Rests the Muse) in the Janskerkhof, which was called the Yellow Castle [37]. Only male students could (and still can) become members of the USC. Naturally, a private community such as the USC required an admission test, by way of initiation. Traditionally, new members (the Greens) were usually subjected to humiliating actions. In Bloembergen's time, the first of these was that their heads were

Fig. 3.3 End of the initiation into the fraternity. From left to right: Jan Caro (friend), Bob Lansberg (friend), and Nico Bloembergen. The shaven heads of the new members are covered with a beret. *Source* Nico Bloembergen

shaved. The Greens, including the Bloembergens, also had to attend a meal where the dessert consisted of a pastry puff filled with green soap. Greens had to pose with the soap puffs on their shaved heads. Outsiders called corps students 'frat boys'.They referred to each other as 'buddies'. After their studies, USC members kept in touch through a *dies*, an annual gathering at the beginning of November, at which they would meet in rotation (Fig. 3.3).

On 21 October, Bloembergen and Jan Caro were registered as faculty members of the USC, Bloembergen as a faculty member for mathematics and physics and Jan Caro at the faculty of law. A characteristic of the student corpora was that members would often remain friends throughout their lives, and sometimes help each other in society later on. In this way, Bloembergen remained in contact with Jan Caro, Jan Kluyver, and Bob Lansberg. Jan Kluyver would also study physics and Bob Lansberg medicine. Jan Caro later becomes a lawyer and Bob Lansberg a physicist specialising in aerospace.

An important pastime was pub crawls. Bloembergen did participate to some extent, but was not much of a barfly. He usually ate once a week in the club. His year in the club was called *The Laughing Cavalier*, after a painting by Frans Hals from 1624 [38, 39].

In Bloembergen's day, twenty to twenty-five first-year students of physics, mathematics, and astronomy would start in Utrecht every year. There would be a few corps members. Physics usually involved the formation of 'pairs', who acted as couples in practice. Bloembergen and Jan Kluyver formed a 'pair' [40].

Bloembergen attended mathematics lectures by Barrau (analytical geometry) and Bockwinkel (algebra and analysis). Barrau had obtained a doctorate in Amsterdam with Diederik Johannes Korteweg in 1907, with a thesis entitled *Contribution to the theory of configurations* [41]. In Chap. 1 we met Korteweg briefly as Ph.D. supervisor of Bloembergen's grandfather Nicolaas Quint. Barrau was appointed in Utrecht in 1928. His opening address was: *Figures and their coordinates.* He had an aversion to any reform in mathematics [42].

Bloembergen studied Barrau's book closely and found analytical geometry in three dimensions interesting. He took his exams at the home of the almost 70-year-old Barrau. Barend Streefland, who arrived in 1941 and passed his candidate exam in 1946, notes that Barrau took excellent care of his lectures. They were somewhat old-fashioned [43].

Bockwinkel was Lorentz's personal assistant for theoretical physics in Leiden and was succeeded by Ornstein. In 1919 Bockwinkel was appointed lecturer in the faculty of mathematics and physics, to teach introductory mathematics. His best-known book was *Lectures on Integration* [44]. There are many anecdotes about Bockwinkel, including the following one from a biology student: *"I only went to two of these lectures. At the start of the first lecture, Dr. Bockwinkel cried out: 'Are there any biologists in the room? Will they raise their hands! If all the biologists want to leave the room now, I know from experience that they won't be interested anyway!' I tried one more lecture [...] but Bockwinkel was right, it didn't interest me"* [45].

According to Maarten Bouman he was a very frustrated man [46].

Bloembergen noticed that Bockwinkel was very unhappy that he would not become a professor. Bloembergen grasped everything about the mathematical concept of limit, and everything else [47]. Kees de Jager (born in 1921), a fellow student and later professor of astronomy, commented as follows on Bockwinkel's lectures: *"The lectures were tough going. Difficult questions could only be answered by Nico. When Nico was sick once, nobody could answer a question. Bockwinkel asked: Does nobody know the answer? Is Bloembergen not around?* [48]. According to Barend Streefland, Bockwinkel was notorious. *"He was a very strange person. His teaching was poor, but what he taught was an eye-opener for me. Existence proofs in mathematics, analysis, differential and integral calculus: we had never heard of this in secondary school. He did not speak very clearly and did not cover much more than his book. He had no social skills, not from any angle."* [49]

Bloembergen attended physics classes by Ornstein and Milatz. Maarten Bouman said he learned practically nothing from Milatz in physics [50]. Barend Streefland did not share that view [51]. Bouman was assistant lecturer to Ornstein and he noticed that Ornstein often did not know a week in advance what he was going to treat [52].

Bloembergen followed Minnaert's teaching methods for physics and got to make simple astronomical observations in the observatory.

He studied organic chemistry at Kögl and did the practical analytical chemistry under the supervision of Strengers in the old veterinary school near the *Museumbrug* (Museum Bridge). Bloembergen had good memories of that.

Van de Hulst was a physics lab assistant and Van der Held a physics lab teacher. Bloembergen did standard experiments, for example, with a torsion balance, or measuring the radiation of a black body. This last experiment was done near a door that kept opening and completely disrupted the measurement by the draught, so that the setup had to be properly shielded. The practical occupied two full afternoons a week. An instrument-making course also had to be followed on Friday afternoon in the basement of the laboratory, and although Bloembergen was not much of a handyman, he succeeded in making a transformer. He first had to master the theory transformers. Then he had to cut slats (50) out of soft iron, after which the holder for the coil had to be sawn out of the slats with a fretsaw. Then the coil was wound and holes drilled in the slats for the wiring. All this under the supervision of the head of the instrument maker, W. Jezeer. As mentioned at the beginning of this chapter, the pre-candidate practice could sometimes go wrong, but Bloembergen found the experimental handiwork as challenging as mathematics and he earned a lasting respect for the essence of matter. The manual for the lab contained a warning that students should not think they were good at physics just because they tinkered with radios. Bloembergen was rather afraid of the opposite: he had never tinkered with anything in his classical education at the grammar school and hoped that everything would go well, despite the fact that he was not much of a handyman. He felt one could learn a lot through perseverance, including sports, for example. He didn't have any special talent for that either, but through perseverance and practice one can still get results, pleasure, and satisfaction. Bloembergen also followed an astronomy course run by Minnaert [53].

The exams were passed without difficulty.

Bloembergen would also play hockey in the USC association Sphaerinda. On 4 November 1938, the seniors played against the first years. The result was 3-1. In the first year, Bloembergen was right winger and Jan Kluyver left. The first-year students were 3-0 behind at half-time. *As was customary every year, that year the hockey match between the representative teams of the seniors and first years took place on the beautifully kept grounds of the S.C.H.C. in Bilthoven. After the break [...], the freshers gritted their teeth and managed to outsmart the seniors' goalkeeper, who had so far kept his goal with great skill, and under loud cries a goal was scored by Bloembergen* [54] (Fig. 3.4).

On 7 January 1939 the inter-university ski championships were held in Roselend in the French Alps. Bloembergen took part, but was not very successful, coming seventh out of eight. He did better in the slalom, where he came third out of eight [55].

Bloembergen also became a member of the rowing club, Triton. In order to be able to row seriously, he had to give up the pub crawls because a lot of training was required. The many canoeing trips he had made with his father were to his great advantage when he began rowing. Bloembergen was part of the lightweight four (stroke) with G. Monsees (bow), A. G. Th. Becking, H. H. F. Hobbel, and Jan Caro (coxswain). Nico's brother Evert was coach! On 19 May 1939, the rowing match was held on the Amsterdamse Bosbaan and the Triton team came first (out of three) [56]. It was a narrow victory [57].

Fig. 3.4 Field hockey club SCHC, October 1938. Nico Bloembergen is the one with the shaved head. *Source* Nico Bloembergen

On 16 June 1939, Bloembergen moved to Schoolplein number 2 [58]. His best friend Jan Caro lived at number 3. At noon, Bloembergen sometimes went to visit his grandmother, who lived in Utrecht a few hundred meters from the Schoolplein, in the Wilhelminapark. To his displeasure, Caro was not allowed to come along. Every Friday, Bloembergen went to Jan Caro's to eat. After dinner he would play billiards with Caro's father and then Bloembergen and Caro would go to the city and the PHRM society or the cinema [59].

Bloembergen continued to go regularly with friends to his parents' house in Loosdrecht. His father Auke often went to the house with friends for a week, and before or after that, Bloembergen and his club could go there for a week. The saying then was that first the old whores went there and then the old gentlemen [60].

Bloembergen became a member of the *Christiaan Huygens* physics society. He also became a member of the Undergraduate Mathematics and Physics Society. In January 1940, he became ab actis (secretary) of this society [61].

Early in May 1940 there was another Varsity rowing match featuring the lightweight four with Bloembergen and Jan Caro among others. However, they were not successful [62].

The situation around the Netherlands became increasingly worrying at the end of the 1930s. Germany had annexed Austria in 1938 and as threatened had taken over part of Czechoslovakia. In September 1939, Germany invaded Poland. The United Kingdom and France then declared war on Germany, and in return, Mussolini's Italy declared war on them. Initially, skirmishes only took place on the border between France and Germany. In April 1940, Denmark and Norway were occupied by the Germans.

For German science, the Nazification was catastrophic: Jewish scientists were no longer allowed to practice their profession and became social pariahs. Quantum mechanics was declared by the Nazis to be Jewish, not 'Aryan' science, and could no longer be practised. Numerous Jewish, socialist, communist, and liberal scientists fled Germany and the areas annexed by Germany for as long as it was still possible. These included Albert Einstein (Swiss, but lived in Germany), Robert Oppenheimer, Otto Frisch, Rudolph Peierls, Hans Bethe, Liese Meitner (Austrian, but lived in Germany), Ludwig Wittgenstein (Austria), Kurt Gödel (Austria), Victor Weisskopf (Austria), Niels Bohr (Denmark), Edward Teller (Hungarian, but lived in Germany), Eugène Wigner (Hungarian, but lived in Germany), Leó Szilárd (Hungarian), John von Neumann (Hungarian), and Enrico Fermi (Italian). These scientists often fled to the United States and the United Kingdom, where they not only strengthened the science in those countries, but also contributed significantly to the war effort, working on the development of radar and the atomic bomb, for example.

At the time, the university was a closed world and within it the USC was a stronghold, in which people did not want to have anything to do with the outside world. Yet the outside world invaded the university and the USC. In the first place, Jewish scientists had disappeared from the German occupied territories, so contacts with German science, which were already more problematic in Nazi regions, were minimized. On the other hand, contacts with the scientists who fled or were deported to the United Kingdom and the United States could be maintained until May 1940.

In the student and university journals there were more and more disturbing reports, all pointing to the approaching war.

On 7 October 1938, two articles relating to the Munich Convention were published in Vox Studiosorum.[3] The first had the headline:

Flight of Intellectuals

It is undoubtedly a daring undertaking to run an article in this magazine which deals with politics, and which even tries to encourage us directly to take a greater interest in this subject. This does not yet look very committed, but apparently the editors did not want to shock their readers too much. It should be kept in mind that until then all politics had been taboo. In the same issue, the editors went one step further:

EDITORIAL Now everything is quiet again. We can continue with our former more or less pleasant life. The people are cheering, there will be no war! […] War is on the doorstep, and it remains there [63].

In 1939, the tone of the articles changed. More and more people were writing about war and there were calls for voluntary service. In *SNIFF,* the mobilization in the Netherlands from September 1939[4] was mentioned [64]. In the issue of 13 October 1939, Evert Bloembergen first appeared as an editor [65]. In the issue of 3

---

[3]The Munich Convention was signed on 30 September at the Munich Conference between France, the United Kingdom, Germany, and Italy on the future of what was then Czechoslovakia. The treaty allowed the Czech Sudetenland, where a minority of Germans lived, to become German without involving Czechoslovakia in any way. This was a reward for the aggressive German politics. The British Prime Minister justified this capitulation under the heading of *Peace for our time* and stated that peace had been guaranteed for ten years until then. The Germans invaded Poland less than a year later.

[4]On 1 September 1939 Germany invaded Poland.

November the war in France was discussed[5]: *Alarm! War in France. We Dutchmen, we are neutral and as such we have nothing to do with this war, which is not ours … But this does not mean that we are not involved and that we will not be jointly responsible for what is to come* [66]. In an issue of 3 November 1939, a call was made for a *VOLUNTARY ANTI-AIRCRAFT DEFENCE CORPS* [67].

In Bilthoven large advertisements appeared in which the Municipal Air Protection Service announced what people had to do to be prepared for attack. Attics had to be tidied up and buckets of sand placed there, because *fire bombs are thrown down from above, so the attic would be the first to be hit.* Absurd government advice is apparently a timeless phenomenon. In every house the windows were to be hidden with *non-transparent material.* In October 1939 an exercise had already been held in the vicinity of the Parklaan, in which the district was completely blacked out. In addition, the municipality constructed shelters. From September 1939, approximately 1500 soldiers were stationed in the municipality of De Bilt. In February there were rumours that parts of the municipality of De Bilt would be evacuated and on 1 March, the municipality published an evacuation plan [68].

In 1940, it became clear that it was only a matter of time before the war reached the Netherlands. On 2 February 1940, Evert Bloembergen became editor-in-chief of Vox and wrote an article about censorship [69].

On 9 February 1940, Vox wrote: *When will it be our turn to join the war?* [70]. Another article appeared on 16 February: *The Netherlands in wartime. Discussion of speeches by F. Muller (Rector Leiden), A. C. Josephus Jitta (State Mediator), J. van Walré de Bordes (Mayor of Middelburg, League of Nations), and J. Linthorst Homan (Queen's Commissioner in Groningen)* [71]. Then on 1 March: *OF WAR AND PEACE* [72]. On 19 April an appeal was published in the Utrechtsch Faculteitenblad to help with harvesting:

Call to all students, male and female, at the University of Utrecht.

We must eat, war or no war, that is clear. Well, the farmers have sown. But as the harvest approaches, they lack hands, now that so many have been mobilized. Those who have time must help. And who has more time than a student in summer?

It may also be possible in the Netherlands to make it compulsory *to help with the harvest. But surely every Dutch student will welcome the possibility to do this work without obligation from above? What is needed for this is clear: You must not only applaud it, but volunteer to do it yourself […]*

The Student Harvest Commission, Utrecht Division

E van Rossum, Chairman

and others [73].

Then on 10 May 1940 the curtain fell for the Netherlands.

The Second World War had major consequences for the university, the teachers, the professors, the staff, and the students. In view of this, we shall discuss the period 1940–1943 separately.

---

[5]This refers to the border control with Germany. For the French this was known as the 'drôle de guerre', for the Americans and British as the 'phoney war', and for the Germans as the 'Sitzkrieg'.

References

1. Greenlaw B. wih.org/blog/2016/01/29/immerse-in-the-dutch-golden-age-of-art-and-science/
2. www.knaw.nl/en/about-us/academy-history/1902-de-tweede-gouden-eeuw
3. Snelders HAM. Inleiding in de geschiedenis van de natuurwetenschappen in ons land (Introduction to the history of the natural sciences in our country). dspace.library.uu.nl/bitstream/handle/1874/21451/c1.pd
4. Kox AJ. Hendrik A. Lorentz. In: Kox AJ (red) Van Stevin tot Lorentz. Portretten van achttien Nederlandse natuurwetenschappers (From Stevin to Lorentz. Portraits of eighteen Dutch natural scientists). Bert Bakker, Amsterdam, 1990
5. Snelders, HAM. Lorentz, Hendrik Antoon (1853–1928). In Biografisch Woordenboek van Nederland. www.inghist.nl/Onderzoek/Projecten/BWN/lemmata/bwnl/lorentz
6. Hendrik Antoon Lorentz. e.citizendium.org/wiki/Hendrik_Antoon_Lorentz
7. www.knaw.nl/cfdata/prijzen/prijzen_detail.cfm?orgid=170
8. Heilbron JL. Max Planck_s compromises on the way to and from the Absolute. In: Evans J, Thorndike AS (eds) Quantum mechanics at the crossroads. Springer, Berlin, 2007
9. Einstein A. H.A. Lorentz als Schöpfer und als Persönlichkeit, Leiden (1953)
10. Vellinga SYA. De uitdaging van crisis en bezetting, de jaren 1936–1946 (The challenge of crisis and occupation, the years 1936–1946). In: Dunk HW von der, Heere WP, Reinink AW (red). Tussen ivoren toren en grootbedrijf. De Utrechtse Universiteit, 1936–1986 (Between ivory tower and large company. The Utrecht University, 1936–1986). Maarssen, Schwartz, 1986
11. Snelders HAM. Julius, Willem Henri (1860–1925). In: Biografisch Woordenboek van Neder-land. www.inghist.nl/Onderzoek/Projecten/BWN/lemmata/bwn2/julius
12. www.nobelprize.org/nobel_prizes/physics/laureates/1911/wien-lecture.html
13. earthobservatory.nasa.gov/Features/Langley/langley_2.php
14. physics.kenyon.edu/EarlyApparatus/Thermodynamics/Radio_Micrometer/Radio_Micrometer.html
15. Einstein A. W.H. Julius, 1860–1925. Astrophysical J 63:196–198, 1926
16. Heijmans HG. Wetenschap tussen universiteit en industrie. De natuurkunde in Utrecht onder W.H. Julius en L.S. Ornstein 1896–1940 (Science between university and industry. Physics in Utrecht under W.H. Julius and L.S. Ornstein 1896–1940). Erasmus Publishing, Rotterdam, 1994
17. Ornstein LS. Toepassing der Statistische Mechanica van Gibbs op molekulair-theoretische vraagstukken. Eduard IJdo, Leiden, 1908
18. plato.stanford.edu/entries/statphys-Boltzmann/
19. www-liphy.ujf-grenoble.fr/pagesperso/bahram/Phys_Stat/.../gibbs_1902.pdf
20. www.brianjford.com/A-BRNRMS.htm
21. Einstein, Albert (1956) [1926]. Investigations on the Theory of the Brownian Movement? (PDF). Dover Publications. Retrieved 2013-12-25
22. Uhlenbeck GE, Ornstein LS. On the theory of the Brownian motion. Phys Rev 23: 823 (1930)
23. Ornstein LS. Problemen der kinetische theorie der stof. In: Jaarboek der Rijks-Universiteit te Utrecht,1914–1915, Oratie 1915
24. Letter Arnold Sommerfeld to Hermann Jordan, President Faculty of mathematics and sciences, 19 June 1929
25. Ornstein LS, Moll WJH, Burger HC. Objective Spektralphometrie. Vieweg, Braunschweig, 1932
26. Dresden M, Kramers HA. Between Tradition and Revolution. Springer, New York, 1987
27. Interview Rob Herber with N Bloembergen, 6 December 2006
28. Interview Rob Herber with M Bouman, 8 March 2007
29. Dresden M, Kramers HA. Between Tradition and Revolution. Springer, New York, 1987
30. O'Connor JJ, Robertson EF. Hendrik Anthony Kramers. www-history.mcs.ac.uk/Biographies/Kramers.html
31. Interview Rob Herber with M Bouman, 8 March 2007

32. Kramers HA. Die Grundlagen der Quantentheorie, 1933
33. Casimir HBG. Het toeval van de werkelijkheid. Een halve eeuw natuurkunde (The coincidence of reality. Half a century of physics). Meulenhoff, Amsterdam, 1983
34. Heijmans HG. Wetenschap tussen universiteit en industrie. De natuurkunde in Utrecht onder W.H. Julius en L.S. Ornstein 1896–1940 (Science between university and industry. Physics in Utrecht under W.H. Julius and L.S. Ornstein 1896–1940). Erasmus Publishing, Rotterdam, 1994
35. Heijmans HG. Wetenschap tussen universiteit en industrie. De natuurkunde in Utrecht onder W.H. Julius en L.S. Ornstein 1896–1940 (Science between university and industry. Physics in Utrecht under W.H. Julius and L.S. Ornstein 1896–1940). Erasmus Publishing, Rotterdam, 1994
36. Vox Studiosorum 74:152, 1938
37. nl.wikipedia.org/wiki/Utrechtsch_Studenten_Corps
38. Interview Rob Herber with J Caro, 22 November 2006
39. Interview Rob Herber with B Lansberg, 29 January 2007
40. Interview Rob Herber with M Bouman, 8 March 2007
41. Michiel Wijers HJ. Proeve van een genealogie van de wiskunde en de informatica in Nederland (Proof of a genealogy of mathematics and computer science in the Netherlands). www.win. tue.nl/~wijers/GeneaMathCSNL.pdf
42. La Bastide-van Gemert S. _Elke positieve Aktie begint met critiek_. Hans Freudenthal en de didactiek van de wiskunde (Every positive Action starts with criticism '. Hans Freudenthal and the didactics of mathematics). Verloren, Hilversum, 2006
43. Interview Rob Herber with B Streefland, 12 February 2007
44. Bockwinkel HBA. Kollege Integraalrekening. Holland, Amsterdam, 1932
45. www.bio.uu.nl
46. Interview Rob Herber with M Bouman, 8 March 2007
47. Interview Rob Herber with N Bloembergen, 8 December 2006
48. Telephone interview Rob Herber with C de Jager, 18 March 2007
49. Interview Rob Herber with B Streefland, 12 February 2007
50. Interview Rob Herber with M Bouman, 8 March 2007
51. Interview Rob Herber with B Streefland, 12 February 2007
52. Interview Rob Herber with M Bouman, 8 March 2007
53. Interview Rob Herber with N Bloembergen, 8 December 2006
54. Vox Studiosorum 74:219, 1938
55. Vox Studiosorum 75:51, 1939
56. Vox Studiosorum 75:116–117, 1939
57. Interview Rob Herber with J Caro, 22 November 2006
58. Utrechtsch Faculteitenblad 6:162, 1939
59. Interview Rob Herber with J Caro, 22 November 2006
60. Interview Rob Herber with B Lansberg, 29 January 2007
61. Utrechtsch Faculteitenblad 7:74, 1940
62. Vox Studiosorum 76:92, 1940
63. Vox Studiosorum 74:179 en 183, 1938
64. Vox Studiosorum 75:179, 1939
65. Vox Studiosorum 75:207, 1939
66. Vox Studiosorum 75:233, 1939
67. Vox Studiosorum 75:245, 1939
68. Brugman JC. Bezet en verzet. De Bilt en Bilthoven in oorlogstijd (Occupation and resistance. De Bilt and Bilthoven in wartime). JC Brugman, Bilthoven, 1993
69. Vox Studiosorum 76:11, 1940
70. Vox Studiosorum 76:19, 1940
71. Vox Studiosorum 76:33, 1940
72. Vox Studiosorum 76:47, 1940
73. Utrechtsch Faculteitenblad 7:119–120, 1940

Chapter 4
BSc and MSc Exams
in Wartime—Infected by Science

In June 1943 Bloembergen took the quantum physics exam with the professor of theoretical physics Léon Rosenfeld. It went well and Rosenfeld wanted to write out an exam form, but Bloembergen said: *"I don't need that anymore, I already have my doctoral degree."* To which Rosenfeld replied good-naturedly: *"Oh, you should have told me that before, then we wouldn't have had to do it at all."* [1]

4.1   The Netherlands at War

The war from 10–15 May 1940

At 3.55 am on 10 May German troops and aircraft crossed the border. The German attack was part of an offensive against the Netherlands, Belgium, Luxembourg, and France to subjugate these countries. The time was set by Hitler himself at 3.55; it would be almost light. The first and most important goal of the bombardment by the German Air Force was the destruction of the Dutch Air Force. Moreover, the defense of the airports Valkenburg, Ockenburg, and Ypenburg near The Hague and Waalhaven, where later German airborne landings were to take place, had to be eliminated as far as possible. The airborne troops in The Hague had to attack the government buildings.

On the evening of 10 May, blackout measures came into force. This would continue to be the case throughout the war.

The Soesterberg airport near Bilthoven was also bombed.

The German army had advanced through North Brabant, but there had been significant delays in the German plans. German troops had also advanced through Gelderland to conquer the west via Utrecht. For this purpose, the Grebbe Line between the IJsselmeer and the Rhine near Rhenen had first to be taken by the Germans. One third of the Dutch army had the task of defending this line to the last. On 11 May the German attack began.

© Springer Nature Switzerland AG 2019
R. Herber, *Nico Bloembergen*, Springer Biographies,
https://doi.org/10.1007/978-3-030-25737-8_4

Dutch soldiers defeated on the Grebbe Line withdrew in groups, one going to Utrecht.

In the municipality of De Bilt (with the villages of De Bilt and Bilthoven), the Bilt police had started arresting members of the NSB, the Dutch fascist movement. From 11 May, people stood in long queues in front of the Post Office to withdraw money from their savings accounts. In some places anti-aircraft batteries were installed to shoot at enemy aircraft [2].

On 13 May, Utrecht was full of Dutch troops who had withdrawn from the Grebbe Line, some soldiers with bleeding feet because they had walked the whole way. Everywhere in the city there were military vehicles. Furthermore, a stream of evacuated residents arrived in the city from the east [3]. The USC premises were opened to the soldiers and they were provided with bread and sausages [4].

In De Bilt, the police had transferred arrested NSB[1] members to Hoorn. The Dutch army had troops with artillery stationed in the forts and fortifications around Utrecht, such as in *Fort op de Voordorpsche Dijk* (or *Fort Voordorp*) between Maartensdijk and De Bilt, *Werk Griftenstein* between Utrecht and De Bilt, and *Fort De Bilt* (then De Bilt, now Utrecht). The inhabitants of De Bilt were evacuated to Bilthoven, further north, due to the expected advance of German troops along the Utrecht road (which ran from Zeist via De Bilt to Utrecht). The inhabitants of the villages of Achttienhoven, Westbroek, Groenekan, and Maartensdijk were evacuated to villages in West Friesland. The meadows around Utrecht were deliberately flooded and the cattle were transported to Woerden and Bodegraven.

The Dutch army, and in particular the 14th Infantry Regiment, was in the aforementioned positions around Utrecht on 14 May, while the German 322nd Regiment of the 207th Division, which had been involved in the fighting at the Battle of the Grebbeberg, was just outside De Bilt. In the Wilhelminapark in Utrecht there was an artillery battery. The Germans had suffered large losses at the Grebbe Line and wanted to avoid a frontal attack in unprotected areas. However, no shot was fired. The German army command opted for terror. Pamphlets were scattered over Utrecht with the following message:

To the Commander in Utrecht.
The Dutch Defense Line at the Grebbe has fallen! The German forces are in the majority and they are ready, from the East, South-West and South, while simultaneously deploying superior armored and air forces (Stuka bombers) to attack the city of Utrecht.
By this message, I would ask the Commander in Utrecht to give up this pointless struggle and surrender the city in order to spare the city and its inhabitants from meeting the same fate as Warsaw.[2]

---

[1] The NSB was a fascist party in the Netherlands which had strong connections with the German Nazi party. In early 1940, they attempted a coup. nltimes.nl/2017/04/27/dutch-nazi-collaborators-attempted-netherlands-coup-german-invasion.

[2] On 24 September 1939, Warsaw was bombed on three consecutive days. This led to many victims and great devastation. The press outside Germany and Italy expressed its indignation and horror.

I would ask you to signal your unconditional surrender (frequency 1102 kHz, call sign: hollow).
Otherwise, to my regret, I will be forced to regard the city of Utrecht as a fortress and to begin the attack using all the military means at my disposal.
The responsibility for all consequences lies solely with you.
The German Supreme Commander. 14 May 1940

The language was warped, but the threat was clear. A German officer with a white flag came with an ultimatum that read the same as in the pamphlets: the city must surrender. The commander of the city of Utrecht, Van Voorst tot Voorst, sent the man back with the message that the ultimatum was rejected.

Shortly afterwards, the bombardment of Rotterdam began. About 1000 people were killed and around 85,000 lost their homes. Only a handful of buildings survived [5]. The Dutch army capitulated on 15 May at four o'clock in the morning.

The evacuated residents of the village of De Bilt were able to return home on the evening of 14 May [6].

Occupation and study
On 15 May the German army entered Utrecht. The Domplein was soon full of military vehicles, and people wondered in great fear how the troops would behave. However, their discipline left nothing to be desired and everything seemed very easy. Soldiers who had overrun the synagogue to demand money were sent to the front in France as punishment [7].

On 16 May, Bloembergen's father Auke left for Rotterdam on a motorcycle to see how the fertilizer factories in Kralingse Veer and Pernis were getting by. Everything looked to be in order. Evert went by car to the house in Loosdrecht. The house was still there, but there was a white cross painted on it, as a sign that it was ready for demolition. There was a driving ban, and a year later Auke Bloembergen's car was claimed by the occupying forces [8].

Utrecht changed during the occupation, becoming another city in another country. In May 1940 the local military commander, the *Ortskommandatur*, was housed in rooms at the town hall. Later he moved to the USC building at Janskerkhof 14, and subsequently to the headquarters of the 88th Army Corps. There were also offices for the *Feldgendarmerie*, *Luftwaffe*, and *Kriegsmarine*, and many offices associated with German civil administration, the various police forces such as the *Ordnungspolizei* (usually abbreviated to *Orpo* and referred to as the *Grüne Polizei*), part of the *Gestapo*, the regular police, and the *Sicherheitspolizei* (*SiPo* or security police, part of the *SS*). Then there was also the Dutch police, the Dutch SS, the *NSB*, and the *WA*. And there were uniformed visits from Germany, such as the party, the NSDAP, the *Waffen SS* (the army unit of the *SS*), German officials in uniform, and so on. The number of different uniforms worn by the Nazis was almost endless. There were signs in all directions indicating the way to their offices, and army and civil service uniforms everywhere. In Bilthoven and De Bilt there were all kinds of offices for the *Lufwaffe* and the *Ortskommandatur*. It was clear that the uniforms of the army units belonged to an occupying force, but all those uniforms of the civil administration gave rise to great unrest. The Netherlands was already being governed, wasn't it? What would all these administrative agencies do with the Netherlands? Some, including the highest German official in the Netherlands, Seyss Inquart, had a skull as an emblem on his cap. That certainly didn't bode well!

In the war years, university business came to a standstill. The non-teaching staff looked after the shop, so to speak [9]. War and occupation had immediate consequences for scientific work at the university. Scientific journals from the United States and the United Kingdom were no longer received and conference visits were initially excluded. Scientists from many countries, but especially the countries that were important for research - the United States and the United Kingdom—could no longer visit. Later it was possible to attend congresses in the countries occupied by Germany and Italy, but with many restrictions. In an era when scientific exchange consisted of publications in journals, conference visits, personal contacts, and correspondence, these restrictions were fatal to the exercise of science.

4.2   Teachers Under the Occupation

Léon Rosenfeld (1904–1974)

After the departure of Uhlenbeck in 1938, the vacancy for a professor of theoretical physics was filled only in 1940. Rosenfeld obtained his doctorate in 1926 at the University of Liège. After that he became postdoc in Paris, Göttingen, and Zurich. During these periods he maintained contacts with Heisenberg and Pauli. From 1930 to 1940 he was a faculty member of the University of Liège and a close associate of Bohr, whom he visited regularly in Copenhagen. Rosenfeld was friends with Bohr [10]. On 3 May 1940, hence seven days before the German invasion, he was appointed professor of theoretical physics. Because of the occupation, his appointment could not be made official until much later (Fig. 4.1).

Prior to his arrival in Utrecht as Visiting Professor, Rosenfeld had visited the Institute for Advanced Study in Princeton for a semester. He was an expert in quantum dynamics. Just before Bohr left for Princeton, Otto Robert Frisch informed Bohr about the recent discovery of nuclear fission by Otto Hahn and Fritz Strassman in Germany. Frisch's niece, Lise Meitner, had produced the theory behind it. Bohr told Rosenfeld about this, but forgot to say that it had to remain a secret until it had been published in *Nature*. So Rosenfeld talked openly about nuclear fission at a meeting in Princeton with I.I. Rabi and Willis Lamb, among others. They in turn talked about this with their colleagues in Columbia University, like Enrico Fermi. The result was that the German research could then be replicated at Columbia [11].

Rosenfeld took office on 15 September 1940. His oration was on 16 February 1941. Rosenfeld was not an inspiring man. According to Maarten Bouman, he was a good theoretical physicist with a great international reputation. His personal relationships with the students were not very intense [12].

Fig. 4.1 Léon Rosenfeld. *Source* University Museum Utrecht 0285-3884, Utrecht University

Leonard Salomon Ornstein

At the beginning of the war Ornstein received an invitation from a friend in the United States, the astronomer Peter van der Kamp, to come to the US with his family. However, he decided to refuse, because he did not want to abandon the laboratory. In September 1940, he was suspended [13].

Johannes Marius Wilhelm (Pim) Milatz (1910–2000) [14]

Milatz studied physics at the University of Utrecht. In 1937 he obtained his Ph.D. with Ornstein with a thesis on neon: *The electronic excitation function of the metastable S_5 level of neon* [15]. On 1 March 1940 he was appointed lecturer in physics. Milatz's public lecture was about Brownian motion. Milatz was trained in Utrecht and received his Ph.D. there as a pupil of Ornstein (Fig. 4.2).

Milatz was initially a lecturer and after Ornstein's forced departure, became his successor as professor of experimental physics and director. Milatz did research on fluctuation phenomena in noise and Brownian motion.

Gerrit Arnoldus Wijnand Rutgers

An assistant to Ornstein who became professor of Applied Physics in Utrecht in 1963.

Pieter Jacobus van Heerden (1915–?)

Assistant to Milatz. He obtained his Ph.D. in 1945 with the thesis *The crystal counter: a new instrument in nuclear physics* [16]. He would go on to the United States, where he became Assistant Professor at Harvard University and then researcher at General Electric and Polaroid, Edwin Land's company which developed the instant camera. Van Heerden worked on three-dimensional holography. In the United States he would still maintain regular contact with Bloembergen (Fig. 4.3).

Fig. 4.2   Pim Milatz. *Source* University Museum Utrecht 0285-4191, Utrecht University

Fig. 4.3   Physics student association Christiaan Huygens. Nico Bloembergen is marked with an arrow. *Source* Nico Bloembergen

4.3   16 May—December 1940—From War to Occupation

The first anti-Jewish measures—dismissal of Jewish teachers

On 16 May, the Queen's Commissioner appeared at the Senate meeting. Bosch van Rosenthal declared that the army had capitulated, but that the queen and government in London remained the legitimate government. The war between the Netherlands and Germany continued, but from 20 May, normal education and examinations were resumed [17].

The following announcement would appear in the Utrecht Faculty Bulletin of 24 May:

Official Communications of the Editorial Board
Now that all the work at our University has been resumed with full force, the Utrecht Faculty Bulletin will also ensure that the student interests of all subscribers are served as well as possible.
By its very nature, it will only be possible for the time being to pass on messages and announcements from the faculties.
May the Faculty magazine in its place cooperate to ensure that study at our University also proceeds in a dignified national spirit full of zeal and dedication! [18]

So, nothing was written about war and capitulation!
In Vox of 1 June Evert Bloembergen wrote:

THE VOX
Other times, other morals. But also another language. See only the newspapers from before and after mid-May. […] Vox is a student magazine and does not interfere in politics [19].

Apparently, Evert wanted to go back to the time before 1938, but he was overtaken by events.

His father Auke Bloembergen was still director of the fertilizer company. *A.S.F./V.C.F.* The production of fertilizer came to a standstill in the course of 1940 due to a lack of raw materials. His co-director, J. D. Waller, was arrested for assisting allied pilots. The head office at Maliebaan was closed and the company moved to the *Meijenhagen* estate on the Groenekanseweg in De Bilt.

In July the premises of the USC society PHRM at the Janskerkhof were claimed by the Germans, and the society moved to the Tivoli Building in the Kruisstraat (demolished in 1955).

In 1995 the journalist Sander van Walsum wrote an interesting book about the activities of the Utrecht University Council during the war: *Even if wounded.* Many of the following quotations can be found there.

On 24 August *General Commissioner* Schmidt announced the beginning of the Jewish action [20]. In August all civil servants had to sign a declaration of abstention, which was a declaration that they abstained from any activity against the German occupier and the German army.

All university officials signed that statement.

Rosenfeld took office on 15 September.

On 16 September, the new rector magnificus, H. R. Kruyt (1882–1959), was faced with a new measure: Jews were no longer eligible for the position of professor. One would expect the university administration, knowing how the Jews in Germany were treated and knowing that the same regime was now taking discriminatory measures against the same population in the Netherlands, to protest and possibly close the university or resign. However, only one protest was received, from history professor Pieter Geijl (1887–1966). In October, a new document was submitted to the attention of the professors: they had to sign a declaration of descent. The church attended by parents and grandparents had to be filled in. Kruyt and with him all the professors had no objection. Only Professor G. J. W. Genité returned the forms with the following note: *the undersigned has reluctantly fulfilled this anti-Dutch assignment.*

Before the war, Utrecht already had an organisation for the central food supply for students, the *Mensa Academica*. In the autumn of 1940, for the first time, there was some food shortages and student kitchens were set up in various locations. On 20 September, the Utrecht Faculty magazine announced this as follows:

Food supply for students during the coming winter months.
It will come as no surprise to anyone that the food supply for students in the coming winter is a cause for concern. Indeed, due to the many distribution measures, it will be difficult in the future to prepare meals, which by expert care and by the right choice of ingredients, are as nutritious as possible [...] they have joined forces to set up a student kitchen, which can provide inexpensive and nutritious hot meals [21].

On 13 November, Kruyt heard rumours that all professors with two 'voljoodsche' (fully jewish) grandparents were to be immediately suspended. On 22 November, Kruyt heard from Victor Jacob Koningsberger (1895–1966, later father-in-law of Evert Bloembergen), a professor of plant physiology who had good contacts with the students, that there was great commotion among the students about the possible dismissal of Jewish professors. On the same day Kruyt heard that three Utrecht professors and four (principal) assistants were to be dismissed '*sofortig*' (immediately). In physics it concerned the professors Wolff and Ornstein and among others the assistant Abraham Pais. On 23 November, all Jewish officials received a letter from the Ministry of Education that they had been relieved of their duties.

Koningsberger gave a speech in the auditorium:

Speech pronounced on Monday, 25 Nov. 1940, in response to
the dismissal of the Jewish professors
My conscience commands me here to remember with deep sorrow and disappointment the dismissal from office of a number of Dutch colleagues, solely for reasons of origin or faith. My great sympathy goes out to them and I feel that we all share in the wrong that they have been done.

After all, since 1579, the year in which the Union of Utrecht[3] *was pronounced in the present Grand Auditorium of our University, it has always been the highest Dutch ideal that nobody should be persecuted for his race or faith. Therefore, in my view, this measure seems to disregard the very nature of the Dutch people, and it should be taken as an insult to the Dutch universities, to Dutch science, and therefore to the Dutch people themselves.*

I know that the vast majority of those before me feel this as much as I do. However, I am not telling you this in order to provoke you into any act, which could be considered illegal or disloyal towards the occupying government. On the contrary, persevere in your admirable self-control and never forget how creditable it would be for all mankind if a people in all circumstances could always behave according to the unwritten and undescribed laws of humanity and decency.

w.g. V.-J. Koningsberger [22].

Thundering applause followed from the auditorium, which had heard the whole speech standing up [23]. Koningsberger left his text on the lectern, thinking that if his words were passed on, they would be correct. However, the paper was burned by a well-meaning spectator who wanted to protect Koningsberger. This is how it came about, according to Van Walsum, that some Utrechters spoke out, while the board was silent.

The paper *Vox Studiosorum* published a statement about the suspension of Jewish professors, signed by the editors [24]:

STUDENT WORLD

A few days ago, for certain reasons, some of our professors were dismissed.

We very much regret this because they, as gifted people and excellent teachers, were of great value to the scientific life at our University.

We may suspect the cause of this resignation to be a political one. Therefore, as Utrecht University Students, we would like to state emphatically and for the sake of completeness that it is customary at our university to keep political and scientific education completely separate.

In the same way, the recently dismissed professors have preceded us in science and only in science.

And that is why we are surprised by this measure. Because it needs to be made clear once again: scientifically, the dismissal is a great loss for our University, and politically, the dismissals seem to be pointless.

If, as is the case today, a dismissal is made, it is both scientifically and politically incomprehensible to us.

---

[3]The Union of Utrecht (*Unie van Utrecht*) was a treaty signed in Utrecht on 23 January 1579, unifying the northern provinces of the Netherlands, until then under the control of Hapsburg Spain. The Union of Utrecht is regarded as the foundation of the Republic of the Seven United Provinces, which was not recognized by the Spanish Empire until the Twelve Year Truce in 1609. The treaty was signed on 23 January by Holland, Zeeland, Utrecht (but not all of Utrecht) and the province (but not the city) of Groningen. The treaty was a reaction of the Protestant provinces to the 1579 Union of Arras (*Unie van Atrecht*), in which the southern provinces decleared their support for Roman Catholic Spain. en.wikipedia.org.

It is well known that the measure is also contrary to our deeply rooted Dutch traditions.
E. BLOEMBERGEN.
J. H. O. INSINGER.
M. C. VAN HARDENBROEK VAN AMMERSTOL
M. W. L. DE GEER.
J. GREIDANUS.
A. M. BERKELBACH VAN DER SPRENKEL.
G. LUBBERHUIZEN

On 4 December 1940, Milatz, a member of the board with the personal title of lecturer, was appointed acting director of the physics laboratory.

Maarten Bouman comments: *In the early days, until we did our doctoral studies, apart from the fact that those Jewish professors were dismissed, Ph.D. students continued relatively normally. There was electricity, so there were still possibilities* [25].

Bloembergen followed organic chemistry lectures by Kögl and also took exams in that subject. He had positive memories of the analytical chemistry lab, which was in the old veterinary school near the Museum Bridge. Practicals were run by Strengers.

When Wolff was suspended, Bloembergen went to mathematics lectures by Nieland. Maarten Bouman thought Nieland had a school-like way of teaching [26]. Ornstein's lectures were taken over by Milatz after his suspension (Fig. 4.4).

Fig. 4.4 Henk Lagendijk.
Source Almanac for
students, 1948

Bloembergen and Jan Kluyver become a 'research couple'. Kluyver (1919–1996) was the son of Albert Jan Kluyver. Together with Ornstein, he was in charge of the Biophysical Laboratory of the Rockefeller Center. Kluyver had been studying in Lausanne for a year and was therefore further ahead than Bloembergen. Jan no longer had to do the everyday practical tests. He was assigned as a 'slave' to a Ph.D. student, Gerrit Rutgers. Kluyver and Bloembergen were thus paired at the lab and had to decide what the latter should do. He could compete with Kluyver and there were some tests he no longer had to do [27]. Kluyver would later become professor of nuclear physics in Amsterdam.

Ornstein met with Rutgers once a fortnight or once a month, when he could talk to Bloembergen and discussed the research with Rutgers. Perhaps even more importantly, Bloembergen and Kluyver were allowed to do research together and thus moved from the reproductive phase into the productive phase of their studies. Bloembergen was completely taken with science. His brother Evert also believed that Nico only became a scientist at university and not before that [28]. The research involved measuring the degree to which alpha-radiation (helium nuclei) from the radioactive element polonium could be stopped by various solids, such as aluminium, nickel, silver, gold, and mica. They discussed their research during the so-called children's colloquium, which was the colloquium before the real one held by the Ph.D. students. Bloembergen found it very interesting. This would be Bloembergen's first publication, even before his doctoral exam [29]! Bloembergen and Kluyver get along very well with Rutgers.

4.4 1941—The Occupier Shows His Grim Character

Numerus clausus and exclusion of Jewish students, BSc examinations, and the Declaration of Loyalty
Within a few months the situation at the universities escalated due to anti-Jewish measures that were causing increasing resistance in the university world.

New anti-Jewish measures were announced on 17 February. There was a numerus clausus for new Jewish students of three percent. The Jewish Council had to deal with implementing this. The Germans actually wanted to refuse any Jewish student, but were afraid of a possible reaction from the students and hoped to prevent this by the 'compromise' of three percent. In Utrecht professors Boeke and Koningsberger protested. The latter gave the following speech:

"After what we experienced together in the last week of November, we have nothing more to say to each other now against the new course measures that have been announced against students of Jewish blood. But precisely because we understood each other well then, I consider it my duty once again – although it may be more often than necessary for many - to call upon your self-discipline. Please persevere in the utmost self-control. Refrain from every demonstration. We shall understand

each other without the need to speak. Our homeland asks this of you, just as it still
asks all of us to remain at our posts undaunted. [...]
Secondly, in the absence of other motives, it will have to be made clear to everyone
that a racial theory cannot be a guideline in a colonial empire, ninety percent of
whose subjects belong to a dozen different races, not to mention the thousands of
Japanese, Chinese, British Indians, Arabs, and Armenians, who have been living in
the Dutch East Indies for centuries. Carry this thought with you wherever you may
go.
And finally, in the Netherlands, a university stands for something quite different to
a cinema or a pub, where one can stay away when access to friends is denied. You
cannot yet fully assess the extent to which the changing circumstances will allow
our University go on, so you must rely on the answer given by your teachers. I,
personally, am convinced that the importance of our homeland still demands of us
that we remain undaunted at our posts." [30]

The students protested vehemently through their magazines.
Vox Studiosorum appeared on 21 February showing the coat of arms and the university
motto *Sol Iustitiae Illustra Nos* on the cover page, with a black border around it as a
sign of mourning [31]. The magazine also published comments by the Rector of the
USC:
(...) Circumstances have now arisen in which intervention in the Dutch student
society is so deep that the USC will have to reflect on what has happened.
Although the NSF has unanimously decided not to proceed to a general strike, the
special circumstances in Delft and Leiden did in fact lead to a spontaneous strike.
This has not failed to set the mood in Utrecht.
However, the Senatus Veranorum has nevertheless taken the position to refrain from
a strike [...]
Nowadays, the opportunity for our Jewish compatriots to study at schools and uni-
versities is limited [32].

The editors (Evert Bloembergen was editor in chief) thought that they should
not remain silent, after Evert had asked them the penetrating question: *"What will*
our descendants say fifty years later, if we remain silent now?" [33]. The magazine
was immediately banned and only reappeared in July 1945. Evert Bloembergen
was called upon to appear before the Gestapo on the Maliebaan. The next day he
had to come back and after a short interrogation he was imprisoned in the House of
Detention in the Gansstraat. Another editor was also imprisoned. In the Bloembergen
household, there was great dejection; nobody knew what would happen. Rie went to
the Maliebaan, where a young and impudent officer blocked her way. On 22 April,
Evert's birthday, his father Auke walked along a path behind the House of Detention
and whistled the family whistle. Evert did indeed hear that. After two months, he
was suddenly released. This early release probably had much to do with the fact that
Evert's future parents-in-law, the Dagevos family, knew the Swiss consul. Through
his intercession, Evert was released earlier. After this incident, Evert felt he had had
enough of student life and left for a job outside Utrecht [34].

From that moment on, Nico Bloembergen became very careful. The magazines *Vox Veritas* (of the association Veritas) and *Vivas Voco* (of the association Unitas) wrote about the ban on the appearance of *Vox Studiosorum* and were then also banned.

On 17 March 1941, Bloembergen took his BSc (*kandidaat*) exam cum laude under the letter D from the academic statute, which meant physics and mathematics with chemistry. After the BSc exam, of the 20–25 students who started studying physics, there were perhaps a quarter left [35]. Bloembergen had survived the transition from school grammar school to the freedom of university exceptionally well and the conditions induced by the ongoing war do not seem to have had any influence on his academic achievements.

Bloembergen continued with SCHC hockey in Bilthoven. He even entered the first team, which played in the big league.

Before the BSc exam, no lectures on quantum mechanics and relativity theory were given in Bloembergen's day. These were emerging fields of expertise and were of great interest to him and he read two books about this 'new physics'. The first was a popular booklet by Albert Einstein and Leopold Infeld called *The Evolution of Physics: The Growth of Ideas from Early Concepts to Relativity and Quanta* (*Physik as Abenteuer der Erkenntnis*) [36]. According to Bloembergen, it was about: "*two men who stand in a falling elevator and feel no gravity*" [37]. The second book, about quantum mechanics, was by Max Born, Nobel prizewinner in 1954. The latter was in fact a series of ten public lectures by Born[4] and it set Bloembergen forever on the trail of atomic physics and quantum mechanics.

After his suspension, Ornstein still came to the lab [38]. Then, in November 1940, he was dismissed. He was so upset about this that he died during the night of 19 to 20 May 1941. His funeral took place on Friday, 23 May, at the Jewish cemetery in Utrecht. Just like the session of mourning by the Senate on the previous Wednesday, it was attended en masse.

Bloembergen was also at the funeral.

After his death, Ornstein's wife and children received protection from former Ph.D. students in Eindhoven [39].

Marcel Minnaert, Professor of Astronomy, wrote a beautiful and moving memoriam in the *Utrechtsch Faculteitenblad*:

IN MEMORIAM Professor Dr. L. S. ORNSTEIN 1880–1941
By Mr. Minnaert
All his work was done in the laboratory, which he had made into a great tool for research and which was his passion and his life. Here he had gathered a large number of staff around him, whom he managed to inspire with his enthusiasm and investigative drive. This laboratory is the most appropriate monument that could have been erected in his honour. We also think of all the work Ornstein has done for the Faculty of Mathematics and Physics and for the university as a whole. We think of his work in the societies of which he was chairman or board member. We think of him as a human being, with his qualities, with his faults, and with his all-pleasing

---

[4]Max Born was director of the Institute for Theoretical Physics in Göttingen in 1933 but he had to leave because he was a Jew. He then accepted a position as a professor in Cambridge.

drive to move forward. We cannot give a proper overview of the great wealth of this life, so filled to the brim, within such a short obituary. Sadness fills us when we recall all he had to suffer in the last few months. But while he was lying on his deathbed, the machines went on working and the electric current pulsated through the thousands of wires in his laboratory. In his deeds he lives. His work cannot perish. We will cherish his memory forever [40].

On 21 June, more than a month after his death, the Senate of the Board of Trustees learned "*that Ornstein cannot become an honorary member.*" [41] A request had been made by the Senate of the Secretary-General for Education in February.

On 4 August, the Faculty wrote in a letter to the Board of Trustees that Milatz' appointment as professor had still not taken place. It was proposed that Milatz be appointed again as temporary manager as of 1 December, and that he be appointed either as of 1 March 1940 (the date of Ornstein's resignation) or—if Milatz objected— as of 20 May 1941 (the day after Ornstein's death).

On 13 November it was reported that from 19 November a Mensa soup kitchen would be installed on the premises of the University Building on Domplein [42].

4.5 1942—Difficulties in the Energy Supply

Lecture times were adapted to get around problems in the gas supply.

On 20 February 1942 Rosenfeld gave his inaugural speech: *Development of the Causality Idea* [43].

In 1942 Bloembergen became friends with Mienke Kesseler. She was a member of the Utrecht Female Students Association along with Bloembergen's sister Diet. After a short time Mienke would break up with Nico. She would later marry the physicist Wim den Boer.

At the beginning of May, Minnaert was arrested and taken to the monastery in Sint Michielsgestel as a prisoner. On July 13th Koningsberger was taken prisoner in Haaren [44].

Bloembergen became a board member of the Mensa, whose administration had been 'simplified' by a machine that Milatz had devised. The device had switches for each item that was for sale. If a certain item was sold, the switch had to be turned over. There was one large switch which could be used to reset everything to zero. Bloembergen spend many hours operating the device [45].

In 1942–1943 the Physical Laboratory also had a field outside Utrecht where vegetables could be grown, including beets and potatoes. Bloembergen went there gardening one afternoon a week. He went more often in autumn when the crops were harvested and more often in spring when when the crops were sown and planted. There is a note, possibly written by the head of the instrument making company, Jezeer, which says that, in the company allotment, on 3 November, Bloembergen was in service in the morning and Bouman in the afternoon.

On 25 September there was a message from the rector, urgently recommending every Utrecht student to register quickly.

On 23 October the *Utrechtsch Faculteitenblad* published the following announcement about Milatz:

The lecturer Dr. JWM Milatz has been appointed full professor in charge of teaching experimental physics at the University of Utrecht and has also been appointed director of the Physical Laboratory [46].

Milatz would not give an oration. This was not considered appropriate in connection with the death of Ornstein [47].

On 16 October 1942, a USC clubmate of Bloembergen, Hendrik (Henk) Langendijk, died in Amsterdam at the age of 22. Langendijk was wearing a yellow Star of David and had been trying to escape from the Gestapo, who were chasing him, when he fell under a tram, whether intentionally or not. Langendijk was a medical student and lived with his parents in The Hague. Bloembergen had been there sometimes to visit and he was very upset about the death of his friend.

In the post-war student almanac, he was commemorated as follows:

IN MEMORIAM
H. LANGENDIJK, Summer 1942
Already at the beginning of the occupation, when many of us had little awareness of the situation our country would find itself in, when most of us had not yet been faced with difficult decisions regarding our attitude towards the occupying forces, he was ready to fight. While life in the Netherlands continued relatively normally and the student world was still virtually untouched, he already sensed that great personal pressure which the occupier would later impose on us all. With many of his peers he tried to resist the measures taken against them. He was the victim of this unequal struggle. All those who knew him saw how much he suffered from all this. Nevertheless, he remained outwardly the same, warm and even cheerful. Thus are our last memories of him as someone who, in spite of everything that was done to him, was always a good and faithful friend, who refused to sit back, but by his attitude, at that time, and perhaps even more so in the years to come, will serve as an incentive to support the cause for which we should all stand up. We are grateful to you, Henk, for giving us this.
Rest in peace.
A. v. d. B.

On 30 October there was more about the heating:

Heating problem.
Thanks to the cooperation of Prof. Kruyt and Prof. Milatz, the Philosophy Faculty has a heated room for evening study. Mondays and Fridays at the Physical Laboratory for 80 people and Tuesdays and Thursdays at the Van 't Hoff Laboratory for 40 people. It is the intention that Wednesday should also be involved, to provide 4 evenings for study opportunity.
A contribution to the costs of f 2,50 florins will pay an employee to open and close the room, and black out the windows, among other things [48].

At the beginning of December, Secretary-General Van Dam informed the Rectores Magnifici that six thousand forced labourers had to be found in the universities and colleges of higher education. In connection with this, he asked for the names of all students enrolled. In the night of 11–12 December, a fire was started in the student administration room of the Academy Building. When Rector Van Vuuren wanted to start immediately with the reconstruction, there was patient cooperation. The perpetrators of the attack were members of the Children's Committee, which included Geert Lubberhuizen and later publisher of the illegal *Bezige Bij* (Busy Bee) [49]. On 13 December, 341 students were still enrolled in Utrecht. For the most part it was NSB members (collaborators) who wanted to force the professors to continue their education [50].

4.6   1943—MSc Examination—The University Closes

Police came to the Bloembergens' door to take Nico with them for the *Arbeitseinsetz*, the forced labor law in Germany. However, Nico was warned and was not at home. In February and March, he went into hiding with an aunt in the provinces [51]. The reason for the raid was betrayal by an NSB official in the SCHC hockey club, who had the addresses of the hockey playing students via a tip from a member of Bloembergen's team. The police first asked about Evert, whereupon the parents said that he was married and no longer lived at home. Then they ask about Nico. The parents said: "*Well, he is a student. We do not know where he is.*" Herbert and Auke were still under sixteen and were not in danger [52].

On 13 March, Secretary-General Van Dam announced the following:

The undersigned […] hereby solemnly declares that he shall observe the laws, ordinances, and other decrees applicable in the occupied Dutch territory to the best of his knowledge and belief, and shall refrain from any action directed against the German Empire, the German Wehrmacht, or the Dutch authorities, as well as any action which, in view of the prevailing circumstances, constitutes a violation of public order in the institutions of higher education.

This was the so-called declaration of loyalty. Students who had signed on 13 April could continue their studies. Non-signatories and graduates were eligible for the *Arbeitseinsatz* in Germany. At the beginning of April it appeared that only 455 students (12.6%) had signed in Utrecht. A real pamphlet war had raged around the declaration of loyalty, and until the end, 'pick-up teams' had been working to convince waverers. In Utrecht they had populated the Domplein in March and April, if necessary on the stairs of the Academy Building, to persuade students who wanted to enter with the apparent intention to sign the statement that they should do otherwise. This failure to sign was an act of great selflessness, because it was completely unclear what the future would bring [53].

Immediately after his BSc examination, Bloembergen started working for the Ph.D. student Piet van Heerden. Bloembergen was developing a new measuring device in the tradition of the Physical Laboratory. Initially, under Moll, the development of electromechanical measuring instruments of great complexity reached the high point of instrumentation. As we have seen, nuclear physics research made different demands. A second point was the recording of spectral lines, which was still done with a photographic plate.

> The use of electron tubes brought with it new possibilities. In 1904, John Ambrose Fleming (1849-1945) invented the electron tube, a diode comprising two electrodes, an anode (a positive plate) and a cathode (a negative filament), that could be used to drive an alternating current. In 1907 Lee de Forest (1873-1961) observed that, if a third electrode was placed between the anode and cathode, the control grid with a negative voltage in relation to the cathode could act as a voltage amplifier. Small voltages obtained in experiments could thus be amplified and measured. This tube was called a triode. In the 1920s the development of electron tubes accelerated to included devices with more and more grids, such as tetrodes, pentodes, and hexodes, double triodes, and so on. These tubes were gradually used in laboratory equipment to amplify signals.

> Another development was the photocell. This electron tube consists of an anode and a cold cathode coated with a metal with low extraction value, such as caesium. The photocell uses the photoelectric effect, discovered by Hertz and explained by Einstein, which can be exploited to convert light intensity into a current that is directly proportional to that light intensity. A further development was the invention in 1934 of the photomultiplier by RCA Corporation. This special photocell combined the photoelectric effect and secondary emission discovered by Westinghouse. With the photomultiplier, amplification factors of 10^6 could be achieved. These developments made it possible to measure much more sensitively. Thus the photomultiplier increasingly replaced the photographic plate.

With the amplification of weak signals another problem arose: that of noise. Noise is caused by the Brownian motions of components in the measurement series.

The measuring device that Bloembergen had to work on was a photoelectric amplifier. The photocell, the detector of light signals, had to be cooled to suppress photothermal noise. The light signal was modulated by an optical chopper, a round disc that had regular recesses and rotated with a fixed frequency. This interrupted the light beam in a regular way. The current coming from the photocell was then detected and amplified by means of phase sensitive detection, synchronous with the interrupted light signals, and read out on a galvanometer. This galvanometer was the only classical electromechanical instrument still present in the setup. The phase sensitive detection resulted in a significant improvement in the signal-to-noise ratio. The research was only published in 1945, due to the circumstances of the war, and proved to be extremely important for Bloembergen's career (see Chap. 6).

Bloembergen attended Rosenfeld's lectures on physics. Rosenfeld did a whole series of lectures for doctoral students, starting with classical mechanics (the mechanics of Newton) and statistical mechanics, and in the second year, electromagnetism. Bloembergen found statistical mechanics very interesting. The third year, Rosenfeld lectured on quantum mechanics. He started with a repetition of classical mechanics and the Poisson bracket. The latter is binary operation defined in Hamiltonian mechanics. In 1833 William Rowan Hamilton reformulated classical mechanics in such a way that it could be used to formulate classical wave equations. But these wave equations could also be used in quantum mechanics. Rosenfeld was to begin his lectures on quantum mechanics at the end of February and he did so, but the closure of the university put an end to that. Because of this, Bloembergen did not have to take an exam in quantum mechanics. The mathematics given in Utrecht were formal and rarely applied. Bloembergen taught himself applied mathematics, mechanics, and electromagnetism from German textbooks.

The physics exams that Bloembergen took for Rosenfeld courses were:

- Theoretical physics.
- Statistical mechanics. He made a mistake during the exam, but did not notice it. Later he discovered the mistake and went to Rosenfeld the next day to tell him. Rosenfeld appreciated that Bloembergen had pointed out his mistake.
- Electromagnetism.
- Quantum mechanics (waived).

According to Bouman, Rosenfeld was a good physicist with a sound international reputation, but his personal relationships with the students were not very intense [54]. Bloembergen liked Rosenfeld very much in statistical mechanics; he gave interesting lectures [55].

At the time, Barend Streefland thought Milatz was the archetype of an excellent professor. He gave extraordinarily good lectures, treating his subjects with great care. But with hindsight he found him "*a bit too much on the classic tour. Virtually no quantum mechanics.*" [56]

Bloembergen followed a special course by Milatz and wrote him a lecture on Brownian motion. The lecture was mimeographed (copied from a stencil) and handed out. He learned a lot from that. Milatz was one of the few professors appointed by the occupying forces. At that time, Milatz was a brilliant organizer and communicator, but he gave students no insight into new developments in physics, such as quantum mechanics. Bloembergen considered Milatz as his teacher in Utrecht, but was of the opinion that Milatz was no longer up to it after the war.

In early April, Bloembergen took his last exam. The mathematics exam for Nieland was taken at the teacher's home in Putten, because the university was considered too dangerous due to the possibility of air raids. The exam was about Fourier analysis and differential equations. Bloembergen did everything in his power to take the MSc exam before the deadline for signing the loyalty declaration. He succeeded on 9 April 1943 and took the exam together with Bouman. Bloembergen got the appreciation "cum laude". Two weeks later the university closed. So, he was no longer a student on the deadline of 13 April!

> The students were summoned on 3 May. Three thousand Dutch students left for Germany via Ommen [57]. About 60% remained in hiding.

Bloembergen promised Rosenfeld to take the quantum mechanics exam anyway and in June the time had come. Rosenfeld and Bloembergen talked together for an hour and Rosenfeld had only one question. Afterwards Rosenfeld felt that Bloembergen had shown sufficient basic knowledge of the subject, but—as already mentioned at the beginning of the chapter—actually taking the exam was not necessary [58].

References

1. Bloembergen N. Encounters in Magnetic Resonances. World Scientific, Singapore (1996)
2. Brugman JC. Bezet en verzet. De Bilt en Bilthoven in oorlogstijd (Occupation and resistance. De Bilt and Bilthoven in wartime). JC Brugman, Bilthoven, 1993
3. Leeuw AJ van der. Utrecht in de oorlogsjaren (Utrecht in wartime). In: Miert J van (ed). Een gewone stad in een bijzondere tijd. Utrecht 1940–1945 (An ordinary city in a special time. Utrecht 1940-1945). Het Spectrum, Utrecht, 1995
4. Walsum JC. Ook al voelt men zich gewond. De Utrechtse universiteit tijdens de Duitse bezetting 1940–1945 (Even though injured. Utrecht University during the German occupation 1940–1945). Universiteit Utrecht, Utrecht, 1995
5. www.warhistoryonline.com/world-war-ii/8-things-need-know-1940-rotterdam-terror-bombing-m.html.
6. Brugman JC. Vergeten bunkers. Forten in en om De Bilt en het Duitse 88^e legerkorps in De Bilt (Forgotten bunkers. Forts in and around De Bilt and the German 88th army corps in De Bilt). JC Brugman, Bilthoven, 2006
7. Leeuw AJ van der. Utrecht in de oorlogsjaren (Utrecht in wartime). In: Miert J van (ed). Een gewone stad in een bijzondere tijd. Utrecht 1940–1945 (An ordinary city in a special time. Utrecht 1940–1945). Het Spectrum, Utrecht, 1995
8. Bloembergen AR. Het gezin van Rie en Auke Bloembergen 1917-1956. Eigen Beheer, Wassenaar, 2003
9. Walsum JC. Ook al voelt men zich gewond. De Utrechtse universiteit tijdens de Duitse bezetting 1940–1945 (Even though injured. Utrecht University during the German occupation 1940–1945). Universiteit Utrecht, Utrecht, 1995
10. fr.wikipedia.org/wiki/Léon_Rosenfeld
11. Mattson J, Simon M. Pioneers of NMR and Magnetic Resonance in Medicine. The Story of MRI. Dean Books, Jericho, 1996
12. Interview Rob Herber with M Bouman, 8 March 2007
13. Walsum JC. Ook al voelt men zich gewond. De Utrechtse universiteit tijdens de Duitse bezetting 1940–1945 (Even though injured. Utrecht University during the German occupation 1940–1945). Universiteit Utrecht, Utrecht, 1995
14. Ritsma RJ. Afscheid van professor Milatz. Sol Iustitae 11:3, 1956
15. dspace.library.uu.nl/handle/1874/323146
16. Heerden PJ van. The Crystal Counter. A New Instrument in Nuclear Physics. Proefschrift, Rijksuniversiteit Utrecht, Utrecht, 1945
17. Vellinga SYA. De uitdaging van crisis en bezetting, de jaren 1936–1946 (The challenge of crisis and occupation, the years 1936–1946). In: Dunk HW von der, Heere WP, Reinink AW (red). Tussen ivoren toren en grootbedrijf. De Utrechtse Universiteit, 1936–1986 (Between ivory tower and large company. Utrecht University 1936–1986). Maarssen, Schwartz, 1986
18. Utrechtsch Faculteitenblad 7:135, 1940
19. Vox Studiosorum 76:111, 1940
20. Leeuw AJ van der. Utrecht in de oorlogsjaren (Utrecht in wartime). In: Miert J van (ed). Een gewone stad in een bijzondere tijd. Utrecht 1940–1945 (An ordinary city in a special time. Utrecht 1940-1945). Het Spectrum, Utrecht, 1995
21. Utrechtsch Faculteitenblad 8:13, 1940
22. www.stichtingvredeswetenschappen.nl
23. Vellinga SYA. De uitdaging van crisis en bezetting, de jaren 1936–1946 (The challenge of crisis and occupation, the years 1936–1946). In: Dunk HW von der, Heere WP, Reinink AW (red). Tussen ivoren toren en grootbedrijf. De Utrechtse Universiteit, 1936–1986 (Between ivory tower and large company. Utrecht University 1936–1946). Maarssen, Schwartz, 1986
24. Vox Studiosorum 76:221, 1940
25. Interview Rob Herber with M Bouman, 8 March 2007
26. Interview Rob Herber with M Bouman, 8 March 2007

27. Mattson J, Simon M. Pioneers of NMR and Magnetic Resonance in Medicine. The Story of MRI. Dean Books, Jericho, 1996
28. Interview Rob Herber with E Bloembergen, 20 November 2006
29. Rutgers GAW, Bloembergen N, Kluyver JC. On the straggling of Po-alpha particles in solid matter. Physica 7:669–672 (1940)
30. Walsum JC. Ook al voelt men zich gewond. De Utrechtse universiteit tijdens de Duitse bezetting 1940–1945 (Even though injured. Utrecht University during the German occupation 1940–1945). Universiteit Utrecht, Utrecht, 1995
31. Vox Studiosorum 77:38, 1941
32. Vox Studiosorum 77:39, 1941
33. Bloembergen AR. Het gezin van Rie en Auke Bloembergen 1917–1956. Eigen Beheer, Wassenaar, 2003
34. Bloembergen AR. Het gezin van Rie en Auke Bloembergen 1917–1956. Eigen Beheer, Wassenaar, 2003
35. Interview Rob Herber with M Bouman, 8 March 2007
36. Einstein A, Infeld L. Physik als Abenteuer der Erkenntnis, Sijthoff, 1938
37. Interview Rob Herber with N Bloembergen, 8 December 2006
38. Interview Rob Herber with B Streefland, 12 February 2007
39. Interview Rob Herber with M Bouman, 8 March 2007
40. Utrechtsch Faculteitenblad 8:152–153, 1941
41. Walsum JC. Ook al voelt men zich gewond. De Utrechtse universiteit tijdens de Duitse bezetting 1940–1945 (Even though injured. Utrecht University during the German occupation 1940–1945). Universiteit Utrecht, Utrecht, 1995
42. Utrechtsch Faculteitenblad 9:52, 1941
43. Utrechtsch Faculteitenblad 9:115–120, 1942
44. Vellinga SYA. De uitdaging van crisis en bezetting, de jaren 1936–1946 (The challenge of crisis and occupation, the years 1936–1946). In: Dunk HW von der, Heere WP, Reinink AW (red). Tussen ivoren toren en grootbedrijf. De Utrechtse Universiteit, 1936–1986 (Between ivory tower and large company. Utrecht University 1936–1946). Maarssen, Schwartz, 1986
45. Interview Rob Herber with N Bloembergen, 8 December 2006
46. Utrechtsch Faculteitenblad 9:45, 1942
47. Interview Rob Herber with B Streefland, 12 February 2007
48. Utrechtsch Faculteitenblad 9:53, 1942
49. Walsum JC. Ook al voelt men zich gewond. De Utrechtse universiteit tijdens de Duitse bezetting 1940–1945 (Even though injured. Utrecht University during the German occupation 1940–1945). Universiteit Utrecht, Utrecht, 1995
50. Vellinga SYA. De uitdaging van crisis en bezetting, de jaren 1936–1946 (The challenge of crisis and occupation, the years 1936–1946). In: Dunk HW von der, Heere WP, Reinink AW (red): Tussen ivoren toren en grootbedrijf. De Utrechtse Universiteit, 1936–1986 (Between ivory tower and large company. Utrecht University 1936–1946). Maarssen, Schwartz, 1986
51. Bloembergen N. Encounters in Magnetic Resonances. World Scientific, Singapore (1996)
52. Mattson J, Simon M. Pioneers of NMR and Magnetic Resonance in Medicine. The Story of MRI. Dean Books, Jericho, 1996
53. Walsum JC. Ook al voelt men zich gewond. De Utrechtse universiteit tijdens de Duitse bezetting 1940–1945 (Even though injured. Utrecht University during the German occupation 1940–1945). Universiteit Utrecht, Utrecht, 1995
54. Interview Rob Herber with M Bouman, 8 March 2007
55. Interview Rob Herber with N Bloembergen, 8 December 2006
56. Interview Rob Herber with B Streefland, 12 February 2007
57. Vellinga SYA. De uitdaging van crisis en bezetting, de jaren 1936–1946 (The challenge of crisis and occupation, the years 1936–1946). In: Dunk HW von der, Heere WP, Reinink AW (red): Tussen ivoren toren en grootbedrijf. De Utrechtse Universiteit, 1936–1986 (Between ivory tower and large company. Utrecht University 1936–1946). Maarssen, Schwartz, 1986
58. Bloembergen N. Encounters in Magnetic Resonances. World Scientific, Singapore, 1996

Chapter 5
Quantum Mechanics and Oil Lamps: Bilthoven in the War

Nico Bloembergen is alone in a small room in the dark. In the flickering light of a piece of cotton soaked in fuel oil, he struggles evening after evening through a textbook on quantum mechanics.

5.1 May 1943: University Closure—A Crystal as Detector for Radioactive Radiation

On 23 November 1940 the dean of the Faculty of Law, R. P. Cleveringa, gave an impressive speech at Leiden University (RUL) in honor of the resignation of his Jewish colleague and teacher Eduard Maurits Meijers. *"It is this Dutchman, this noble and true son of our people, this man, this student father, this scholar, who has been relieved from his function by the hostile foreigner who dominates us today! I told you I would not speak of my feelings; I will contain them, even though they threaten to burst like boiling lava through the crevices which I feel at times, at their insistence, will open in my head and heart."* [1, 2]. The Leiden students then went on strike and the RUL was closed.

The students of Leiden had largely come to Amsterdam and Utrecht to continue their studies there. Now that the measures taken by the occupying forces affected the entire student population (see Chap. 4), there was less and less talk of education, and in April 1943 education at Utrecht University had come to a virtual standstill. When attending in lectures, practicals, and exams, any student could be arrested. Most students stayed away. By the beginning of May, no more education was provided. Unlike other universities, exams were still taken. The teachers stayed on, some reluctantly, and in September the new academic year started with a shrinking number of students. In December, the whole education effort was clinically dead. Only 285 students were still enrolled in Utrecht on 1 December [3].

© Springer Nature Switzerland AG 2019
R. Herber, *Nico Bloembergen*, Springer Biographies,
https://doi.org/10.1007/978-3-030-25737-8_5

In the house at Schoolplein 3 in Utrecht, Jan Caro's father, who was involved in air protection, was hiding Jewish people in the basement. An older sister of Jan started a newspaper. Jan himself lay on his bed with his legs in plaster for months [4].

Bloembergen was appointed as a university firefighter after his doctoral exam. That was also written on his identity card and therefore he was temporarily protected from being drafted into the *Arbeitseisatz* [5]. Moreover, that meant he could still gain access to the lab and work there.

Pieter van Heerden was working on solid-state detectors for nuclear physics, using silver chloride and diamond. All kinds of ionizing radiation could be measured, such as alpha-radiation (helium nuclei), beta-radiation (electrons), and gamma-radiation (electromagnetic radiation). The detector was based on a voltage pulse that was produced when the crystal was struck by radioactive radiation [6]. The crystal was cooled with liquid air (-77 K, -196 °C) to suppress leakage currents (ions). This subject was very new and there would be a lot of interest in it in Harvard after the war, as we shall see in this chapter. Nowadays such a detector is called a scintillation detector. Bloembergen was working on the internal, secondary emission from such a detector. When the ionising radiation hits the detector, some of the electrons remain 'hanging' in the crystal, so to speak. He would measure the motion of these secondary electrons through the crystal. The research results appeared in a publication, so Bloembergen already had three publications by the end of the war [7]. This publication also only appeared after the war just like his first research on alpha-radiation.

In the laboratory there was electricity, but no heating. The equipment was outdated, because nothing new could be purchased since May 1940 [8]. Bloembergen continued working on the publication on the scintillation detector until 1945. In October 1943 Werner Heisenberg came to visit the laboratory. He was on a trip through the Netherlands to get acquainted with the state of affairs in Dutch physics in the framework of a cultural exchange. Heisenberg was in the Netherlands at the invitation of Kramers, Kronig, and Casimir, who after years of isolation like many "*in the large family of physicists craved high level scientific discussions*" (in order not to arouse suspicion and give the visit an informal appearance, Heisenberg slept at Kramers' home) [9]. Heisenberg was an important man in quantum mechanics and, after the forced departure of Jewish scientists in the service of the Nazi regime, he was busy using science outside Germany for the German cause. Heisenberg was fascinated by the possibility of splitting the atom and was the right-hand many of the rocket builder Wernher von Braun. The Allies considered Heisenberg such a big threat that they were making plans to kill him [10]. However, Heisenberg made efforts in the Netherlands to prevent the claims of the German occupiers. Involving foreign scientists for the German regime "*is not an option,*" Heisenberg stated, and added: "*That would have a huge impact in the scientific world, and to the detriment of Germany from a propagandist point of view.*" [11].

Maarten Bouman was still an assistant at the laboratory and therefore had the opportunity to continue working on his Ph.D. Bloembergen did not have this advantage and so could only continue after May 1945 [12]. Later we shall see whether this really would turn out to be a handicap. Jan Kluyver, Bloembergen's buddy at the lab, hadn't graduated yet in April 1943 and had to work in Germany for two years. He only graduated in 1948. When Bloembergen was not in the lab, he was busy with the vegetable garden [13].

In the middle of the year Reinhard, commander of the German 88th Army Corps,[1] decided that the command center housed in the former PHRM society at the Janskerkhof in Utrecht would be transferred to Bilthoven. The command center itself came to the Bilderdijklaan, the logistic part in and around the town hall Jagtlust (which was evacuated) at Soestdijkseweg South, the liaison department at the Gerard Doulaan and the Van Dijcklaan (in the Van Dijck school, where the Bloembergen children had been), and the commander of the artillery corps at the Beethovenlaan, in the corner of Soestdijkseweg North. For the construction of the command centre located between Bilderdijklaan, Hasebroeklaan, and Tollenslaan, about 300 m from the Bloembergens' home, a narrow-gauge railway line was built from the station, on which tipping carts could be pulled by a diesel locomotive. They were in such a hurry that the work continued uninterrupted day and night. On the corner of the Bilderdijklaan and the Soestdijkseweg, a special guard post was set up with a hand siren: every time there was air raid warning, the siren would sound. Twelve bunkers were built on this complex. The largest bunker was 13 by 36 m. The villas on the Bilderdijklaan and Hasebroeklaan were also cleared for the accommodation of the officers. When the complex was finished, it was surrounded by barbed wire. The entrances were guarded by sentry stations equipped with machine guns [14].

On the list of personnel from the fire brigade of the Physics Laboratory of the Rijksuniversiteit, N. Bloembergen appeared as a fireman physicist and M. A. Bouman as assistant firefighter. This list was drawn up on 8 July and signed by J. M. W. Milatz, professor and director.

On a note from Jezeer dated 3 December 1943, it is mentioned that the Physical Laboratory supplied a total of 350 half days of farm work. The staff, including students and scientific staff, therefore worked 175 days in the laboratory's own vegetable gardens.

5.2 1944: Raids and Bombing—Back in Bilthoven—The Hunger Winter

Milatz wrote a letter to the Board of Trustees on 18 April 1944 [15] in which he stated that the previous day he had been interviewed by an official of the *Security Service* who reproached him for the following:

Milatz had wrongly allowed the assistants P. J. van Heerden, Th. Schut, and N. Bloembergen to be exempted from the *Arbeitseinsatz*. Milatz was ordered to deny these gentlemen access to the Physical Laboratory, dismiss them, and order them to

---

[1] The 88th Army Corps (LXXXVIII A.K.) was stationed throughout the Netherlands, with the exception of parts of Zeeland, i.e., Noord Beveland, Zuid Beveland, Walcheren, and Zeeuws Vlaanderen, and comprised about six divisions with a total of 40,000 men.

report to the employment office. Milatz was made personally responsible for carrying this out.

Milatz stated that he lacked the authority to dismiss Van Heerden, Bloembergen, and Schut. The official then declared that he would report the matter to the Department in Apeldoorn and the Board of Trustees. The official did not want to mention his name.

Milatz then summoned all the workers in the laboratory and instructed them to follow the orders of the *Security Service*. The following day the *Rector Magnificus* approached the *SS-Untersturmführer* M. H. Heyting. He assured him that this was not the working method of the *SD*, so Milatz could just ignore the official's requests, and the whole affair ended well.

Bilthoven

Soesterberg was an important airbase for the Germans. Bombers were stationed there to attack the United Kingdom. Soesterberg is just a few kilometers from Bilthoven. To protect the airbase, the Germans set up five pieces of defense artillery in the Heidepark. To this end, open bunkers were built, surrounded by smaller bunkers for ammunition. In the middle was a command bunker. In total more than 40 bunkers were built [16]. Heidepark was located at the end of the Van Ostadelaan (where the Bloembergens lived) and the heavy blasts of the artillery must have been extremely loud there. The searchlights were clearly visible at night. The exclusion zone, *Sperrgebiet*, started about 100 m away from the Bloembergen's house. In 1943/1944 the anti-aircraft defence was moved to an area between the Lassuslaan in Bilthoven and the Pleineslaan in Den Dolder, closer to Soesterberg.

On 15 August 1944 a massive attack took place on Soesterberg airfield by about 300 B17 bombers. People were sitting on the rooftops in Bilthoven North to watch the attack and the tremendous noise of the bombs could be heard everywhere in De Bilt and Bilthoven. Moreover, in the area between Bilthoven and Soesterberg, a lot of damage was done and there were casualties [17].

> 3 September 1944: liberation of Brussels. Antwerp was liberated on 4 September, and on the same day, the London government had incorrectly announced that the allied armies had crossed the Dutch border. There were also claims that Breda would be liberated, too. In the south, German soldiers were on the run everywhere, and many Dutch people were waiting to greet the Allies, up as far as The Hague and Amsterdam. On 5 September, the NSB and Germans panicked and fled en masse to Germany: *Dolle Dinsdag* (Mad Tuesday).

Chaos reigned in the municipality of De Bilt, where the soldiers of the 88th Army Corps command began to leave in great haste. The soldiers evacuated the former town hall of Jagtlust and camouflaged their cars with man-sized tobacco plants that they pulled out of the town hall gardens [18].

> On 17 September Operation Market Garden started, an allied offensive from the southern Dutch border from Weert to Arnhem. The London government called for a general railway strike, upon which the 30,000 staff stopped work. As intended, this strike obstructed the German war machine, but at the same time brought food supplies to a standstill. This led to the *Hunger Winter* in the west of the Netherlands. As many as 20,000 died due to the famine [19].

Bloembergen wanted to travel by train to Utrecht on 18 September to work in the lab. He came back from the station with the message that there were no trains, because the railways were on strike [20].

After the railway strike every man between 16 and 50 years of age ran the risk of being arrested for the *Arbeitseinsatz*, so Bloembergen decided to move back to his parents' house in Bilthoven and go into hiding there. There were far fewer raids in Bilthoven than in Utrecht. Father Auke also created primitive hiding places in the house on the Van Ostadelaan and later in the temporary housing. Evert and Diet had left the house, so Nico found himself there with two brothers and one sister (Fig. 5.1).

At half past two in the afternoon of 24 September, an officer from the *Wehrmacht* would tell the Bloembergens in the Van Ostadelaan that they had four hours to clear the house. The house was to be requisitioned. Auke junior was sent out to recruit a few friends. Fortunately, the mother of a distant cousin, Herman van Aalderen, told Rie that the Bloembergens could stay with her. At half past four a German soldier came to tell them that that they would have 24 h of respite. Most of the furniture did not have to go with them to Van Aalderen's house; they could take it to the office of Herman van Aalderen's father, Herman senior, who was arrested in August and would not return from Germany.

In their temporary home, Herbert Bloembergen developed pleurisy. Herbert spent six months in bed, usually with a fever. At first the room was heated with the scarce coal then available, but later on the specialist said the room had to be cold.

Fig. 5.1 Nico Bloembergen's family house, Van Ostadelaan 9, Bilthoven. He lived here from 1925 till 1946, with an interruption during the Second World War. *Source* Rob Herber, 2014

Nico Bloembergen got jaundice, presumably hepatitis A.[2]

A month later the Bloembergens and Van Aalderens had to leave the house: this was also requested by the Germans. The house would be used as a *Wehrmachtsheim*, a military home [21]. The Bloembergens were offered a rather small furnished house in the Nachtegaallaan. They stayed there until the end of May 1945. The sick brothers Nico and Herbert were carried one by one by Auke senior to the Nachtegaallaan, a distance of only 500 meters, on two bicycle wheels with a tarpaulin cover borrowed from the Green Cross.[3] Herbert was there for six months until May 1945. The family did not go hungry in the Hunger Winter. There was grain in the house to bake their own bread, and the rape oil and coal were also moved to the Nachtegaallaan. Cooking was done on a coal stove. In the evening, however, there was no light, but there were petrol lamps. There was fuel oil for the lamps, but they didn't burn very well and didn't provide much light [22].

> From September onwards, the *Wehrmacht,* at first organized in the west, later become completely disorganized, commandeering everything they could use. Bicycles, clothing, blankets, all means of transport, radios, batteries, petrol, oil, vacuum cleaners, sewing machines, iron bolts, and everything that was edible were claimed [23].

> The police chief Rauter made the following announcement:

> The German SS and Police Forces, including the Arbeitskontrolldienst (Labour Control Service) and the entire Dutch police force, have been ordered to arrest, without delay, in cities and villages, on streets and squares, all 16 to 50 year old Dutch people hanging around there, who apparently have nothing to occupy them and spend their time idly, and to transfer them to the employment camp in Amersfoort. From there they will be made available to the Labour Offices in Germany [24].

On 9 October a raid took place in the center of Bilthoven [25].

There is a folder in the archives showing that the Physical Laboratory received extra nutrition from the Mass Food Division of the Wartime Food Supply Office. For example, there is an invoice for hot food supplied on 19 August 1944 by the Rijkskeuken in the Begoniastraat for 140 L at 30 cts, making a total of 42 Dutch florins. In September 1944, the amount had to be paid in cash.

The Personnel Administration in the folder included the general staff, academic staff, and students. The students had an entrance fee of f 0.50 to f 1.00 per week. Number 13 was N. Bloembergen.

On 26 and 28 November the railway line was bombed (the Germans were still using military trains) and so was the station in Bilthoven. Bombs fell on Vinkenlaan and Emmaplein, where a number of shops and homes were hit [26]. During one of these attacks Herbert Bloembergen could see from his bed in the Nachtegaallaan,

---

[2]Hepatitis A is caused by the hepatitis A virus. This is a less severe form of hepatitis, which occurs mainly in children. The symptoms come after an incubation period of two to six weeks and are generally fatigue, mild fever, sometimes pain in the upper abdomen, and nausea. In adults, hepatitis A usually accompanies jaundice, which is a yellowing of the skin and the whites of the eyes. There is no possible treatment with medication. The disease must be left to take its course, and in most cases lasts no longer than 6 weeks.

[3]At that time, the Green Cross (*Groene Kruis*) was an association that was actively involved in nursing the sick and improving hygiene conditions.

which is near the Vinkenlaan, that a fighter-bomber was coming over. It dropped a bomb, but Herbert knew so much about mechanics from the technical school that he could predict that the bomb wouldn't hit their house. All the windows were broken due to the impact of the bomb. His mother Rie was in his room and stood between the window and his bed at the most dangerous moments to protect him [27]. Nico, who was also in the room, chided her: *Mother, you shouldn't do that. It does not help in any way and it is dangerous.*

The *Hunger Winter* from 1944–1945 was a major disaster for the west of the Netherlands. At eight o'clock in the evening people were no longer allowed on the street because there was a curfew. There was no gas left to cook with. There was no electricity and therefore no light. There was no more coal, so houses could no longer be heated. An emergency stove, usually placed on the normal stove, might serve as a cooker. In it wood chips and twigs would be burnt. A Philips handheld dynamo with light bulb could be used as a light source - these were known as squeeze cats because of the spinning sound the dynamo made. Candles usually served as a light source, but when there were no longer any left, people switched to oil lamps. They went to bed early in the evening on an empty stomach in a cold, dark house. Food consisted of what could be provided by the soup kitchens, usually a watery sugar beet stew or a pinkish-purple potato peel soup. Animal feed was considered suitable for human consumption. Everything that moved was hunted down, including gulls, cats, and dogs. Bread was no longer made from flour, but from all kinds of surrogates (such as pea flour). It was sticky, grey, and only 4 cm high. A slice of bread would covered with a slice of sugar beet. For these foods, one had to queue up for hours at the soup kitchens. The men who were not in danger of being arrested for labor service often went on hunger trips to the north of Noord-Holland, Gelderland, or Overijssel, using bicycles with wooden tires, to exchange valuables such as jewelry for food.

Twenty thousand Dutch people died of hunger and cold during that winter.

What about the food supply in the Bloembergen House? Before September there were ration voucher that guaranteed a sufficient, but nevertheless meager food supply. Without scruples, Auke senior also bought all kinds of food such as cheeses and even a pig (slaughtered in the kitchen) on the black market. In September it became clear to him that the supply would dry up. He calculated that, until the new harvest in summer 1945, a few "muds" of potatoes would be needed (one mud was about 70 kg). He went on his bike with his trailer over the river Lek near Vianen and bought a few muds from a farmer. Auke senior was very caring when it came to his family, and helped others too, when he could. He also knew someone who could deliver coal bought on the black market [28].

According to Maarten Bouman: *"During the time the University was closed, a permanent staff had to be present, including instrument makers and scientific officials. They had to keep business ticking over. I had a job as a custodian. The glassblowers blew glass for the Christmas decorations. The instrument makers made an oil press. This could be used to press rapeseed oil, which could then be used for food. The Physical Laboratory had a license to listen to the BBC. Zernike from Groningen had made a crystal clock with a tuning fork to accurately measure time and that device was kept in the Physical Laboratory. Zernike regularly came to Utrecht and then the BBC was switched on, because they broadcast 'boom, boom, boom, boom'[4] as a time signal to check the clock. Then once there was a kind of robbery. The whole lab was inspected and they [the Germans, RH] came across the radio and turned it on: 'boom, boom, boom, boeom' started it. There were no collaborators in the faculty.*

People had to spend a lot of time for the survival of their family. The people in the lab went to get potatoes. For example, they would take a cargo bike to the Vollenhoven estate in De Bilt. The university also had large fields of potatoes near the Meern. One afternoon a week would be spent gardening there. In the last year [1944, RH] all those potatoes had been stolen. [29]

On 29 December, there was an air raid on the command center of the 88th army corps in Bilderdijklaan. Many villas were also affected, although the Germans had already left. The hotel restaurant Heidepark (see Chap. 3) and Het Nieuwe Lyceum [the New School] were also hit [30].

5.3 Quantum Mechanics and Oil Lamps

Bloembergen did not sit around doing nothing during the Hunger Winter. After the doctoral examination he realized that he had to master quantum mechanics on his own. It was clear to him that quantum mechanics was the future, also in chemistry. He went carefully through *The Quantum Theory of Electrons and Radiation* [31] by Kramers (see Chap. 3 for more on Kramers). Bloembergen did not find Kramers' book easy or even the most useful book to learn quantum mechanics from. He also had lecture notes by Uhlenbeck. The oil lamp, which burns with rolling fuel oil, is used as lighting. Every hour the wick had to be snuffed out and the glass had to be cleaned.

Since quantum mechanics would play a dominant role in Bloembergen's work, there follows a short introduction to this remarkable physics of the very small, as far as Nico would have known about it in 1944.

For the introduction, use was made of *Thirty years that shook physics* from George Gamow [32] and *Quantum generations. A history of physics in the twentieth century* of Helge Kragh [33]. Other material is drawn from *Quantum mechanics at the crossroads,* edited by James Evans and Allan S. Thorndike [34].

Applied physics and technology in the industrial revolution showed great vitality at the beginning of the nineteenth century. Invention after invention came on the scene. The material properties that engineers had to deal with led to developments in physics. Faster travel also entailed a standardization of time and the need to communicate to run train networks and control station lighting. Electricity and magnetism were two new branches of physics that received a lot of attention. Heinrich Hertz (see

---

[4]In wartime, the BBC started their broadcast with the first four notes (GGGE) of Ludwig van Beethoven's Fifth Symphony in C minor, opus 67. These four notes were very important for the occupied countries, because that was the sound of the free world.

also Chap. 2 on the high school period and the Hertz experiment) could generate electromagnetic radiation and in 1895 Wilhelm Röntgen discovered the radiation named after him. Antoine Becquerel discovered radioactivity in 1896, and in the same year Pieter Zeeman showed that emitted atomic radiation (for example the yellow light of sodium in the well-known sodium lamps) was split into components with different wavelengths under the influence of a magnetic field. In 1897 J. J. Thomson and Pieter Zeeman, in collaboration with Hendrik Lorentz, demonstrated the existence of the electron (see also Chap. 3). This gave a new impulse to experimental research and also led to theoretical reflection on atomic models.

In physics in the 1930s, as described in Chap. 4, the photocell was used for spectroscopy research. This electron tube is based on the photoelectric effect, which was described by Albert Einstein in 1905. Einstein wondered whether light would behave in emission and absorption processes as if it were an energy quantum. He predicted a linear relationship between the energy of the electrons released by incident light and the frequency of light. Only in 1915 was this confirmed experimentally.

In 1897 J. J. Thompson showed that negatively charged particles, electrons, could be released from the atom. A positively charged residue remained, called an ion. Thomson thought the atom consisted of a swarm of thousands of electrons circling around in an enclosed and positively charged space. However, the spectra observed when an element was heated could not be explained in this way. In 1909, a former student of Thomson's, Ernest Rutherford, working with two of his own students, Hans Geiger and Ernest Marsden, fired alpha-particles (helium nuclei) from radioactive radium via a diaphragm at a thin metal foil. Almost all the alpha-particles flew straight through, but some were deflected and a very small fraction of them (1 in 8000!) even rebounded, as if they had collided with something. As Rutherford said: *It was quite the most incredible event that ever happened to me in my life. It was almost as incredible as if you had fired a 15-inch shell at a piece of tissue paper and it came back and hit you.*

What the alpha-particles had crashed into were the massive, positively charged nuclei of the atoms. The electrons moved around this nucleus, according to Rutherford. The problem was that the dimensions of an individual atom could take all possible values.

Niels Bohr solved this problem in 1913. He fixed the electron orbits by means of a limiting condition, or a quantum rule. The resulting 'stationary' state of the electron was still described by classical mechanics, but the classical radiation theory was abandoned. For example, according to Bohr, an electron circling in a stationary state would not emit radiation. Such radiation could only occur in a transition to another state with a lower energy level. When radiation was absorbed, the electron would 'jump' to a stationary state with a higher energy level. The energy difference was directly proportional to the frequency of the radiation, the proportionality factor being Planck's constant.

Bohr applied his theory to hydrogen, the simplest atom. He assumed that it only had one electron. The spectral lines of atomic hydrogen had previously been measured and they corresponded well with his theory—as did those of ionized helium.

Until then, no one had been able to shed any light on the harmonic relationships between the spectral lines of specific elements.

The Institute of Theoretical Physics at Copenhagen University run by Bohr became the leading institute for quantum mechanics in the world. Many worked at the institute for one or two years, including Paul Dirac, Nevill Mott, Hans Kramers, Hendrik Casimir, Wolfgang Pauli, Werner Heisenberg, Max Delbrück, Carl von Weizsacker, Léon Rosenfeld, George Gamow, Lev Landau, John Slater, Robert Oppenheimer, and others.

Bohr's model did not give a good description of the other, more complex elements. Arnold Sommerfeld expanded Bohr's theory in 1915 with elliptical orbits and additional quantum rules. This enabled him to explain the so-called fine structure of the hydrogen spectrum: with very precise determination, the spectral lines of hydrogen appear to consist of a few very close lines. Sommerfeld also succeeded in explaining, in terms of Bohr's theory, the splitting of spectral lines in a magnetic field as previously measured by Zeeman.

In 1925, Wolfgang Pauli made the postulate that two electrons from the same atom cannot be in the same quantum state.

Also, in 1925, Samuel Goudsmit (1902–1978) and George Uhlenbeck (1900–1988) stated that the electron also spins around its own axis, just as the planets rotate around the Sun and around their own axis.

In 1925, Heisenberg made a complete break with the classical description of electron motion in the stationary states of the atom. These orbits could not be observed in any way other than through incoming and outgoing radiation. He drew up a number of calculation rules for the relationships between the absorbed and emitted radiation and presented them as a new quantum theory of the atom. The calculation rules he proposed were soon recognized as those for certain mathematical objects, called matrices.

In 1926, Erwin Schrödinger, building on an idea by the Frenchman Louis de Broglie, came up with a wave equation for the description of a quantum mechanical system. This is now known as the Schrödinger equation. This equation does not indicate where an electron happens to be at any given time, but just the probability of finding it in a certain place in the atom. It soon became clear that Schrodinger's wave theory was broadly in line with Heisenberg's matrix theory. Werner Heisenberg published his uncertainty relation in 1927. This stated that, at a quantum mechanical level, one could never determine exactly the momentum and position of a particle like an electron at the same time: either the position was well known and then the momentum was uncertain, or the momentum was known and then the position was uncertain.

Paul Dirac united the theories of Heisenberg and Schrödinger and also derived the Dirac equation in 1928. This equation gave a relativistic description of the electron. Dirac thus succeeded in reconciling quantum mechanics with the theory of relativity.

When Bloembergen started studying in 1938, the development of quantum mechanics was still in full swing. However, according to Gamow's booklet (written in 1966), developments were slower: according to him, there were many more developments in the period 1900–1930 than in the period 1930–1960.

5.4 1945: More Raids—Food Drops and Liberation

Bloembergen's sister Diet stayed with her future parents-in-law in Roermond in September 1944; Roermond remained occupied territory. In January 1945 the Germans evacuated the population through Germany to the east and north of the Netherlands. Diet went by train (run by the *Wehrmacht*) to Amersfoort and from there back home to Bilthoven.

Due to a lack of other food, flower bulbs were eaten in the Bloembergen household. They were cooked for hours without much result and Bloembergen still remembers the bitter taste.

Auke senior went with Diet and To on a hunger hunt to the Veluwe in the eastern part of the country, and later with people from the office to Overijssel. Nico, Herbert, and Auke couldn't come along, because they were in danger of being arrested for the *Arbeitseinsatz*. And anyway, Nico and Herbert were sick in bed [35].

> At the end of February the first relief goods were delivered by the Swedish Red Cross. This was followed by five further shipments from the Swiss and International Red Cross. The town services provided free distribution. The first distribution took place between 6 and 9 March [36]. In Dutch bakeries, the real flour that was supplied was used to bake Swedish white bread. For the population in the starving regions, Swedish white bread and a packet of margarine were an unforgettable experience: after years of eating the surrogate agents and the government bread made from potato flour, after years of terror, occupation, and starvation, people now tasted real bread, the taste and smell of which they had completely forgotten [37].

Jan de Vries, mathematics student and USC clubmate of Bloembergen, did candidate exam A on 3 February 3 1941 (mathematics and physics with astronomy). Bloembergen went with friends Jan Caro, Bob Lansberg, and another USC clubmate on a steam train to the engagement of Jan de Vries with Clara Brink. Bloembergen also met Clara's father, Hendrik Gerard Brink (1875–1944). We shall meet Clara Brink (1921–2009) again in Chap. 8. Jan de Vries was taken prisoner in Amersfoort and shot in Varsseveld on 2 March 1945.

In the post-war student almanac, he was commemorated as follows (Fig. 5.2):
IN MEMORIAM
J. DE VRIES, 2 March 1945
When we think back to Jan de Vries, we see him as a good friend, always open and honest with us. He did not like formalities and ceremony. His dealings with men were direct and uncomplicated; he was young, and possessed all the beautiful qualities of youth. Jan was a hard worker. He loved to take action: not talk, not weigh things up at length, but do. He could do an unbelievable amount of work in a short time. His focus was mainly on his studies: mathematics and physics, in accordance with the tradition in the de Vries family. Two mathematical articles will be published from his last years. But he also managed to find time for his many hobbies. He was especially attracted by radio technology and photography. He was a good cellist. In him, the philosophy faculty loses a great organizer. In particular, he showed his gifts in this area at the congress in 1942. In the years 1943 and 44 he made himself useful by publishing lecture notes for his fellow students, scattered over such a wide area. In

Fig. 5.2 Jan de Vries.
Source Almanac for
students, 1948

the spring of 1945, he made contact with an illegal newspaper: he joined in, was
caught, and fell on 2 March 1945 in Varsseveld. He is one of those among the fallen,
of whom we will think often and for a long time to come.
A. G. Th. B.

On 2 April, five resistance fighters of the Bilthoven *KP* (a resistance group) carried
out an attack near the Prinsenburg campsite on the Soestdijkerweg between Bilthoven
and Soest. Two Germans happened to pass by on their bicycles and shots were
exchanged. One German soldier was shot dead. As a reprisal, on 4 April ten people
were taken from the prison on Wolvenplein in Utrecht and shot at the site of the
attack. There is a memorial to this [38].

In April, the *Jagdkommando* (a special operations group) sent Andries Jan Pieters
from the Dutch SS to Loosdrecht. Forty people in hiding and illegal workers were
subjected to all kinds of atrocities. The body of a Jewish person in hiding was thrown
into the Loosdrechtse Plassen after being tortured to death. Pieters raped the wife
of one of the detainees. His group also plundered on a large scale. On 3 May 1945
they were arrested by the German police, on behalf of none other than the notorious
Willy Lages,[5] who was not even a minor war criminal himself. Even according to the

---

[5]Willy Paul Franz Lages (October 5, 1901–April 2, 1971) was the German chief of the *Sicher-*
heitsdienst in Amsterdam during the Second World War. From March 1941 he led the so-called
Zentralstelle für jüdische Auswanderung (Central Bureau for the Jewish Emigration). As such he

standards of the occupying forces, Pieters went way beyond any reasonable behavior and he was executed on 21 March 1952.

> In April and May food drops took place in the west of the Netherlands (operation *Manna)* and thus brought relief [39]. Within nine days, more than 800 aircraft brought almost 12,000 tons of 'manna' [40].

On 2 May there were drops near Utrecht, among others on the meadows near the *Lage Weide* (Fig. 5.3). More than 5000 packages dropped by 58 *B-17 Flying Fortress* bombers were assembled by special teams. Quite a lot got into private hands, and in the afternoon, cans of milk were being sold for eight guilders on the black market [41].

> On 4 May, the *Wehrmacht* capitulated in Northwest Europe.

> On the evening of 4 May, the (illegal) radio announced that Norway, Denmark, and the Netherlands had been liberated.

Many inhabitants of Utrecht took to the streets, but the German soldiers drove them back and shot at them [42].

> The commander of the German troops in the Netherlands did not recognize the capitulation. On 6 May the Second World War ended for the Netherlands. On that day the first food ships arrived in Rotterdam.

Along the Utrechtseweg (the road between Utrecht and Amersfoort/Zeist that runs through De Bilt) (Fig. 5.4) and Hessenweg in De Bilt, many people were there to welcome the Canadians. They sang loudly 'Oranje Boven', a patriotic song, meaning

Fig. 5.3 Food drops above Utrecht, made by 'Flying Fortresses' of the USAF, May 1945. *Source* Frans Beenen

---

was responsible for the deportation of Dutch Jews to the concentration camps in Germany and occupied Poland. en.wikipedia.org.

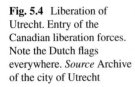

Fig. 5.4 Liberation of Utrecht. Entry of the Canadian liberation forces. Note the Dutch flags everywhere. *Source* Archive of the city of Utrecht

something like 'Orange High'. The word 'Orange' referred of course to the Dutch Royal family [43].

Most Dutch people tried to sit out the war as best they could. There was certainly little sympathy for the Germans, but they tried to adapt as well as possible to daily life. Most students were no exception. Physics students also tried to study as well as possible, take exams, and take part in lab activities. Bloembergen did the same, just like his fellow student Maarten Bouman and later Barend Streefland. Initially, in 1941, his brother Evert was still inclined to resist the banning of Jewish students, which meant that he spent almost two months in prison. His youngest brother Auke writes about this in the family chronicle: *The family was happy to stick with just one prisoner of war. That is also understandable, because no one from the family really went into resistance later on. I shall stop there* [44].

Bloembergen soldiered on with his studies. He felt it was a very bad thing to have lost his (Jewish) friend Henk Langendijk in 1942, but when asked how he got through the war he answered: *You have to go on.*

What did the study of physics do for Bloembergen? He passed the candidate and doctoral exams cum laude. He got a good grounding in physics, albeit largely classical. He supplemented this by a study of quantum mechanics using Kramers' book. Practical and theoretical physics were taught at a high level in Utrecht; the development of measuring equipment was a specialty there from which Bloembergen would benefit a lot in his future career.

Finally, Bloembergen already had three publications to his name, although due to the circumstances of the war, not everything had yet appeared in print.

References

1. vandaagindegeschiedenis.nl/26-november/
2. loe.niod.knaw.nl/grijswaarden/De-Jong_Koninkrijk_deel-04_tweede-helft_zw.pdf
3. Walsum JC. Ook al voelt men zich gewond. De Utrechtse universiteit tijdens de Duitse bezetting 1940–1945 (Even though injured. Utrecht University during the German occupation 1940-1945). Universiteit Utrecht, Utrecht, 1995
4. Interview Rob Herber with J Caro, 22 November 2006
5. Bloembergen N. Encounters in Magnetic Resonances. World Scientific, Singapore, 1996
6. Heerden PJ van. The Crystal Counter. A New Instrument in Nuclear Physics. Proefschrift, Rijksuniversiteit Utrecht, Utrecht, 1945
7. Bloembergen N. Note on the internal secondary emission and the influence of surface states. Physica 11:343–344, 1945
8. Bloembergen N. Encounters in Magnetic Resonances. World Scientific, Singapore, 1996
9. www.delftintegraal.tudelft.nl
10. Powers T. Heisenberg's war: the secret history of the German bomb. Jonathan Cape, London, 1993
11. Delft D van. Tegen de roof: het Kamerlingh Onnes laboratorium in oorlogstijd. Gewina 30:247–264, 2007
12. Interview Rob Herber with M Bouman, 8 March 2007
13. Mattson J, Simon M. Pioneers of NMR and Magnetic Resonance in Medicine. The Story of MRI. Dean Books, Jericho, 1996
14. Brugman JC. Vergeten Bunkers. Forten in en om De Bilt en het Duitse 88ᵉ legerkorps in De Bilt (Forgotten Bunkers. Forts in and around De Bilt and the German 88th army corps in De Bilt). JC Brugman, Bilthoven, 2006
15. Letter from JMW Milatz to the Board of Trustees of the University of Utrecht, 18 April 1944
16. Brugman JH. Vergeten bunkers. Forten in en om De Bilt en het Duitse 88ᵉ legerkorps in De Bilt (Forgotten Bunkers. Forts in and around De Bilt and the German 88th army corps in De Bilt). JC Brugman, Bilthoven, 2006
17. Brugman JC. Bezet en verzet. De Bilt en Bilthoven in oorlogstijd (Occupation and Resistance. De Bilt and Bilthoven in wartime). JC Brugman, Bilthoven, 1993
18. Brugman JC. Vergeten Bunkers. Forten in en om De Bilt en het Duitse 88ᵉ legerkorps in De Bilt (Forgotten Bunkers. Forts in and around De Bilt and the German 88th army corps in De Bilt). JC Brugman, Bilthoven, 2006
19. en.wikipedia.org
20. Bloembergen AR. Het gezin van Rie en Auke Bloembergen 1917–1956. Eigen Beheer, Wassenaar, 2003
21. Mattson J, Simon M. Pioneers of NMR and Magnetic Resonance in Medicine. The story of MRI. Dean Books, Jericho, 1996
22. Bloembergen AR. Het gezin van Rie en Auke Bloembergen 1917–1956. Eigen Beheer, Wassenaar, 2003
23. Jong L de. Het Koninkrijk der Nederlanden in de Tweede Wereldoorlog. Deel 10b. Het Laatste Jaar II. Eerste Helft (The Kingdom of the Netherlands in the Second World War. Part 10b. The Last Year II. First half). Staatsuitgeverij, 's Gravenhage, 1981

24. Whiting C, Trees W. Van Dolle Dinsdag tot Bevrijding. Het langste oorlogsjaar (From Mad Tuesday till liberation. The longest war year). Unieboek, Bussum, 1977
25. Brugman JC. Bezet en verzet. De Bilt en Bilthoven in oorlogstijd (Occupation and Resistance. De Bilt and Bilthoven in wartime). JC Brugman, Bilthoven, 1993
26. Brugman JC. Bezet en verzet. De Bilt en Bilthoven in oorlogstijd (Occupation and Resistance. De Bilt and Bilthoven in wartime). JC Brugman, Bilthoven, 1993
27. Bloembergen AR. Het gezin van Rie en Auke Bloembergen 1917–1956. Eigen Beheer, Wassenaar, 2003
28. Bloembergen AR. Het gezin van Rie en Auke Bloembergen 1917–1956. Eigen Beheer, Wassenaar, 2003
29. Interview Rob Herber with M Bouman, 8 March 2007
30. Brugman JC. Vergeten Bunkers. Forten in en om De Bilt en het Duitse 88^e legerkorps in De Bilt (Forgotten Bunkers. Forts in and around De Bilt and the German 88th army corps in De Bilt). JC Brugman, Bilthoven, 2006
31. Kramers H A. Quantum Theorie des Elektrons und der Strahlung (Quantum theory of electrons and radiation). In: Hand und Jahrbuch der Chemischen Physik Leipzig, 1938
32. Gamow G. Thirty years that shook physics. Dover, New York, NY, 1966
33. Kragh H. Quantum generations: a history of physics in the twentieth century. Princeton University Press, Princeton, NJ, 1999
34. Evans J, Thorndike AS (eds). Quantum mechanics at the crossroads, Springer, 2007
35. Bloembergen AR. Het gezin van Rie en Auke Bloembergen 1917–1956. Eigen Beheer, Wassenaar, 2003
36. Miert J van. 'En toen kwam het er eigenlijk op aan.' September 1944-september 1945 ('And then it really matters.' September 1944-September 1945). In: Miert J van (red). Een gewone stad in een bijzondere tijd. Utrecht 1940–1945 (An ordinary city in a special time. Utrecht 1940-1945). Spectrum, Utrecht, 1995
37. www.wikipedia.nl/Zweeds_wittebrood
38. Brugman JC. Vergeten Bunkers. Forten in en om De Bilt en het Duitse 88^e legerkorps in De Bilt (Forgotten Bunkers. Forts in and around De Bilt and the German 88th army corps in De Bilt). JC Brugman, Bilthoven, 2006
39. Jong L de. Het Koninkrijk der Nederlanden in de Tweede Wereldoorlog. Deel 10b. Het Laatste Jaar II. Tweede Helft (The Kingdom of the Netherlands in the Second World War. Part 10b. The Last Year II. Second half). Staatsuitgeverij, 's Gravenhage, 1982
40. Whiting C, Trees W. Van Dolle Dinsdag tot Bevrijding. Het langste oorlogsjaar (From Mad Tuesday till liberation. The longest war year). Unieboek, Bussum, 1977
41. Miert J van. 'En toen kwam het er eigenlijk op aan.' September 1944-september 1945 ('And then it really matters.' September 1944-September 1945). In: Miert J van (red). Een gewone stad in een bijzondere tijd. Utrecht 1940–1945 (An ordinary city in a special time. Utrecht 1940-1945). Spectrum, Utrecht, 1995
42. www.hetutrechtsarchief.nl
43. Brugman JC. Vergeten Bunkers. Forten in en om De Bilt en het Duitse 88^e legerkorps in De Bilt (Forgotten Bunkers. Forts in and around De Bilt and the German 88th army corps in De Bilt). JC Brugman, Bilthoven, 2006
44. Bloembergen AR. Het gezin van Rie en Auke Bloembergen 1917–1956. Eigen Beheer, Wassenaar, 2003

Part III
From Atoms to Humans:
Nuclear Magnetic Resonance and MRI

Chapter 6
A Leap into the Deep

Bloembergen sails on a Liberty ship through the Channel on its way to the United States. Liberty ships have no mine detectors and they only have a primitive warning system: a ship's boy stands on the bow and looks out for mines. If he sees something, he gestures and shouts wildly at the bridge.

6.1 Utrecht and Bilthoven: 5 May 1945–Mid January 1946

On 5 May Herbert Bloembergen talked to people on the street from the window in his room and saw flags. At nine o'clock in the evening he called from his room: "*We are free!*" The following days were confusing, because although the capitulation was now a fact, there were still German troops everywhere. On one of those days, probably 7 May, the great bells of the Dom tower rang out. This was an impressive tower, 112 m high, built in the thirteenth century, and the sound could be heard in Bilthoven. Later in May, the Bloembergen family returned jubilantly to their home on the Van Ostadelaan, which had been occupied since 1944. It had been released before May and would be cleaned and tidied up. It was still in reasonable condition [1].

On 5 May, hidden flags, orange eyelashes, and portraits of the royal family were hung in the streets of Utrecht. Dutch flags and orange banners were forbidden by the Germans, because the Dutch flags represented the national identity and the orange banners represented the Dutch Royal family. At 12.30 pm three Canadians, members of the Red Cross, rode into the city on motorcycles. However, the *Ortskommandant* (local commander) remained at his post and the *SS* was still patrolling the streets. On 6 May the situation changed a little. After negotiations between the city commander

© Springer Nature Switzerland AG 2019
R. Herber, *Nico Bloembergen*, Springer Biographies,
https://doi.org/10.1007/978-3-030-25737-8_6

of the *Domestic Armed Forces (BS)*[1] and the German authorities, about 250 prisoners were released from Wolvenplein prison [2].

On 6 May, General Blaskowitz signed the capitulation order for German troops in the Netherlands in the auditorium of the Wageningen University of Applied Sciences.

For Utrecht, 7 May was Liberation Day. Flags were hung from the Dom tower and the *BS* commander installed himself on the Domplein. Around noon there was a skirmish in the district of Oudwijk, in which ten members of the *BS* were killed. Another member of the *BS* was shot down on the Chartroise road. There was also fighting on the Marnixlaan and the Van Hoornekade with further victims. The Mayor Van Pelkwijk returned in the afternoon. At about three o'clock in the afternoon, Canadian and British troops travelled through the city [3]. A cheering crowd had gathered; Viestraat, Neude, and Potterstraat were full of cheering people.

On 8 May came the unconditional surrender of Germany. All fighting officially had to end at 23.01 h.

On 8 May the *BS* began arresting German soldiers in Utrecht. They were taken out of their barracks and assembled on the Paardenveld. From there they went on to encampments outside the city.

On 9 May Auke cycled to De Bilt, where the Canadians had entered. Here, too, there were cheering people everywhere. Herbert took his first steps outside the door in more than six months and watched the Allies entering the Soestdijkseweg for the first time [4].

Ernst Strubbe tells us: "*I saw the first Canadians entering the Holle Bilt. One part went* via *the Utrechtseweg to Utrecht and another part* via *the Soestdijkseweg to Baarn.*" [5].

On 10 May, a dramatic thanksgiving service was held by the Jewish community in Utrecht. Religious education and services in the synagogues were resumed [6].

Only on 13 May did Canadian soldiers come to Bilthoven to collect weapons. This happened near the bombed hotel Heidepark. After that the Germans had to go to the bunker complex with Gerard Doulaan [7].

In the service schedule of the Physical Laboratory, it says that N. Bloembergen entered the premises out of office hours from 16 to 23 May, presumably to continue his research there.

The Mayor of Utrecht felt that there was an increased level of debauchery after the liberation and wrote:

The men sought relaxation and entertainment and tried to earn money through the black market. Jeeps and trucks were driven wildly and there were many accidents. [...] There was dancing, flirting, and open petting. The latter gave offense. Measures against this, e.g., to stop people lying down along the canals and in the meadows, were difficult to implement and met with resistance [8].

The food shortages were certainly not over after the liberation and persisted well into the summer [9]. The power stations and central kitchens were not yet operational. On 12 May,

---

[1] The BS or *domestic armed forces* were created on 5 September 1944 and stemmed from three important opposition groups: the *Order Service*, the *Rural Fighters*, and the *Council of Resistance*. At the time, the BS consisted of no more than 10,000 men. The commander was Prince Bernhard.

a week after the liberation, the newspaper reported that 10,000 inhabitants of Amsterdam were on the verge of starvation. The number of sufferers of starvation edema[2] in Amsterdam was estimated at 20,000 to 30,000. The same was true for other cities in the west. Thousands were dying as a result of malnutrition.

On 22 May an official of the Red Cross observed that the nutritional situation in Utrecht had been very bad for the past three weeks. The average calorie intake had dropped to 800 kcal in the last weeks of the Hunger Winter,[3] and from 12 April to 12 May, it was only 600 kcal. The following week, however, it rose to 1400 kcal. In Utrecht, 4000 cases of edema were reported. The municipality distributed hot food and clothing. The food situation improved in the second half of June. All references to the occupation were quickly removed.

On 24 May, the German military stationed around Utrecht began to retreat via Loosdrecht, Aalsmeer, IJmuiden, Castricum, and Petten to the Afsluitdijk and then on to Germany. Most of them had to walk, but some went by bike or horse and cart, doing an average of 30 km a day [10].

In the weeks following the liberation, many Dutch people returned home, workers, soldiers, and a group of 400 Jews. The latter often had no home to go to and no work; moreover, many had suffered from the loss of family members. Little or nothing was done for these compatriots, because the London government had decided that it could not discriminate. In 1930, 1218 Jews lived in Utrecht and the surrounding area; during the war, more than a thousand were murdered [11]. Thousands of children were still living in the north and east, and they returned in June or later.

In June to July, children also left Utrecht for North Brabant and even Switzerland to regain strength [12].

The trams started to run again, the garbage was collected, the harbors were repaired, and the rubble of Schiphol airport was cleared. The trains could not run yet, because the rolling stock had been taken away, the rails damaged, the signal boxes sabotaged, and the bridges destroyed. Extensive regions of the country lay in a state of ruin. The floodplains of the Meuse and Rhine were full of mines. Cities were uninhabitable because of the war. Some 107,000 Jews were deported from the Netherlands, of whom an estimated 90,000 were murdered in the concentration camps [13]. In addition, 16,000 non-Dutch Jews who were residing in the Netherlands were murdered [14]. That was the deplorable state of the Netherlands after the liberation [15]. For the first months a *Military Authority*[4] was responsible for

---

[2]Edema is a swelling of body parts due accumulation of fluids. It can occur for various diseases and is caused by a disruption of the water intake. Hunger edema is due to a protein deficit brought about by a limited range or amount of food. Any loss of protein from the blood has an impact on the water metabolism due to the disappearance of the colloid osmotic pressure, and that means that no water can be extracted from the tissues. In the case of a hunger edema, water tends to accumulate in the abdomen (hunger belly). From: www.healthsquare.nl.

[3]Normally, a mature woman 31–50 years old needs 2400 kcal, and a mature man 31–50 years old needs 3100 kcal. Teenagers and people in their twenties need more, the elderly less.

[4]The Dutch government in London faced huge administrative problems after the liberation of the Netherlands. That is why administration was legally entrusted to a military government that was set up for this purpose. The legally trained Major-general Mr. H. J. Kruls was chief of staff of the Military Authority. This authority was founded in 1944 and dissolved in 1946. Kruls remained in charge throughout this period nl.wikipedia.org.

government. A major problem was the trial of 120,000 Dutch people who had collaborated with the occupying forces. Purification committees were set up everywhere [16]. These matters would drag on for years to come.

That summer was a summer like none before.

The Netherlands found itself divided. On the one hand there was a frenzy over the liberation. This was acted out in countless neighborhood dinner parties, in which our liberators also participated. The Canadians acquainted us with the music that had been forbidden in the previous years: "In the mood" by Glenn Miller was number one [...] It was one endless party, that summer. On the other hand, many underwent unfathomable grief over absent relatives or other horrors they had experienced during the war [17].

After the liberation Bloembergen went back to his room at the Schoolplein again [18]. Before the war he had already decided that he wanted to do a doctorate, but in the United Kingdom, France, or Germany [19]. After May 1945 he did consider possible topics in Utrecht, but found nothing suitable [20]. He had no contacts in Sweden, where there might have been opportunities because the country remained neutral and had not suffered destruction during the war. Switzerland was also a possibility, but they did not accept foreign students. He had become quite depressed, because he didn't know what to do [21]. His brother Evert said: *"Why don't you go to America?"* It seems that Nico took this to be good advice and he wrote to three American universities in June: the University of Chicago, the University of California at Berkeley, and Harvard University [22]. It took him days to draft the letters and address the handwritten letters to the chairmen of the physics department at each university. He knew very little about the United States [23]. His choice of universities was based on the physics journals he had seen, such as *Physical Review* in the library of the Physical Laboratory in Utrecht, periodicals that had been received until May 1940. Any developments over the previous five years were completely absent. The periodicals usually presented studies that Bloembergen did not understand. He had heard of Enrico Fermi at the University of Chicago and E. O. Lawrence in Berkeley [24]. Bloembergen chose several subjects more or less at random. He didn't care which university, as long as good research could be done there.

In July or August, Bloembergen received a message from the University of California (Berkeley), stating that the university was unfortunately not allowed to admit foreign students to their campus. Federal government regulations stipulated that no foreigners could work or study on campus during the war. Bloembergen was astonished by the letter. He knew that the war was now over in Europe and had assumed there was peace everywhere, so he was surprised to discover that the war between the Allies and Japan was still under way in Asia. He thus became aware that the world was bigger than Europe. The University of Chicago did not answer. What Bloembergen didn't know was that both Berkeley and Chicago were involved in the Manhattan Project to develop an atomic bomb. Years later, these universities would offer him a place as a professor!

Bloembergen received a positive response from Harvard University. The chairman (dean) of the Harvard Physics Department, Otto Oldenberg, asked Bloembergen to send copies of letters of recommendation, proof of identity, and diplomas obtained. Bloembergen sent everything in and he was promptly admitted to the Graduate School of Arts and Sciences for the academic year 1945–1946. A graduate school is a (university) school that awards academic titles to students who have obtained a bachelor's degree. In Bloembergen's case, the bachelor's degree corresponded to what was then the Dutch candidate examination. The Dutch doctoral exam was not recognised as such in the United States.

The former Dutch academic system consisted of three steps. The first was the candidate exam and consisted of theoretical and practical study material in science, mostly made up of mathematics, physics, chemistry, and biology, depending on the course. This can be compared to the bachelor's degree. The second step, referred to as the doctoral exam, involved more specialized theoretical and practical study, and in Bloembergen's case, physics. This step could be compared to some extent to a master's degree, but it was more intensive because the student had to carry out some of his own research. The third step was the dissertation, and consisted of research, only, for the doctor's degree.

Meanwhile, Bloembergen was still cycling around with messages from the former underground in the area. He eventually stopped, considering that it was no irrelevant.

Vox Studiosorum, the student magazine, would be published again in July. The front page of the first number carried an *In Memoriam* for the students who died in the war. On page two it said that the Utrecht Student Corps had been resurrected [25].

On 30 July Pieter van Heerden obtained his Ph.D. with the thesis: *The crystal counter. A new instrument in nuclear physics.* His supervisor was Milatz. In the introduction Van Heerden explained that the well-known Geiger-Müller counters were always used to measure radioactive radiation. These counters counted every particle of alpha-radiation, but also cosmic radiation and UV light. The disadvantage with these counters was that they could not distinguish the different types of particles and radiation and that the energy of the radiation or the particles could not be determined. The subject of the thesis was the development of a detector that did not have these defects of the Geiger-Müller counter. Van Heerden found out that, if crystals such as silver chloride were ionized by radioactive radiation, a voltage pulse would be emitted. The height of that pulse was proportional to the energy of the particle, so the energy spectrum of weak radioactive radiation could be determined. Other crystals tested were diamond, zincblende, and silver and thallium halides [26].

The August edition of *Vox Studiosorum* stated that the Supervisory Board of the 'Mensa' foundation had set up a Committee of Governors of the University House (at the Lepelenburg) with N. Bloembergen as ab actis (secretary) [27]. Evert Bloembergen was still praeses of *Vox* until September (Fig. 6.1).

On 6 August, an atomic bomb with a strength of 15 kilotons of trinitrotoluene (TNT) had been dropped on the Japanese city of Hiroshima. Some 78,000 people were killed almost instantaneously. In the days that followed, many more were killed by the radioactive fallout. On 9 August an atomic bomb was dropped on Nagasaki, killing another 40,000 people. Japanese capitulation was signed on 2 September.

In September, Bloembergen knew that he had been admitted to Harvard University. But how could he pay the study costs, not to mention the travel and accommodation

Fig. 6.1 The board of the Mensa Academica, August 1945. Bloembergen standing on the right. *Source* Nico Bloembergen

costs? He was on a shortlist of twenty to thirty promising students drawn up by the (Dutch) Ministry of Education, Arts, and Sciences (OK&W) for scholarships to the United States. Those students who had studied Chinese literature or American history, for example, were given priority in the allocation of scholarships (by the Rockefeller Foundation). The places available in the United States were reserved for the humanities, and certainly not for a science like physics, which was sensitive in times of war. The two physics students on the list, Bram Pais and Bloembergen, had to wait until dollars become available. In Bloembergen's case, as in his choice to enroll in American universities, his family offered a solution. His father Auke was willing to withdraw the necessary amount in guilders. Pais had no family to support him and had to wait another year before he could go to the United States. But even with the Dutch money from his father, Bloembergen had not yet solved all his problems [28].

> During the war, an enormous amount of money had been put into circulation, especially by the German occupiers. But there were few goods available to buy with this in the looted country. The Finance Minister Piet Lieftinck started a financial sanitation operation in June. From 26 September to 2 October, everyone could have 300 guilders in their possession and that amount had to be exchanged for new money during that same week. For one week, until 2 October 1945 there was no valid paper money in the country with a single exception. Everyone with a distribution card could change ten guilders in withdrawn notes for new money. This action was known as *Lieftinck's tenner*. All banks had to declare assets in

excess of 900 guilders to the Tax and Customs Administration. All funds were verified and investigations could be initiated. It was almost impossible to exchange guilders for dollars because there was practically no foreign money present in the Netherlands after the war and no goods were exported, so for example no dollars entered the country.

Bloembergen went to the Ministry of Education, Arts, and Sciences and explained that he was on the list of students for scholarships and that he had also been accepted by Harvard University, a first-class university. Furthermore, he added that his father could make the money available in guilders. Bloembergen's stubbornness convinced the officials and he got authorization to exchange guilders for dollars. He needed $1800, which included $400 for tuition fees and $1400 for livings. At the exchange rate at that time, that came to 5000 guilders.[5]

After obtaining the authorization, Bloembergen went to Amsterdam to exchange money at De Nederlandsche Bank (the Dutch Central Bank), then located at the Turfmarkt. There were only a few trains a day and it takes two hours in a goods train, because the railways were still in a deplorable state. Bloembergen showed the bank official his license to purchase $1,800, and the poor man nearly fell off his chair, declaring that he had never seen so much money. He asked Bloembergen who he was, who could wish to exchange so much money. The official went on to say that the only ones who usually received such licenses were bulb traders and professors, and then the amounts involved were never more than $500. The bulb dealers wanted to make orders to be able to deliver in the spring of 1946 and the professors to maintain their international contacts. Bloembergen explained to him that those people only needed money for three to six weeks, whereas he had to live on it for a whole year.

There was also a passport and a visa to be applied for. All in all, it took much longer in the paralyzed Dutch bureaucracy than it would have under normal circumstances. Bloembergen stood by getting more and more annoyed, because the date for the start of the lectures in Harvard was getting closer and eventually went by.

The journey itself was also a problem. It was not possible to fly, because there were no civil aircraft flying to the United States. Ocean steamers were not yet available, but there were cargo ships, the so-called 'Liberty' ships, that were mainly used for supplies during the war. After the war they were used to provide the Netherlands with food supplies from the United States. The ships went back empty with seawater as ballast and could then carry just over ten passengers from the Netherlands.

Bloembergen could only travel to the United States in mid-January 1946. He thus missed four months owing to all the bureaucratic delays. He didn't know that, thanks to that delay, he would arrive at Harvard University at exactly the right time [29, 30].

Field hockey, which was important for Bloembergen, restarted in the Netherlands on 14 October with a regional competition [31]. Bloembergen continued to play in the SCHC, the Bilthoven club.

---

[5]After a few years, Bloembergen was able to pay the amount back to his father at a much more favorable rate, so that he didn't lose out on balance.

6.2 1946: From Utrecht to Cambridge

Bloembergen's father Auke took him to Rotterdam in mid-January in the family car. On the way they got two flat tires (the quality of the cars and especially the rubber of the tires left much to be desired) and they couldn't go any further. Bloembergen got a lift from a passer-by, hugged his parents along the side of the road, and then went on to Rotterdam.

The ship sailed along the Hook of Holland and via the English Channel to the Atlantic Ocean. The Channel was still full of mines, laid by both the British and the Germans. Both countries had laid thousands of these, and although international treaties prohibited this, the Germans had also laid magnetic mines that exploded whenever an (iron) ship approached. The mines had been laid by submarines, surface vessels, and aircraft [32]. Many Liberty ships suffered from cracks in the hull and deck and some sank due to these defects. The captain of Bloembergen's ship decided to avoid the storms of the North Atlantic and takes a long detour via the Azores. That was a good decision; along the way, they had two emergency signals from ships that had broken in two in Iceland. Yet they still ended up in a storm, during which the captain reduced the speed to slightly over 3 km an hour for 24 h!

Once they had safely crossed the Atlantic, they were unable to moor as intended in the port of New York, because a dockers' strike was under way there. They had to steam south along the New Jersey coast and then up the Delaware to Philadelphia, where they finally moored in early February. The crossing had lasted eighteen days.

Why did Bloembergen go to Harvard?

The education and the environment in Bilthoven had formed a rather sheltered environment. The primary school was close to home, just like the hockey club. The gymnasium in Utrecht was less than 10 km by bicycle and the university was 10 km away from Bloembergen's family home. His parents had always supported him and Bloembergen had only ever needed to take small steps and make small decisions. His parents, brothers and sisters, Bilthoven, the hockey club, they had always been somewhere nearby.

Bloembergen's decision to go to Harvard was thus a major break from all this. His brother Evert had shown the way and his father Auke had supported him financially, but he was the one who had taken the decision to go to Harvard- then already a renowned university- and to continue his education there, especially with the idea of doing good research there. This was a huge step. Bloembergen did not know Harvard University, he did not know the customs of American universities, and he did not know the town of Cambridge.

After his first research, before his candidate (BSc) exam at Utrecht University, Bloembergen had increasingly been considering research as his main aim in life. In the post-war Netherlands, which were lying in ruins, things would have been very difficult. Countries like France and the United Kingdom had also suffered enormously from the war. Germany, the most important country for physics before the Second World War, was obviously not even an option. An American University like Harvard had thus become the logical destination.

6.3 Physics in the United States

Education

According to a letter Bloembergen sent to Milatz on 27 February 1946, at the time of his arrival in Cambridge, secondary education and the first years of a university education looked like this:

Secondary education is at high school until the age of 17. Very little is learned there. "Some cannot add 1/3 + 1/4".

This is followed by a four-year course at the university (or at a college), during which a main subject and subsidiary subjects are followed. After four years, one becomes bachelor and can go on to graduate school. The academic year consists of three teaching periods, or terms. In each term one must enroll for four courses. Each has three hours of lectures per week, but one is expected to work ten hours per week at home on the compulsory literature. In the middle and at the end, each course is examined. The student has very little freedom and can also take little initiative. It is a school system that aims at competition and high marks. Some courses have experimental practice, e.g., about atomic physics, rectifier tubes or microwaves. Only after two years of experimental courses may one register for research courses, both theoretical and experimental. Compared to the Dutch system, it is a school system and students only learn to work independently at a later stage.

According to Bloembergen, this was even too late for experimental research. Theoretically, students were taught better, but according to Bloembergen that did not mean that they actually achieved better results. Bloembergen met up with Uhlenbeck who considered that the American students had the same level as the Dutch ones, but that they worked much, much harder in the US [33].

Research

The first American physicist to receive the Nobel Prize was Albert Michelson in 1907. Michelson was a real experimenter and did important work in optics. He determined the speed of light with great accuracy. He also designed an interferometer, a measuring instrument with which the properties of two or more light beams could be determined. This was what he used with Edward Morley to show that the speed of light does not increase or decrease when the orientation of the beam of light changes with respect to the direction of motion of the Earth. Josiah Willard Gibbs (1839–1903) worked at several universities, also in Europe, but remained as professor at Yale University in New Haven, Connecticut from 1869. Gibbs is considered to be the father of chemical thermodynamics. He also published on theoretical thermodynamics and statistical mechanics. In 1901, before the Nobel Prize was established in 1901, he received the Copley Medal, the most prestigious award at the time [34]. Ornstein's dissertation *Application of the Statistical Mechanics of Gibbs to molecular and theoretical questions* [35] dealt with Gibbs' theories. Bloembergen had always had a particular interest in statistical mechanics and applied it in his work.

Research in industry
The research laboratories of AT&T (Bell Labs), General Electric (GE), and Radio Corporation of America (RCA) contributed a lot to the development of electron tubes such as diodes, triodes, photocells, and photomultipliers [36]. However, more fundamental research was also carried out in these laboratories, similar to what was being done in the Philips Physics Laboratory.

As the research in Utrecht had already shown, the development of pre-war electronics was of great importance for the future development of physics. More sensitive measurements could be made and Brownian movement and the noise it caused could be taken into account.

Physics in the United States, from underprivileged child to world leader
In the United States before the Second World War, research in physics was almost non-existent and its practitioners relatively unknown. If people in the United States had said in family circles that they were 'physicists', nobody would have known what that was, or they might have thought it was a 'physician' [37]. In the Netherlands, Lorentz and Kamerlingh Onnes were celebrities who made the newspapers.

The Second World War turned physics in the United States on its head.

Science in the service of defense can be taken as the main driving force behind physics research in the United States from 1940 to 1960. The budget for military research roses from a total of 50 million in 1935 to 100 million in 1940 and 3000 million in 1945 (in $ from 1964). Later, the budget gradually increased to 6000 million in 1960 [38]. In June 1940, the National Defense Research Committee was established to coordinate defense research. In May 1941, this was replaced by the Office of Scientific Research and Development (OSRD), which was only given an advisory function after the war. The OSRD had almost inexhaustible resources at its disposal. Two major projects are explained briefly below. The OSRD was dissolved in December 1947. Under the leadership of Vannevar Bush, former director of the OSRD, and president Karl Compton of MIT, a strictly military Research and Development Board (RDB) was established in 1947 [39] (Fig. 6.2).

Manhattan Project
Since the 1930s, research into the possibility of controlled nuclear reactions had been carried out in Germany. In the United States, there was a fear that a controlled nuclear reaction might be developed there, and in 1939 a small research group was set up in Chicago to investigate the use of nuclear reactions as a weapon. Later this was continued in three main locations: the research site in Los Alamos, New Mexico, the uranium enrichment plant in Oak Ridge, Tennessee, and the production plant at the Hanford Site, Washington. Eventually, in 1946, more than 130,000 people were working on this project, including thousands of physicists. In Los Alamos, J. Robert Oppenheimer was director and research leader. Well-known physicists and mathematicians working on the project were Enrico Fermi, Richard Feynman, John von Neumann, and Edward Teller. The developed atomic bomb was put together in Los Alamos [40]. On 6 August 1945 the first atomic bomb was dropped on Hiroshima. At the end of 1945, this explosion had claimed 140,000 lives, some of which were due

Fig. 6.2 Vannevar Bush with a differential analyzer, a mechanical analog computer for solving differential equations. *Source* wiki.sjs.org

to the radioactivity released during the nuclear explosion. On 9 August a plutonium bomb was dropped on Nagasaki and this claimed 65,000 victims in the short term. Later, the project leader, Oppenheimer, strongly opposed further development of such a barbaric weapon.

Radar (Radio Detection and Ranging)

The history of radar began in the nineteenth century after Maxwell and Hertz had formulated the basic equations of electromagnetism. Christian Hülsmeyer developed a simple ship detection device in Germany in 1904. He demonstrated this device in Germany and the Netherlands using radio waves, the echoes of which could detect (metal) ships [41]. The device consisted of a spark generator, whose signal was directed to the antenna (a dipole, like those used for TV and FM radio reception) with a funnel-shaped reflector. If a signal was received by a similar antenna, which was connected to it, a bell sounded. Hülsmeyer called his device the *Telemobiloscope* because it could detect (metal) ships up to a distance of 3 km. It was not put into production due to lack of interest [42]. From 1920 to 1930, work continued in various countries, made possible by the use of radio tubes and other components developed for radio technology. In particular, a lecture by Marconi in 1922 aroused interest in radar:

*It [now] seems to me that it should be possible to design apparatus by means of which a
ship could radiate or project a divergent beam of these rays in any desired direction, and
these rays, if they came across a metallic object, such as another steamer or ship, would
be reflected back to a receiver screened from the local transmitter on the sending ship, and
thereby immediately reveal the presence and bearing of the other ship in fog or bad weather*
[43].

From 1934 to 1939, eight countries worked on radar under great secrecy. In the
early years research mainly focused on radio waves. Soon it became clear that,
by working with shorter wavelengths, much sharper resolution could be achieved.
However, techniques for using these shorter waves, microwaves,[6] were still in their
infancy.

The Netherlands was at the forefront of these developments. In 1927 the Dutch
government commissioned an investigation into whether 'lethal radiation' could
be used for military purposes [44]. Around 1930, Von Weiler and Gratema had
developed communication equipment (the walkie-talkie) with a wavelength of 1.4
m for the purpose of directing field artillery.[7] During a test, communication between
sender and receiver was disturbed by seagulls flying over. When an aircraft flew over,
communication was restored. In a test set-up—in which a transmitter was installed
in Scheveningen and a receiver in Waalsdorp, both in The Hague—it was established
that an aircraft could reflect the signal. Von Weiler had developed the principle of
radar without being aware of foreign developments. Then, in 1936, work began on
a method of using radio waves to detect aircraft [45]. A wavelength of 70 cm was
used, made possible by specially developed electron tubes called 'apple tubes'. Only
one antenna was needed. On an oscilloscope one could then see the 'bleep', a dot
that lit up on the screen, when an echo came back from an airplane. On the day
of the German invasion in May 1940, four aircraft were ready, one of which was
waiting on the Maliebaan in The Hague. The plane worked well, but there was no
artillery to shoot with. That was very frustrating for Von Weiler! The latter fled to
the United Kingdom with the builder of the aircraft, Staal. There, scientists were
amazed and also jealous to discover the Dutch radar system that worked so well.
The Dutch know-how was quickly integrated into the British systems [46]. They had
already developed one with separate transmission and reception antennas, using a
wavelength of 1.5 m.

The United Kingdom was looking for a system where wavelengths of 10 cm
could be used. Microwaves, however, are not so easy to generate. With radio waves,
resonance can be achieved by connecting a coil and capacitor, the so-called LC
circuit. However, this is not possible for microwaves. A second problem is the transport of microwaves. For radio waves, transport can be done with copper wires; for
microwaves so-called wave pipes, or metal tubes, are required. The technique is more
like plumbing with water pipes than radio wave electronics. Randall and Boot at the
University of Birmingham succeeded in February 1940 in designing a new type of
microwave, the microwave generator. This microwave consisted of eight so-called

---

[6]Wavelength 1 m to 1 mm, frequency 300 MHz to 300 GHz.

[7]Wavelength 10 m to 1 m, ultrashort wave or nowadays VHF, from 20 to 300 MHz.

vibration cavities [47]. In a vibration cavity, a vortex is created by applying a strong magnetic field and a high voltage to the cathode. The anode then begins to resonate in the same way as a tuning fork. The first microwave oven was produced by GEC in the United Kingdom in July 1940.

Robert Buderi describes in his book *The Invention That Changed the World* [48] that the invention of microwave radar was of greater significance for winning the war than the atomic bomb.

Forget Oppenheimer and Teller; radar researchers at MIT [...] won World War II [49].

In September 1940, when the United Kingdom was being constantly bombarded by the Germans and a German invasion threatened, the British inventions for military purposes were transported in the deepest secrecy in a black metal box by a Canadian ocean steamer to the neutral United States. One of those secrets was the microwave device with its vibrating cavity. Its application not only resulted in greatly improved resolution and therefore more detailed observations, but also in 1000 times higher sensitivity. This allowed enemy aircraft and ships to be observed from a much greater distance. The limited size of the system meant it could be used on board aircraft.

Radar also contributed to planetary astronomy as early as 1946, when it was used to examine the surface of the moon to some extent. Nowadays, radar is used in meteorology (rain radar), ocean wave measurement, aviation control, and speed measurements (for the police, for example).

In the *Radiation Laboratory* (*Rad Lab)* which was set up at the Massachusetts Institute of Technology (MIT) in Cambridge, Massachusetts, and which functioned from October 1940 to 1 January 1 1946, as many as two thousand physicists [50] were working on the development of radar for defense purposes, according to Bloembergen. That is probably an exaggeration, because Forman indicates that there were about five hundred working on it at MIT [51]. At the end of the war (in August 1945), MIT's staff was twice as numerous as in 1940 and its research budget was ten times as large [52]. According to Bloembergen, there were seventy-hour working weeks and holidays were an unknown phenomenon for years [53]. Harvard University, like MIT, is located in Cambridge and only three metro stops away. Harvard was also home to many scientists who received defense subsidies, were involved in the Rad Lab, or otherwise worked with MIT (Fig. 6.3). There was a lot of cooperation and consultation between the two universities. On 1 August 1946, the Office for Naval Research (ONR) was established with the aim of promoting scientific research for the benefit of the fleet and the marines. This ONR would become of great importance to Bloembergen after 1948 as a grant provider.

Cambridge and Harvard University
In 1630, seven hundred Puritans sailed from England to Boston to establish a church community there. They founded a number of villages around Massachusetts Bay, but could not agree on a capital. Then, 8 km from Boston, on the banks of a side stretch of the Charles River, the village of Newtowne was established on a hill. In 1636 Harvard College was founded and became one of the first universities in the United States, just like the University of Utrecht in the Netherlands [54]. In 1638,

Fig. 6.3   The construction of a microwave cavity in the Rad Lab at MIT in 1945. *Source* engineer-inghistory.tumbir.com

the university was named after John Harvard, a pastor who, at his death, had left his library and half of his estate to the new institute. In 1779, John Adams dedicated a whole paragraph in the Constitution of Massachusetts to the *University at Cambridge,* the first time Harvard was mentioned as a university [55]. In 1846, the three villages of Old Cambridge, Cambridgeport, and East Cambridge were merged into the city of Cambridge. Harvard University was located in Old Cambridge on Harvard Square, the oldest part of the city. Charles W. Eliot, president from 1869 to 1909, transformed the small institution into a contemporary university. Before the Second World War, Harvard, like other American universities, remained an institution with a mainly regional function. It was not until the 1950s that this changed. William Bender, 'dean of admissions', stipulated that students should be admitted independently of their financial position. Students were interviewed and not admitted only on the basis of grades [56].

In 2010 Cambridge had 105,000 inhabitants.

Jefferson Physical Laboratory [57]
The first building in the United States dedicated to physics was the Jefferson Physical Laboratory of Harvard University, which is still in use. The Jefferson Lab was built in 1884 for teaching and research in physics. In (Fig. 6.4) this building and the later additions named after Pierce and Lyman, many scientific discoveries have been made. Confining to examples in quantum mechanics and its applications, we may mention:

- The Lyman lines in hydrogen.

Fig. 6.4 Jefferson Laboratory, Harvard University. *Source* en.wikipedia.org

- The equivalence principle and quantum field theory.
- The Russell-Saunders coupling or LS coupling between spin and orbit in light atoms.
- Nuclear magnetic resonance (NMR), discussed further in the following chapters.

In the next chapter we will see how Bloembergen would begin to feel at home in this laboratory and how his first important scientific work would see the light of day.

References

1. Bloembergen AR. Het gezin van Rie en Auke Bloembergen 1917–1956. Eigen Beheer, Wassenaar, 2003
2. Miert J van. 'En toen kwam het er eigenlijk op aan.' September 1944-september 1945 ("And then it really matters". September 1944–September 1945). In: Miert J van (eds). Een gewone stad in een bijzondere tijd. Utrecht 1940–1945 (An ordinary city in a special time. Utrecht 1940–1945). Spectrum, Utrecht, 1995.
3. Miert J van. 'En toen kwam het er eigenlijk op aan.' September 1944-september 1945 ("And then it really matters". September 1944–September 1945). In: Miert J van (eds). Een gewone stad in een bijzondere tijd. Utrecht 1940–1945 (An ordinary city in a special time. Utrecht 1940–1945). Spectrum, Utrecht, 1995.
4. Bloembergen AR. Het gezin van Rie en Auke Bloembergen 1917–1956. Eigen Beheer, Wassenaar, 2003

5. www.historischekringdebilt.nl.
6. Miert J van. 'En toen kwam het er eigenlijk op aan.' September 1944–September 1945 ("And then it really matters". September 1944–September 1945). In: Miert J van (eds). Een gewone stad in een bijzondere tijd. Utrecht 1940–1945 (An ordinary city in a special time. Utrecht 1940–1945). Spectrum, Utrecht, 1995.
7. Brugman JC. Vergeten Bunkers. Forten in en om De Bilt en het Duitse 88ᵉ legerkorps in De Bilt (Forgotten Bunkers. Forts in and around De Bilt and the German 88th army corps in De Bilt). JC Brugman, Bilthoven, 2006.
8. www.hetutrechtsarchief.nl.
9. Miert J van. Oorlog en bezetting. Nederland 1940–1945 in woord, beeld en geluid (War and occupation in the Netherlands 1948–1945, in word, image, and sound). Kosmos-Z&K, Utrecht/Antwerpen, 1994.
10. Brugman JC. Vergeten Bunkers. Forten in en om De Bilt en het Duitse 88ᵉ legerkorps in De Bilt (Forgotten Bunkers. Forts in and around De Bilt and the German 88th army corps in De Bilt). JC Brugman, Bilthoven, 2006.
11. www.jhm.nl.
12. Miert J van. 'En toen kwam het er eigenlijk op aan.' September 1944–September 1945 ("And then it really matters". September 1944–September 1945). In: Miert J van (eds). Een gewone stad in een bijzondere tijd. Utrecht 1940–1945 (An ordinary city in a special time. Utrecht 1940–1945). Spectrum, Utrecht, 1995.
13. Jong L de. Het Koninkrijk der Nederlanden in de Tweede Wereldoorlog. Deel 7. Mei '43–juni '44. Eerste Helft (The Kingdom of the Netherlands in the Second World War. Part 7. May '43–June 44. First half). Staatsuitgeverij, 's Gravenhage, 1976.
14. Presser J. Ondergang. De Vervolging en Verdelging van het Nederlandse Jodendom. 1940–1945 (The Demise. Persecution and Destruction of Dutch Judaism 1940–1945). Tweede Deel. Staatsuitgeverij, 's Gravenhage, 1965.
15. Werkman E, Keizer M de, Setten GJ van. Dat kan ons niet gebeuren… (We cannot do that …) De Bezige Bij, Amsterdam, 1980.
16. Miert J van. 'En toen kwam het er eigenlijk op aan.' September 1944–September 1945 ("And then it really matters". September 1944–September 1945). In: Miert J van (eds). Een gewone stad in een bijzondere tijd. Utrecht 1940–1945 (An ordinary city in a special time. Utrecht 1940–1945). Spectrum, Utrecht, 1995.
17. Groot MJ de. Utrechtse kinderen op transport naar boeren (Farming out Utrecht children). De Oud-Utrechter, 17 April 2018.
18. Bloembergen AR. Het gezin van Rie en Auke Bloembergen 1917-1956. Eigen Beheer, Wassenaar, 2003
19. Mattson J, Simon M. Pioneers of NMR and Magnetic Resonance in Medicine. The Story of MRI. Dean Books, Jericho, 1996
20. Interview Rob Herber with M Bouman, 8 March 2007.
21. Mattson J, Simon M. Pioneers of NMR and Magnetic Resonance in Medicine. The Story of MRI. Dean Books, Jericho, 1996
22. Bloembergen N. (ed). Encounters in Magnetic Resonances. World Scientific, Singapore, 1996
23. Mattson J, Simon M. Pioneers of NMR and Magnetic Resonance in Medicine. The story of MRI. Dean Books, Jericho, 1996
24. Mattson J, Simon M. Pioneers of NMR and Magnetic Resonance in Medicine. The Story of MRI. Dean Books, Jericho, 1996
25. Vox Studiosorum 77-81 voorblad (1945).
26. Heerden PJ van. The Crystal Counter. A New Instrument in Nuclear Physics. Proefschrift, Rijksuniversiteit Utrecht, Utrecht, 1945.
27. Vox Studiosorum 77-81: 3 (1945).
28. Bloembergen N. (ed) Encounters in Magnetic Resonances. World Scientific, Singapore, 1996
29. Bloembergen N. (ed) Encounters in Magnetic Resonances. World Scientific, Singapore, 1996
30. Mattson J, Simon M. Pioneers of NMR and Magnetic Resonance in Medicine. The Story of MRI. Dean Books, Jericho (1996)

31. De Sportwereld. 6 Aug 1945.
32. Mattson J, Simon M. Pioneers of NMR and Magnetic Resonance in Medicine. The Story of MRI. Dean Books, Jericho, 1996
33. Letter from N. Bloembergen to JMW Milatz, 27 February 1946.
34. Moyer AE. American Physics in Transition. A History of Conceptual Change in the Late Nineteenth Century. Tomash, Los Angeles, 1986
35. Ornstein LS. Toepassing der Statistische Mechanica van Gibbs op molekulair-theoretische vraagstukken (Application of the Statistical Mechanics of Gibbs to Molecular and Theoretical Issues). Eduard IJdo, Leiden, 1908
36. Kragh H. Quantum generations: A history of physics in the twentieth century. Princeton University Press, Princeton, NJ, 1999
37. Interview Rob Herber with C Townes, 13 November 2008.
38. Forman P. Behind quantum electronics: National security as basis for physical research in the United States, 1940–1960. Hist Stud Phys Biol Sc 18:149–229, 1987
39. en.wikipedia.org.
40. en.wikipedia.org.
41. en.wikipedia.org.
42. www.radarworld.org/huelsmeyer.html.
43. Marconi G Radio Telegraphy. Proc Inst Radio Eng 10:215–238, 1922.
44. Pair C le. In: www.celepair.net/radar-web.htm.
45. Bruin R de, Grimm P, Hoiting H, Luijten P, Vliet J van, Zwaalf L. Illusies en incidenten. De Militaire Luchtvaart en de neutraliteitshandhaving tot 10 mei 1940 (Illusions and incidents. Military aviation and neutrality enforcement up to 10 May 1940). Koninklijke Luchtmacht, Den Haag, 1988.
46. Staal M. Hoe de radar naar Hengelo kwam (How the radar came to Hengelo). In: www.maxstaal.com.
47. histru.bournemouth.ac.uk.
48. Buderi R. The Invention that Changed the World. How a Small Group of Radar Pioneers Won the Second World War and Launched a Technological Revolution. Simon and Schuster, 1997.
49. www.nytimes.com/books/.
50. Letter from N. Bloembergen to JMW Milatz, 27 February 1946.
51. Forman P. Behind quantum electronics: National security as basis for physical research in the United States, 1940–1960. Hist Stud Phys Biol Sc 18:149–229, 1987
52. Forman P. Behind quantum electronics: National security as basis for physical research in the United States, 1940–1960. Hist Stud Phys Biol Sc 18:149–229, 1987
53. Letter from N. Bloembergen to JMW Milatz, 27 February 1946.
54. www.cambridgema.gov.
55. www.hno.harvard.edu.
56. Rosovsky H. The Harvard Community. In: Twaalfhoven BWM (ed). Harvard and Holland. Indivers, Hilversum, 1986
57. Holton G. How the Jefferson Physical Laboratory came to be. Physics Today. December 1984:32–37, 1984

Chapter 7
To Harvard: Tintin in America (1946–1947)

April 1946 Isodor Rabi comes to visit the Lyman Lab, where Bloembergen is working. For Bloembergen (aged 25), Rabi, now 47 years old, is the leading figure in nuclear magnetic resonance. Edward Purcell, Bloembergen's group leader, fails to show Rabi the nuclear magnetic resonance signal, because it is not easy to balance the detector. Purcell is quite disillusioned about this. After their departure, Bloembergen sorts everything out and within an hour he gets a signal. He runs from the basement upstairs, where Rabi is still present and shows the delighted Rabi and Purcell the signal.

Tintin in America

In Philadelphia, Bloembergen took the train to Baltimore to visit Robert Dicke of Johns Hopkins University [1]. During the war he had worked for the Rad Lab on photomultipliers with secondary emission amplification by a factor of 10^{11}. What was particularly important for Bloembergen's work in Harvard was that Dicke had developed a phase sensitive detector. In contrast to Bloembergen's detector, which worked with a galvanometer, Dicke's was completely electronic.

Bloembergen travelled on to New York to see some of the city and made an unannounced visit to Columbia University's Physics Department [2]. European scientists were an attraction and George Pregram, faculty president and also secretary and treasurer of the American Physical Society, found Bloembergen important enough to take him to the faculty club for lunch. In the afternoon, John Dunning (1907–1975) took him to his laboratory to show him a cyclotron.[1] Dunning was working on nuclear fission and showed Bloembergen his experiments. He had been the first

---

[1] A cyclotron is a circular particle accelerator used to accelerate electrically charged atomic particles. These particles can then be made to strike a plate, for example, to release neutrons. Nowadays, cyclotrons are used primarily to produce isotopes for medical research, as a source of radiation to treat cancer, or to produce (positrons) isotopes for a positron emission tomography (PET) scan (a functional imaging technique used to observe metabolic processes in the body). en.wikipedia.org.

© Springer Nature Switzerland AG 2019
R. Herber, *Nico Bloembergen*, Springer Biographies,
https://doi.org/10.1007/978-3-030-25737-8_7

person to achieve nuclear fission in the United States (see the discussion of Rosenfeld in Chap. 4). Dunning also had a special neutron spectrometer that used the time-of-flight method.[2] Bloembergen was quite astonished; he hardly understood what it was all about.

Bloembergen was widely received as though he were a visiting professor, someone of importance, while in his own eyes he was "*nothing*", just a student who had not even finished his studies. After the long years of war, the three days in Baltimore and New York were extremely exciting for him, and seemed filled with both material and spiritual promise [3].

Indeed, this visit to Baltimore and New York was rather reminiscent of *Tintin in America*. In the comic strip by Hergé, the hero of the story, the journalist Tintin, is celebrated after all kinds of adventures in an official reception. But the humble Tintin lowers his eyes before all the praise [4].

To Harvard: ten days to decide
Bloembergen continued his journey by train and travelled from New York to Boston. He was travelling in the evening and got into conversation with a fellow passenger on the train. When his fellow passenger heard that Bloembergen was going to study physics at Harvard, he was very interested and mentioned that he was a nephew of Percy Bridgman, professor of physics at Harvard. Bridgman (1882–1962) was awarded the Nobel Prize for high-pressure research in 1946 (Bloembergen would meet him later). The man was worried when he heard that Bloembergen would arrive at Boston South Station late in the evening without having a clue where he would go. He invited Bloembergen to spend the night in his house and they get off the train in New Haven. Bloembergen meets the man's wife and 14-year-old daughter. He was particularly struck by this spontaneous hospitality, which he later remarked was typical of the United States.

The next morning the host took Bloembergen to the station in New Haven and put him on the train to Boston. His host also told him how to get to Harvard by metro.

That same afternoon, the administration of the Graduate School of Arts and Sciences at Harvard University helped him to find a room in the Weld Hall student house at Harvard Yard. The administration tells him that he was too late to register for the spring period ('spring term'), but they agreed to waive the fine he was due. Bloembergen only had to pay the $200 tuition fee for the spring period.

The next day Bloembergen's trunk arrives at the student house by Rail Express. In the afternoon he lunched with the chairman of Physics Department, Otto Oldenberg. Oldenberg had studied as a German at the California Institute of Technology (CalTech) in Pasadena in 1930 and eventually ended up in Harvard.

Since Bloembergen had already passed his exams in Utrecht and had done some research, he was one of the few students who could start research immediately. Although there were already many American students starting in Harvard in the spring after their military service, almost no one had any experience in research.

---

[2]Time-of-flight (TOF) is a collective name for a number of methods that measures the time that an object, particle, or electromagnetic wave takes to travel through a medium.

They therefore had to start by attending lectures and doing practicals. Bloembergen had a big advantage over these students and felt he had been lucky about this.

Bloembergen asked Oldenberg what research he could do. Oldenberg said he should have a look around and then decide with whom he would like to do research. For Bloembergen, the contrast between the University of Utrecht in wartime and Harvard University as he found it then was tremendous and he was overwhelmed by all the possibilities. The faculty members were extremely helpful and sympathetic, which helped Bloembergen to make his choice. Oldenberg suggested talking to J. Curry Street and Edward Purcell because they were both in need of research staff. Within ten days Bloembergen had to choose what he wanted to do!

J. Curry Street

Street had played an important role in MIT's Rad Lab, which was responsible for the development of defense radar (see Chap. 6). Street was not only on the steering committee of Rad Lab, but had also worked in Rad Lab's Division 10, Ground and Ship. In early 1946, after five years of radar research, Street was able to pick up his research on cosmic rays again and hoped to find a student for that.

Bloembergen with Edward Purcell

Purcell had also worked at the Rad Lab. Six weeks before Bloembergen's arrival, he had discovered nuclear magnetic resonance (NMR) in solid matter, for which he would receive the Nobel Prize in 1952. Incidentally, Purcell had used Street's electromagnet.

Bloembergen knew a little about nuclear magnetism. He had heard Cor Gorter talk about NMR at a meeting of the Dutch Physics Association during the war. Bloembergen believed that radar and related microwave research offered great opportunities for spectroscopic research. He wanted to do cross-disciplinary research and according to him Purcell's research gave such an opportunity. Purcell told him that his (NMR) research was related to what Gorter had done, and Bloembergen commented: "*Well, that's a connection.*" [5]

Purcell also needed a student, but the question was whether he would accept Bloembergen. The latter had a 'carbon copy'[3] of his publication about the phase sensitive detector with galvanometer he had worked on in Utrecht. However, it would only be published in the course of 1946 [6], and there was just one copy. Yet this was sufficient for Purcell: they also had a phase sensitive detector in Harvard, albeit a more advanced version than Bloembergen's. Purcell was therefore convinced that Bloembergen was well enough equipped to conduct research on NMR. In an interview, when asked [RH]: "*If you had done something else…*", Bloembergen added: "*Then Purcell might not have accepted it.*" [7] Purcell suggested that Bloembergen should talk to two others in order to check whether he really wanted to work for him (Purcell) and he referred him to Samuel Goudsmit and George Uhlenbeck, who

---

[3]Before the development of photographic copiers, a carbon copy—not to be confused with the carbon print family of photographic reproduction processes—was the under-copy of a typed or written document placed over carbon paper and the under-copy sheet itself. en.wikipedia.org.

Fig. 7.1 George Uhlenbeck.
Source www.
nationalmedals.org

both worked in MIT's Rad Lab. Bloembergen met them at the front door of the Rad Lab—he was not allowed to go in because the research being carried out there was secret. Goudsmit (Figs. 7.1 and 7.2) felt that Cambridge would be far too expensive for Bloembergen and told him that he would be better off going to a cheaper university, like Michigan (where Goudsmit himself worked); there the cost of living was only half what it was in Cambridge. Uhlenbeck made another remark: "*Well, with Purcell you can take it easy.*"

In the end Bloembergen decided to start NMR research with Purcell. "*Purcell was a very friendly, modest man. He was six years older than me.*" [8]

Peter Pershan, student and later colleague of Bloembergen, shared this view: "*Purcell was a true gentleman, an elegant man. He was very different from Nico. Nico looked for problems and had a large research group. Purcell liked to work alone with one or two people.*" [9]

George E. Uhlenbeck (1900–1988) [10] and Samuel Goudsmit (1902–1978) [11]
Uhlenbeck studied chemical technology at the Technical University in Delft in 1918 and physics at the University of Leiden in 1919. He passed the bachelor's degree in 1920 and was then a pupil of Ehrenfest. In 1923 he took his doctoral degree in physics in Leiden. In his Ph.D. research he collaborated with Samuel Goudsmit.

Fig. 7.2 Samuel Goudsmit.
Source www.physicsforme.
com

Goudsmit graduated under Paul Ehrenfest in Leiden in 1921. By 1925, Goudsmit had already published ten papers in Dutch, English, and German journals.

Uhlenbeck knew a lot about statistical mechanics and Goudsmit a lot about spectra, and together they began to investigate anomalous spectral lines in hydrogen and helium. In 1925 they suggested that the fourth quantum number of an electron proposed by Wolfgang Pauli (*a two-valued quantity with no classical counterpart*) had to be the spin of the electron around its axis, in the way a spinning top rotates around its axis [12]. Ralph Kronig had described the same thing in the United States six months earlier, but Wolfgang Pauli had opposed this idea and Kronig had not published. The consequence was that no Nobel Prize was ever given for the discovery of the rotation of the electron, because it was not clear who had thought of it first. In 1927 Uhlenbeck was able to complete his thesis with a scholarship in Copenhagen and Göttingen. In 1927 he obtained his Ph.D. with Goudsmit under Ehrenfest. Uhlenbeck's thesis title read: *On statistical methods in the theory of quanta*. Also in 1927, he and Goudsmit left for the University of Michigan, Ann Arbor, where he became an instructor. In 1929 Uhlenbeck became Assistant Professor and in 1930 Associate Professor. Goudsmit spent nine years at the University of Michigan, working there with Linus Pauling on the famous book *The Structure of Line Spectra* (1930).

In 1935 Uhlenbeck returned to the Netherlands after strong pressure from Kramers, who had himself become professor in Leiden, and he became professor of theoretical physics in Utrecht. The title of his oration read: *The principle of energy conservation*. In 1939 he left again for Ann Arbor, where he became Full Professor. According to Casimir, there was a connection between Uhlenbeck's decision to go back to Ann Arbor and Goudsmit's decision to reject an appointment in Amsterdam to succeed Zeeman [13].

Goudsmit left around 1940 for a temporary job in Harvard. Goudsmit was very much against Nazism and wanted to do something to contribute to the war effort against Nazi Germany. In 1941 he was appointed to the Radiation Laboratory of the Massachusetts Institute of Technology in Cambridge. Uhlenbeck also worked there. Goudsmit played an important role in the application of the radar developed by the British. Furthermore, he became the head of the documentation department so that he was able to keep up with all technical developments, and this later turned out to be a key factor in his becoming editor in chief of *Physical Review*. Towards the end of the war, Goudsmit became the scientific leader of the *ALSOS* project, which was set up early in 1945 to investigate laboratories in liberated Europe. The military were particularly interested in the progress the Nazis had made on nuclear energy. Goudsmit knew practically every well-known physicist in Europe. He also visited the Netherlands and his former parental home in The Hague, which had been plundered. Goudsmit's parents had been murdered in a concentration camp. "*Sam never did recover the very light touch that he had before the war, but gradually he recovered a fair measure of his old buoyancy.*" [14]

In 1964, Uhlenbeck and Goudsmit received the Max Planck medal for discovering the spin of the electron. Ilya Progogine made the following remark about Uhlenbeck: "*One of the greatest Dutch physicists was George Uhlenbeck.*" [15] Martinus Veltman also stated: "*He was a great man, Uhlenbeck. That was someone I had a lot of respect for. He worked with Goudsmit. They always gave each other every possible praise. After the death of Goudsmit, Uhlenbeck received the Wolff Prize in Israel, I believe 100,000 dollars. Uhlenbeck then gave half of the prize to Goudsmit's widow. That was the kind of man he was.*" [16]

Back to college
At Harvard there were about ten professors of physics and the same number for mathematics. In Utrecht there were only two for each discipline. In addition, there was close cooperation with MIT, which had a similar number of professors. Due to the strong development of radio technology and radar during the war, many special lectures were given on radio tube technology at Harvard. There were also lectures about antennas, wave pipes, and electronic clocks for radar.

Bloembergen first followed a general lecture series and a series on nuclear physics. The latter was given together with MIT. He attended lectures by Purcell and a lecture by Stanford University's William Hansen (1909–1949) on nuclear moments. Bloembergen told Milatz in a letter of 27 February 1946 that Gorter's experiment in Harvard was in the spotlight and that Purcell et al. had obtained positive results

[17]. Although Bloembergen formally met the requirements to follow the graduate student program (doctoral training), Purcell advised him to follow two series of courses, namely statistical mechanics by Edwin Kemble and electromagnetism by Julian Schwinger.

Edwin Kemble (1889–1984)
Edwin Kemble had studied in Harvard and obtained his Ph.D. in physics in 1917. After the First World War he was appointed in Harvard to set up theoretical physics there. Later in his career he focused on education. From 1940 to 1945 he was head of the faculty of physics [18]. During the Second World War he was also associated with the Rad Lab and he had an important task in the ALSOS, then led by Goudsmit.

Julian Schwinger (1918–1994) [19], Genius by Profession
Julian Schwinger was only two years older than Bloembergen and was already considered a genius by Bloembergen and many others in 1946. Today, Schwinger is considered one of the greatest physicists of the twentieth century. He studied at the City College of New York, where he published his first physics publication at the age of sixteen. At (Fig. 7.3) the age of nineteen he had already written his dissertation, but because he had neglected his mathematics courses, he did not receive a doctorate. Isodor Rabi sent him to the lectures by Uhlenbeck at Columbia University, but those were too early in the morning and Schwinger never attended them. Uhlenbeck then set an exam for the sake of form—Schwinger knew as much as Uhlenbeck himself—and Schwinger got an 'A', the highest rating [20]. From 1939 to 1941 he worked at Berkeley, first with a grant from the National Research Council and later as an assistant to Robert Oppenheimer.

During the war he went to the Rad Lab. He worked in Uhlenbeck's group and provided the mathematical support for further development of radar. Schwinger started working at four o'clock in the afternoon and then continued until deep into the night. He was associated with Harvard from 1945 to 1974. When Bloembergen attended his lectures in 1946, he was still Associate Professor; in 1947 he would become Full Professor. Schwinger was one of the founders of renormalization theory during the 1940s.

In 1965 Schwinger, together with Richard Feynman, was awarded the Nobel Prize for quantum electrodynamics. This was just one of the many awards and prizes he would receive.

No lectures were given at Harvard in the afternoon. Schwinger chose to teach from 12.00 to 13.00. However, he would never show up until 12.30 pm. Bloembergen got an A for Kemble's statistical mechanics course and an A for Schwinger's electromagnetism course. The latter was very sparing in handing out the highest rating: he distributed an A a total of four times and B thirty times (along with seven or eight lower ratings). Faculty chairman John Van Vleck told Schwinger: *"Julian, at Harvard we have more A students than that."* Bloembergen got to know many other well-known physicists, such as Walter Kohn, C. Luttinger, R. J. Glauber, J. Blatt, R. Karplus, and F. Frohlich.

Fig. 7.3 Julian Schwinger. *Source* www. schwingerfoundation.org

The lectures at Harvard and MIT helped Bloembergen to expand his English vocabulary, but he never managed to understand the English of Niels Bohr who was also teaching there!

John Hasbrouck Van Vleck (1899–1980): of Dutch Origin, Father of Magnetism
Van Vleck had been called the most successful American physicist since Gibbs [21] and *"the father of magnetism."* [22] Van Vleck's family originally came from the Netherlands and he always had a weakness for that country. He thought it was a rather weird country and always called Bloembergen *"my fellow Dutchman."* [23] Van Vleck followed courses given by Bridgman and Kemble at Harvard, because his father had taught there, and he obtained his Ph.D. with the latter in 1922. Van Vleck kept close contact with the European quantum physicists; in 1923 he visited Kramers and Bohr. From 1923 to 1928, he worked at the University of Minnesota. After (Fig. 7.4) that he got a permanent appointment in Wisconsin until 1934. It was there in 1929 that the British physicist Paul Dirac (1902–1984) came to visit. Dirac had just developed his famous wave equations which united the theory of relativity

Fig. 7.4 John Van Hasbrouck 1930. 'Van between two fans.' *Source* John Comstock

and quantum mechanics. *For Dirac, Van Vleck was the perfect walking companion, uninterested in small talk and quite content to say nothing for hours* [24]. Van Vleck had a passion for railways and knew many train timetables in the United States and Europe by heart. In the summer of 1930, he attended the University of Zurich and completed his book there: *The Theory of Electric and Magnetic Susceptibilities*[4] [25]. Van Vleck maintained a special friendship with Kramers and the work in Leiden on paramagnetic[5] relaxation[6] was of great interest to him. In 1934, Van Vleck became Full Professor at Harvard, and in 1943, head of Harvard's Radio Research Laboratory, an institution closely associated with MIT's Rad Lab. In 1946 he was still connected with Harvard as a Full Professor.

---

[4]Magnetic susceptibility is the extent to which a material becomes magnetic by magnetization (e.g., moving a magnet over the material) or due to exposure to a magnetic field.

[5]Paramagnetism is a form of magnetism, with no spontaneous magnetism occurring in the absence of an external field. If a magnetic field is applied, then the spins of the atoms align themselves with the field lines of the magnetic field. At higher temperatures, the effect gets smaller as the vibrations due to the higher temperatures tend to disturb the arrangement.

[6]Relaxation refers to the return to the equilibrium state. For example, if a magnetic field is applied (at low temperature) and this is suddenly switched off, the return of the spins of the atoms of the non-equilibrium state during magnetization to the equilibrium state without magnetization is then called relaxation.

Research conditions

After the 'courses', Bloembergen only did research in the laboratory. Purcell gives him a quarter (!) assistantship, so he was a bit less hard up financially speaking. Every time he ran out of money, he had to ask for an extension.

Cambridge was a small city, but Boston could be reached by metro within fifteen minutes. Cambridge was not the kind of place that could provide much distraction, but rather a small town, and it soon felt familiar to Bloembergen. His room was on the university campus, very close to the lab. He could eat in the university canteen or in the numerous restaurants near the campus, so he hardly left the university grounds at first.

According to Bloembergen, there was an exceptionally good research climate in the faculty of physics. There were numerous lectures, symposia, exchange programs with MIT, guests from the other universities in Massachusetts (Boston University, Tufts University), New York (Columbia, New York), Connecticut (Yale), and so on. They worked much harder than in Utrecht.

In his own words, Bloembergen had ended up among researchers who would later receive the Nobel Prize for their work. First of all, of course, Purcell himself. Then there was Norman Ramsey, Van Vleck, and Schwinger. As early as 14 March, Bloembergen would give a lecture at a colloquium in MIT, in which he presented Piet van Heerden's crystal counter at Purcell's request. This counter was discussed when Bloembergen talked about his work in Utrecht. There was great interest in this detector in Harvard. As mentioned earlier, it was used to detect radioactive radiation. Bloembergen was so successful with the lecture that Piet van Heerden was asked to become Assistant Professor in Harvard.

Bloembergen told Milatz in a letter that he had no major problems with English, but he did have to read the formulas in English. He gives as an example: $1/\sqrt{N}.e^{h\nu/kT}$.

The cost of living was *"fantastically high"*, but the wages *"idem"*. As assistant, Bloembergen had to set up apparatus [26].

7.1   The Story of Nuclear Magnetic Resonance (NMR). Was NMR a Dutch Invention?

What is NMR?

According to a simple model, an atom consists of a nucleus surrounded by a (cloud of) electrons. This nucleus consists of protons, which are positively charged, and neutrons, which have no charge. Every electrical charge that moves causes a magnetic field, and so therefore do the protons of a rotating atomic nucleus. That field is proportional to the spin. When an external magnetic field is applied, two energy levels are created, one low level where the direction of the rotation axis of the nucleus is parallel to the magnetic field, and one higher level where the rotation axis lies opposite to the magnetic field. The very small difference between the two energy levels is proportional to the strength of the magnetic field. Indeed, strong magnetic fields are needed to make the difference visible at all; with modern NMR spectrometers, such a field would be 1–20 T.[7] If a radio frequency of (typically) 60 MHz is now applied, which corresponds exactly to the energy difference between the two levels, the nuclei at the low energy levels will be shifted to the high levels. If the radio frequency is subsequently slowly increased (field sweep), the radiation will first be absorbed when the nuclei are excited; this is called nuclear magnetic resonance. Then, when the nuclei return to their old energy level, emissions will occur.

The story of NMR

Isodor Rabi (1898–1988), 'Mr. NMR'

Rabi obtained his doctorate with Wills in 1927 on a subject in theoretical physics that he had devised himself. He studied the magnetic susceptibility of a group of crystals. Rabi went to Göttingen, the Mecca of quantum physics, in 1927, and then did theoretical physics in Hamburg with Pauli. Otto Stern, also in Hamburg, studied molecular beams and Rabi was very interested in this. Heisenberg recommended Rabi to Columbia University and Rabi was accepted there in 1929 and in the end would never leave. In Hamburg the magnetic moment of the whole atom could be determined for a number of chemical elements. Rabi wanted to do the same for the nucleus, but the magnetic moment was 2000 times smaller than that of the whole atom. In 1933 Rabi and his student Victor Cohen succeeded in using improved equipment and measurement methods to demonstrate the spin of the nucleus in sodium vapor (Fig. 7.5). The Rabi group determined the magnitude of the magnetic moment of a number of other elements. Stern had shown in 1932 and 1933 that if the magnetic field on a molecular beam was suddenly reversed, the spins of the atoms would also be 'reversed'. After 1935, Rabi started to investigate the same thing for atomic nuclei. To this end, he applied a rotating magnetic field. After extensive theoretical considerations and the development of complex equipment, the research reached a dead end, until 1937. Then Gorter visited Rabi!

Cor Gorter (1907–1980)

Gorter [27, 28] studied physics in Leiden. During his studies he was, according to him, strongly influenced by Paul Ehrenfest [29]. Ehrenfest found that although he did not understand much of the wonderful new techniques developed in the radio industry, there were great opportunities for applications in scientific research. Gorter took this into account in his thesis, in which he suggested spectroscopic research with short radio waves. In 1932 Gorter obtained his doctorate with Wander de Haas (1878–1960), the successor of Heike Kamerlingh Onnes. His dissertation had the title: *Paramagnetic properties of Salts*. In this thesis, Gorter summarized what was known about paramagnetism at the time: a statistical treatment of the contributions of atomic magnetic moments, the size of which could be calculated for each atom or ion using the theory of atoms. From 1931 to 1936 Gorter was curator of the Teylers Museum. There he became familiar with radio techniques. From 1936 to 1940 he was

---

[7]For comparison, the Earth's magnetic field at the surface is about 50 μT.

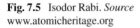

Fig. 7.5 Isodor Rabi. *Source* www.atomicheritage.org

a lecturer at the University of Groningen. After his dissertation, Gorters focused on the inertia of the magnetization and in particular the way the directions of the atomic magnetic moments, as determined by quantum theory, are distributed statistically. These are processes which are determined by the interaction of the ion in question with its environment and which are influenced by the characteristics of the crystal structure on an atomic scale. He was the first to carry out systematic research on these issues.

As early as 1934 he developed a method to determine the magnetic moments of atomic nuclei in a magnetic field at low temperatures. He proposed to create a radio frequency field that could 'flip' the spins (Fig. 7.6).

Gorter was the first to describe paramagnetic relaxation in 1936 [30]. He failed to demonstrate resonance. However, Gorter had only a small group in Groningen. He later showed by calculation that the method they had used, called calorimetry,[8] was not sensitive enough to reveal any effect.

Rabi and Gorter, a Nobel Prize, and a footnote

In the summer of 1937, Gorter left for Columbia University, where Rabi's group was struggling with resonance. Gorter knew Rabi and thought that his molecular

---

[8]Calorimetry is a method for measuring the heat capacity of a substance or the heat of a chemical reaction on the basis of energy absorbed.

Fig. 7.6 Cor Gorter. *Source* Museum Boerhaave, Leiden

beam technique could avoid the problems he himself had had. Gorter was shown around by one of Rabi's Ph.D. students, Norman Ramsey. He showed Gorter the arrangement with the molecular beam that was deflected by a rotating magnetic field. Gorter realized that if the rotating magnetic field was replaced by a constant magnetic field with a radio frequency field across it, resonance could be achieved. Gorter presented Rabi with his calculations, but Rabi did not seem interested. When he returned to the Netherlands, Gorter was expecting Rabi to meet him in Groningen and continue the discussion. But when Rabi heard Gorter use the word "oscillator" for the radio frequency field, he quit the research of his two research groups the Monday following Gorter's visit on Saturday and changed the equipment as Gorter had suggested. According to Ramsey, Rabi himself had already thought of something like this, but it was only through Gorter's discussions and proposals that Rabi became aware of the potential of the method Gorter was proposing [31].

A few months later a Letter to the Editor by Rabi appeared in Physical Review, announcing the discovery of NMR in lithium [32]. The method used was Gorter's. The latter was thanked by Rabi for his advice in a footnote: *We are very much indebted to Dr. Gorter who, when visiting our laboratory in September 1937 drew our attention to his stimulating experiments in which he attempted to measure nuclear moments by observing the rise in temperature of solids placed in a constant magnetic field on which an oscillating field was superimposed.* Rabi was awarded the Nobel Prize in 1944 *for his resonance method for recording the magnetic properties of atomic nuclei.* Gorter wrote in a review article that covered his failed attempts to demonstrate, among other things, NMR: "*I cannot deny that I felt some pride, mixed with the feeling that my contribution had been somewhat undervalued though my advice was acknowledged in the Letter.*" [33]

In a letter dated 5 March 1938, Gorter wrote in clear language: "*I was very glad to see today in the Phys. Rev. that you have realised the experiment I proposed to you last summer. When you told me you were going to try it I was convinced that you would succeed. But it is certainly worth a congratulation that you and your collaborators succeeded so soon and with such good precision.*" [34] Rabi did not write back regarding Gorter's proposal, but he did give praise for all the results achieved [35].

According to Bloembergen, if Gorter had been staying in the United States at the time, he would have shared the Nobel Prize.

After his discovery, Rabi already predicted that, although he had only found NMR in isolated molecules, this phenomenon would also occur in complex matter such as human tissue and steel.

Gorter in Amsterdam and Leiden

Gorter became professor at the University of Amsterdam in 1940, succeeding Pieter Zeeman. In his oration in June 1940 he described once again the way nuclear magnets in external magnetic fields give rise to extremely sharp "*spectral lines*", stating that this method would lead to accurate and reliable results for determining the strength of such magnets [36]. In 1942 he tried to demonstrate NMR in lithium chloride and potassium fluoride, again in Leiden, working with Bert Broer (professor of theoretical physics, 1916–1991), again without success. According to Bloembergen, Gorter was a bad experimenter and the measuring setup used was rather primitive. That may explain why he found nothing. Bloembergen did find NMR in the same monsters after the war.

In the Hunger Winter, Gorter wrote a standard work on paramagnetic relaxation in the Zeeman Laboratory in Amsterdam, but due to the circumstances of war and the subsequent liberation, it was not published until 1947 [37].

Felix Bloch (1905–1983), research up to 1946

Independently of Rabi and Gorter, Bloch had succeeded in demonstrating nuclear magnetic resonance [38]. In the atomic nucleus, next to the positively charged proton, there is the uncharged neutron. Bloch was fascinated by the idea that the non-electrically charged neutron might have a magnetic moment. In 1937 during Fermi's visit to Stanford, Bloch came up with the idea of using magnetic resonance to determine the magnetic moment of the neutron (Fig. 7.7). In 1938 he tried to generate neutrons using a radioactive radium source. Rabi came to take a look every morning, but it didn't work. In the autumn of 1938, in collaboration with Luis Alvarez (1911–1988), Bloch began experiments in Berkeley with the cyclotron there. The first effect with a neutron beam between iron plates that were either magnetized or not was found on 3 May 1939. As Rabi had done, the beam passed between the poles of a magnet and a radio frequency field was applied. Bloch continued the experiments with a larger cyclotron of his own until the summer of 1942, but was then recruited by Oppenheimer for the Manhattan atomic bomb project.

Fig. 7.7 Felix Bloch. *Source* www.phys.ethz.ch/the-department/history.html

Norman Ramsey (1915–2011), the first step toward the NMR of compounds

Ramsey studied at Columbia University and became a student under Rabi. He did his research on molecular beams. Ramsey became interested in magnetic resonance, and in the summer of 1937 he contacted Rabi. He advised Ramsey to choose a different subject, because the research on molecular beams was now being sufficiently exploited. Two months later, everything was different, because by then Gorter had made his September visit and Rabi had demonstrated magnetic resonance. Ramsey was Rabi's only Ph.D. student and would be the first to complete a thesis on NMR. In molecular hydrogen measurements, Ramsey had found six different very sharp peaks instead of the expected single peak [39]. This was the birth of radio frequency spectroscopy. Atomic hydrogen contains only one proton. Molecular hydrogen, H_2, however, has two protons, whose magnetic moments interact not only with each other and the external magnetic field, but also with the internal structure of the molecule (Fig. 7.8). The molecule itself also has a magnetic moment. Research on hydrogen deuteride HD pointed to something else. This is the compound of atomic hydrogen (H), whose nucleus contains just one proton, and deuterium (D), whose nucleus contains one proton and one neutron. It turned out that HD has not only a magnetic dipole, but also an electric quadrupole moment. The molecule does not look like a sphere, but more like an American football, spinning around its axis. Since Ramsey made the discovery with equipment that was built with the help of others, he was not allowed to include it in his thesis and had to look for a different subject. Ramsey completed his thesis in 1939 and left for Washington DC as a Carnegie

Institution Fellow. He obtained his Ph.D. in 1940 on the radio frequency spectrum of the rotating magnetic moment of the molecules H_2, D_2 (deuterium), and HD (hydrogen deuteride). In September 1940, Ramsey accepted a position as associate at the University of Illinois in Urbana-Champaign. This was a position which could possibly lead to an 'assistant professorship'. After only a few weeks Ramsey went on various missions to carry out research for the war effort. In January 1946 he returned to Columbia University to resume his research with Rabi on radio frequency spectroscopy of molecular beams. Later he did research on particle accelerators and together with Rabi became one of the founders of Brookhaven National Laboratory, the national accelerator laboratory [40]. In 1947 he became professor at Harvard University and mentor of the many Ph.D. students who worked on NMR.

Edward Mills Purcell (1912–1997) [41] and his group: Robert Pound (1919–2010), Henry C Torrey—NMR in paraffin

In 1929, Purcell went to Purdue University, Indiana, to study electrical engineering. Throughout his scientific life, he would retain his love for the engineering profession and look for applications of his research. At one-point Purcell shifted his attention from engineering to physics (Fig. 7.9). He managed to get a scholarship for Harvard and stayed there for the rest of his career from 1934. In 1936 he attended a lecture series by Van Vleck in Harvard on electrical and magnetic susceptibility, which according to his own words was decisive for his training in physics. Malcolm Webb, a theoretical physics student who had come from Wisconsin to Harvard, like Van Vleck, also contributed a great deal to Purcell's knowledge of quantum mechanics. Webb and Purcell published an article on magnetic susceptibility anomalies at temperatures below 1 K (−272 °C), previously recorded experimentally in 1926 by Debye and

Fig. 7.8 Norman Ramsey. *Source* www.independent.co.uk

Fig. 7.9 Edward Mills Purcell. *Source* www. sciencephoto.com

others. Purcell's dissertation was about mass spectrography,[9] a technique introduced by Aston in 1919. He obtained his Ph.D. at Bainbridge in 1938 and remained at Harvard as an 'instructor'.

Like many others, Purcell was recruited for the Rad Lab at MIT in the Second World War. In 1942 he succeeded Ramsey as head of the Fundamental Development Group, which was part of Rabi's Research Division. Robert Dicke also worked for Purcell in the group.

Purcell came across the 'water problem' at the Rad Lab. In the effort to work with radar of shorter wavelengths, a wavelength of 1.25 cm was chosen for further research. However, this turned out to be a bad choice: compared to the 3 cm radar used so far, the results were not better, but worse. Water vapor in the atmosphere absorbed all radar radiation of 1.25 cm and this wavelength therefore turned out to be unusable. Then Purcell had the idea of investigating magnetic field absorption by nuclei.

In early 1945 it became clear that the Second World War was coming to an end and the physicists at the Rad Lab began to think about returning to their respective

---

[9]Mass spectrography, currently called mass spectrometry, is used to separate and identify molecules. To this end, a sample is ionized and the ions formed are accelerated by an accurately controlled electric field. After that, the ions follow a circular path through an applied magnetic field. The ions are spatially separated different their different mass to charge ratios and subsequently detected.

universities. During a lunch, Purcell discussed a certain idea with two colleagues from the Rad Lab, Henry Torrey and Robert Pound. *Briefly, it involved measuring the forces at work inside atoms, taking advantage of the fact that most nuclei act as if they are extremely small, rapidly spinning bar magnets. The only difficulty to be solved was that no one knew whether the extremely feeble effect of these magnets in ordinary substances could actually be detected* [42]. Torrey came from Rutgers University, New Jersey, and Pound was a Harvard Junior Fellow. The necessary equipment had to be borrowed and a sufficiently strong magnet was found in a shed at the Lyman Laboratory (Harvard). That magnet was inherited from J. Curry Street (see Chap. 6), who had used it for his measurements of cosmic radiation. Experiments were carried out at every free moment, in the evening, at night, and on Sundays (all three were still working in the Rad Lab). Measurements were performed in paraffin. Pound designed the measurement electronics; he was one of the key figures at the Rad Lab regarding noise problems and became known as the electronic wizard. He later became Full Professor, although he never got his Ph.D. The (Fig. 7.10) measurements were made using a bridge, as was customary in such measurement techniques. An example of such a bridge is the Wheatstone bridge, where an unknown resistance or an unknown voltage can be measured very accurately by setting a kind of balance (formerly a galvanometer) that is kept at 0 V. They had no idea how long it would take for the

Fig. 7.10 Robert Pound.
Source www.nytimes.com

protons to relax again after they had reached the higher energy level. In a calculation by Ivar Waller for electron spin resonance in crystals, he found that it could take a long time before relaxation occurred [43]. Torrey calculated the relaxation time for NMR and found that it could be anything between a few and many hours. In December 1945 the equipment was ready for the measurement to be made. Purcell, Pound, and Torrey stayed at work until four o'clock in the morning and when they went home in a snowstorm, they had a working setup. It took a few days before they had time to continue working, but one Saturday Purcell went to the barn, warmed up the equipment, and waited for the others. They continued together and the experiment worked: for the first time NMR had been demonstrated in condensed matter. The Street electromagnet had stayed on for ten hours because of the uncertainty about the relaxation time. However, that time turned out to be only 0.01 s. The publication of the work was received on 24 December and appeared as a publication by the Rad Lab at MIT [44]. There were three references in the article, in which earlier work by Rabi, Gorter, and Waller was mentioned.

In 1952 Purcell would share the Nobel Prize for NMR work with Bloch.

Bloch in 1946 [45], NMR in water

Bloch had returned to Stanford after the Manhattan project. As described earlier, Gorter and Broer's attempt to obtain nuclear magnetic resonance in 1942 had failed. Bloch was interested to find out whether, with a stronger radio frequency field than that used by the previous groups, magnetic induction[10] of atomic nuclei could be achieved. They recorded the signal from the second coil, which was induced by the first coil. Therefore, Bloch's group called this method magnetic induction. Bloch used a magnetic field of about 0.18 T and a radio frequency field of 7.76 MHz. The test object was an ampoule containing 0.1–1 g of water. Shortly after Christmas 1945, the experiments did succeed, two weeks after the successful experiments by Purcell's group. Bloch had no idea beforehand how long he had to wait until he saw relaxation occur. Purcell had obtained 0.01 s for paraffin, but Bloch found that it was much longer for water, namely 2 s [46]. In October 1946 Bloch's most important article appeared, in which he discussed the two relaxation times he had found: the thermal or longitudinal relaxation time T_1 and the transverse relaxation time T_2 [47]. From then on, Bloch's research focused mainly on NMR in liquids (Fig. 7.11).

Bloch and Purcell

The difference between the working method used by the Purcell group and the one used by Bloch can be described as follows. It is possible to bring a (suspension) bridge to resonance. The Tacoma Narrows Bridge was a suspension bridge in Washington state that was opened for traffic in July 1940. But on 7 November 1940 the enormous suspension bridge started to swing under the influence of the wind, until finally the middle part broke in pieces; the bridge was referred to as *Galloping Gertie*. The wind (67 km per hour) provided an external frequency that corresponded to the bridge's

---

[10]Magnetic induction occurs in nuclei when a strong radio frequency field is applied across a magnetic field. The effect occurs across the radio frequency field and can be measured with a coil.

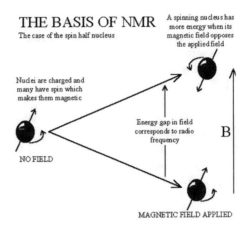

Fig. 7.11 The principle of nuclear magnetic resonance (NMR). The nuclei of the atoms are charged and rotate around their axis, creating a magnetic field. This is the situation if there is no external magnetic field. If a strong magnetic field B is applied, a part of the cores will stand parallel to the magnetic field (the lower ball) and a part will stand transversely to the magnetic field (the upper ball). These last balls have a higher energy level. The higher energy level can be achieved by applying a radio frequency signal with the correct frequency. *Source* www.chem.ch.hujiac.il

own natural frequency. This natural frequency of a bridge can also be determined by attaching a small vibrating device to the bridge and measuring whether resonance occurs further on using a sensitive measuring device. This was Purcell's method, made difficult by the fact that highly sensitive measuring equipment was required. Bloch's method was similar to the wind blowing against the bridge; if the power was sufficient, the resonance would become clearly visible without the need for sensitive equipment. There was one drawback: the resonance frequency distribution had to be extremely narrow, otherwise nothing would resonate. Purcell used one coil with sensitive detection equipment, while Bloch used two perpendicular coils.

One more fundamental difference, according to Rigden, was that Bloch assumed more classical physics with concepts like precession,[11] torque,[12] and electromagnetic induction, which relate to the spatial distribution and reorientation of magnetic moments relative to an external magnetic field. Rigden called this the dynamic approach. According to Rigden, Purcell assumed the quantum mechanical transitions of the nuclear spins from one discrete energy level to a higher one due to energy absorption from an oscillating magnetic field. Rigden calls this the spectroscopic approach [48].

Before Bloch and Purcell saw each other for the first time at a meeting in honor of Rabi's Nobel Prize (1944) in Cambridge from 22 to 24 April 1946, Bloch did not

---

[11] Precession is a motion of the axis of rotation of a rotating object under the influence of an external force, as seen, for example, in a spinning top.

[12] The torque is a measure of the rotational effect of a force, such as the force needed to loosen or tighten a bolt.

believe that Purcell had also demonstrated nuclear magnetic resonance. Only later did Bloch become convinced.

It is clear that Purcell and Bloch succeeded in demonstrating NMR in condensed matter using totally different approaches. Rabi could have demonstrated NMR in individual molecules in a molecular beam. To detect NMR in lithium 7 (^{7}Li), Rabi had to vaporize molecules of lithium chloride and bring them into a vacuum. Purcell and Bloch demonstrated NMR in paraffin and water, respectively, while the material remained intact. What Purcell and Bloch showed, namely NMR of protons in intact material, would prove to be of great significance for research in physics and chemistry, and later for applications in chemistry and biology.

7.2 Bloembergen and NMR, a Happy Combination

Independent research on NMR, from barn to cellar

When Bloembergen was about to start his experiments, Purcell and Pound were still working at the Rad Lab. Purcell and Pound were involved in a huge project to record all the activities of the Rad Lab during the Second World War [49]. This work, which is paid for by the United States government, would eventually fill 28 (!) volumes; the authors took six months to write it [50].

Bloembergen was thrown in at the deep end, because Purcell and Pound had no time for him at the beginning. His first assignment was to remove the paraffin from the resonant cavity in the existing equipment in the shed near the Lyman Lab, which had been used in December 1945 by Purcell, Pond, and Torrey in the 30 MHz resonant circuit. Then he had make a kind of mud from a mixture of mineral oil and calcium fluoride powder and fill the 30 MHz cavity with it. The intention was to create a suspension in which the calcium fluoride would float, as it were, in the oil. The approximately 1 L of this 'mud' had to be pressed through a narrow opening in the vibration cavity (just like a toothpaste tube). Pound adjusted the electronics, Purcell controlled the magnetic current, and Bloembergen read the resonance signals. Within a few days, Bloembergen was able to demonstrate resonance of both protons and fluorine 19 (^{19}Fl). So just two weeks after his arrival in Harvard, Bloembergen was already working on groundbreaking research (Fig. 7.12).

Purcell decided that a new NMR setup was needed. Street's magnet was not only a monster, but also produced a magnetic field with 'monster-sized' inhomogeneities. An ancient electromagnet was obtained from the Société Genevoise, dating back to 1905, hence older than the people who worked with it. Bloembergen had to make a control circuit for the magnet, because at the frequency of 60 Hz at which the magnet was supplied, there was a large 'ripple' at 1 Hz. Bloembergen could not get the wrinkle out. Pound and Purcell didn't believe him until Pound started looking at it himself: two months of work by Bloembergen for nothing. The strange deviation was caused by an error in a winding. The problem was solved by feeding the magnet, not with AC voltage, but with DC voltage from batteries [51].

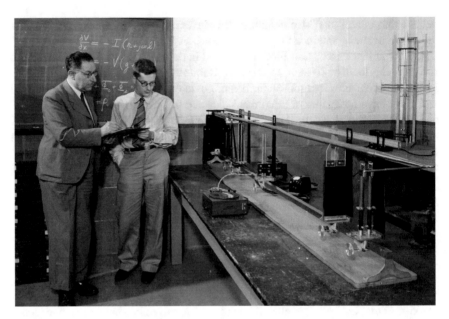

Fig. 7.12 Bloembergen (right) in the laboratory of Professor Le Corbeiller in Harvard, December 1946. Photo for a booklet. *Source* Nico Bloembergen, Cruft Laboratory, Harvard University

Bloembergen then had to make a phase sensitive detector. The bridge circuit previously used by Purcell, Pound, and Torrey was certainly sensitive, but the noise in the measurement setup was also amplified and very low voltages required a different measurement setup.

Robert Dicke had designed a microwave 'radiometer' in the Rad Lab's Fundamental Development Group. When the problems with the 1.25 cm radar appeared, Dicke had helped determine the absorption line of water vapor in the atmosphere. Part of that radiometer was a phase sensitive detector. Bloembergen's experiments in Utrecht with a phase sensitive detector and his knowledge of the Brownian motion responsible for the noise now came in handy. The Utrecht detector had had a galvanometer as measuring device, while Dicke had designed a so-called lock-in-amplifier that worked entirely electronically. Later this detector would become commercially available from Princeton Applied Research (Fig. 7.13).

Bloembergen built Dicke's detector.

In an interview about that time, Pound tells us: *And I oversaw most of the work—I educated him [Bloembergen, RH] in the electronics aspect of doing this, and we decided to go for a full modulated lock-in amplifier detection system and so forth* [52].

Bloembergen went to the instrument shop for that, but there was no money to have it made there and he tinkered with all kinds of materials to build a generator that would generate radio frequency signals. Bloembergen tells us later in an interview: "*We didn't even have money to buy a big resistor to control the current of the electromagnet*

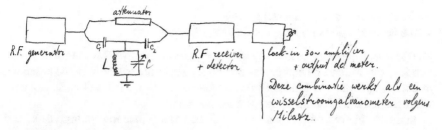

Fig. 7.13 Phase sensitive detector (lock-in-amplifier) as designed by Dicke and adjusted by Bloembergen. *Source* Letter 29 June 1946 from Bloembergen to Gorter. The Dutch text translates as follows: This combination works according to Milatz as an AC galvanometer

[…] We had to make it ourselves with a nickel silver band, which I then mounted with a clamp and a little oil in-between to control the current." [53] The setup was rather primitive; the basis was the generator, which generated a frequency of 30 MHz. There was also a so-called LC (coil-capacitor) circuit to detect the radio frequency, with a tunable capacitor. This was similar to former radio receivers, in which the dial to tune to different radio stations was also connected to a tunable capacitor. The generator coil had 7 turns, was 2 cm long and 7 cm in diameter, and was placed in a constant magnetic field. Two auxiliary coils were used to superimpose a 30 Hz alternating field of 100 μT. The 30 Hz alternating voltage signal was detected by a radio receiver, and then a phase sensitive detector. The signal was read on a voltmeter [54]. Bloembergen told Mattson: *"my craftmanship was marginal"*. The setup ran for eighteen months, the whole time that Bloembergen worked on NMR under the Purcell's leadership [55].

Until then Purcell had required about 1 L of sample to determine NMR. In Bloembergen's setup, thin-walled tubes were used, for which only 1 mL of sample was needed, i.e., a factor of 1000 less. The value of the constant magnetic field was slowly varied until the strength of the signal changed. At that point, resonance was reached.

In Utrecht, Philips radio tubes had been used for the measurement setups and Bloembergen now had to familiarize itself with RCA radio tubes. Furthermore, he had to learn technical English and deal with American components such as capacitors and resistors. He also had to familiarize himself with instrument-making procedures. However, he got plenty of help with all this, especially from Gordon Voorhis, one of Curry Street's Ph.D. students.

The measurement setup showed how much the work fitted in with the Utrecht tradition of instrument design, in particular with regard to Brownian motion (noise) and the phase sensitive detector. The principle of this measurement setup was used in all subsequent research until the 1950s.

Purcell, Pound, and Torrey had formulated a number of goals for NMR research in their first NMR publication:

- The determination of magnetic moments for other atomic nuclei.
- The analysis of spin-lattice coupling.

- The standardization of magnetic fields.
- The determination of the sign of the magnetic moments.

The first measurements are made by Bloembergen with paraffin (for control), calcium fluoride, and water. The equipment was no longer in the barn, but in the basement of the Lyman Lab, and Bloembergen had long working days. He often stayed until midnight and the lab almost became a second home.

In the Spring of 1946, Bloembergen's father Auke came to visit him. Auke had come to set up a sales office in the United States for the fertilizer company ASF/VCF [56].

Purcell and Pound were still working most of the time at MIT's Rad Lab; it was only in the summer of 1946 that they returned to the Lyman Lab and could spend more time on experiments.

At one point, Bloembergen told Pound that the strength of the signal changed when the sample was put back between the magnetic poles. Pound immediately and systematically passed the sample through the opening and they showed that the signal was strongest in one particular place. This was due to inhomogeneities in the magnetic material. Therefore, the actual line width could not be determined, since it varied with the strength of the magnetic field. However, the relaxation time T_1 could be determined. Somehow, the motion of molecules in the liquid was producing inhomogeneities in the magnetic field. Because Bloembergen had followed lectures on Brownian motion in Utrecht in 1942, he was familiar with the frequency distribution of random variables. Purcell suggested that he read Debye's book *Polar Molecules* [57].

After reading Debye's book, Bloembergen realized that, in polar fluids[13] such as water, due to the random rotation of molecules, a loss of permittivity may occur. The explanation lies in the fact that the Brownian motion of molecules in a fluid can be divided into two types of movements. Translation takes place when a molecule is pushed along by another molecule. Rotation occurs when a molecule starts to rotate around its axis due to a collision. In a polar fluid such as water (H_2O), the charge on the molecule is unevenly distributed. The oxygen atom O has a double negative charge and the hydrogen atoms H each have one positive charge. The molecule then forms an electric dipole. Due to the random rotation, the water becomes less polar, as it were. When the rotation comes to a halt, as in NMR, it is called relaxation.

In autumn 1946, Bloembergen calculated the dipole-dipole interactions between the spins (rotation around the axis) of two protons. He determined the spectral distribution of collisions between water molecules: one molecule gets a harder collision than another and the spectral distribution can be expressed as a function of the collision impulse.

---

[13]Polar fluids are good current conductors. Examples are water and ethanol (alcohol). A-polar liquids do not conduct, or hardly conduct at all, e.g., toluene, turpentine, chloroform, and alkanes such as paraffin. The permittivity (formerly the dielectric constant) is the tendency of a material to polarize in the presence of an electric field. Water has a high permittivity, that is, it can be easily polarized, whence an electric current can flow.

By means of a mathematical operation called the Fourier transformation, a signal can be transformed from the time domain where the measurement takes place (e.g., signal strength against time) to the frequency domain (signal strength against frequency). In this way it often becomes clear which are the most important frequencies that contribute to the signal.

From the Fourier analysis, it was clear to Bloembergen that, after the collision leading to rotation, the longitudinal relaxation T_1 (see Bloch) depended on the angular velocity (how fast the molecule was rotating around its axis after the collision). Regarding the transverse relaxation T_2, an important contribution could be seen near zero frequency. Bloembergen tested these assumptions experimentally by selecting liquids with different viscosities, such as alkanes from light to heavy oil and a mixture of water and glycerol. He showed that the more viscous the liquid was, the faster the relaxation would be. He also measured the relaxation of water. Bloch had done that before and found 2 s. Bloembergen found 2.3 s and that is the value that still stands today.

For crystals, the NMR signal is measured to have a broad line. The group led by Purcell realized that the line width should have a minimum if measured diagonally on the crystal lattice. Bloembergen got to work and in a few days had made a cylinder of fluorite (the calcium fluoride of previous experiments) which could be rotated in the field. The crystal was green, probably due to impurities in the crystal that shortened the relaxation time. Purcell calculated the minimum value for the crystal lattice on an old-fashioned mechanical calculator (the make was Marchant), and the theoretical and measurement results came out in a joint publication [58].

Purcell also had discussions with Bloch about their research. According to Bloembergen it was clear that there was competition, but they remained 'friendly' to each other. In July 1946 Bloch sent the theoretical explanations for his experiments to *Physical Review.* These were published in October 1946 and included a set of formulas, later known as the 'Bloch equations', for T_1 and T_2.

Bloembergen also studied the influence of increased gas pressure on the line width, using hydrogen as an example, and found that, contrary to the line broadening observed in optical spectroscopy, line narrowing is observed in NMR, and he gave a theoretical explanation for this [59].

As a final experiment he determined the relaxation time of deuterium and hydrogen in a 50/50 mixture of heavy water (D_2O) and plain water. He then measured the quadrupole moment[14] of deuterium in the liquid state, where Rabi and Ramsey had only done so for molecular beams.

In autumn 1946 Purcell and Pound had little time for research due to educational obligations and Bloembergen was largely alone once again. However, Purcell, Pound, and Bloembergen continued to have very intensive work meetings.

At the end of 1946, Bloembergen's brother Evert, who then lived in Quebec, Canada, came to visit him. Evert worked for a subsidiary of the fertilizer company in

---

[14]A quadrupole is one of a sequence of configurations of, e.g., electric charge or current, or gravitational mass that can exist in an ideal form, but is usually just part of a multipole expansion of a more complex structure reflecting various orders of complexity.

which his father Auke was co-director [60]. Evert is astonished to see Nico's working conditions in the basement without daylight and in the stench of acetone and sulfuric acid [61].

Around Christmas and the New Year, Bloembergen had to stay in his room because of a flu attack and he calculated T_1 for water.

All these experiments and theoretical considerations occupied Bloembergen over the last three months of 1946. During this period, Bloembergen progressed from a laboratory assistant to a fully-fledged research assistant.

Around this time, Purcell commented to Bloembergen: *"You were very well prepared"*. He meant that, in Utrecht, Bloembergen had had a good education as a scientist [62].

In January 1947, Bloembergen had to give his first conference talk. This took place at the annual meeting of the American Physical Society at Columbia University in New York City for an audience of perhaps a thousand physicists. Among those present were Rabi and Charles Townes. Here Bloembergen would present the results of his research on NMR relaxation in liquids. He had just ten minutes to shine his light on T_1 in liquids. Thanks to a rehearsal before Purcell and Pound in Harvard, he managed to finish his talk, in English but with a strong accent, just a few seconds before he was asked to stop by the chairman. After the talk, Rabi approached him and said: *"Let's all get out of here and discuss things privately."* Bloembergen missed the rest of the lectures due to the discussions with Rabi and others [63].

In his letter to Gorter on 2 March 1947, Bloembergen explained that Purcell's method had been applied in Chicago for the accurate determination of the neutron and deuteron moments of tritium [64]. The investigation was secret because it was being carried out for the US Navy. In his letter, he says: *"It's a pity that Purcell missed out on this tritium, but there is no way of competing against these people. We would have difficulty getting hold of any tritium."*

In spring 1947, Bloembergen had completed his experiments. The number of intensive work meetings with Purcell and Pounds was gradually decreasing. Purcell and Pound went on to engage in other NMR research and allowed Bloembergen complete freedom in his research on fluids. Bloembergen felt happy in his research room in the basement, although the stink of acetone and sulfuric acid vapors was still very present. He used acetone and 'dry ice' (solid carbon dioxide, CO_2) to cool to $-78\,°C$. The sulfuric acid vapors were caused by the nightly charging of the truck batteries, which were also kept in the basement, and without any extraction facility. Occupational hygiene had not yet been invented!

In an interview with Mattson, he commented on this:

My plodding study during the grim war years in Holland started to pay off. The traditional Dutch system had given me a very solid background in the fundamentals of physics. I was well prepared to catch up quickly with the advances made in the free countries during the war. So I plunged into a room in the basement of Lyman Laboratory, never noticing the lack of daylight [65].

He discussed things regularly with George Pake (1924–2004). Pake was Purcell's second Ph.D. student working on NMR in the adjacent cellar; they spoke to each other through a hole in the wall. Pake was using a magnet developed by Purcell. This magnet was moved to the Physics Lecture Hall, where Bloembergen gave a public lecture in the Harvard Physics Colloquium series in April 1947. The subject was the same as in the previous presentation, but this time it was combined with relaxation theory in time-varying internal magnetic fields. Commenting on Bloembergen to Mattson, Pake said: *"Bloembergen was almost a second professor for me, although he was a graduate student. He was older and had a lot more experience. When I was starting out, I knew nothing about magnetic resonance and I would shout through the hole in the wall to Bloembergen who would give me tips as to how to get out of my particular dilemmas. And I really liked him, because his experimental operation was way ahead of mine."* [66]

George Pake and Bloembergen lost sight of each other until 2001. Then Pake came to live in Tucson, Arizona, three houses down the road from Bloembergen. They visited each other a lot, but Pake passed away in 2004 [67].

Before Bloembergen started writing his thesis, he made a journey by train and bus to Denver, Colorado, where he met the parents of his roommate in Perkins Dormitory, Herbert Broida. The prairies and the Rocky Mountains made an overwhelming impression on Bloembergen. On his way back to Cambridge, he gave a lecture at the University of Chicago, where he met Herbert Anderson, who was doing NMR research on (radioactive) ^3H tritium. Fermi was also working there at the time. He was in hospital, but wanted to know how the NMR study in Harvard was progressing. Bloembergen visited Fermi in hospital and informed him.

Back in Harvard, Bloembergen met Gorter for the first time.

In the summer of 1947, Van Vleck, a close acquaintance of Gorter, invited the latter to give a course at Harvard. There had been an exchange of letters between Gorter and Van Vleck since 1932. On 6 August 1945, Van Vleck wrote to Gorter:

"This is just a line to say how happy I was to get your card and to learn that you and your family had successfully weathered the barbarian invasion. [...] Practically all the scientists who write from the occupied countries speak particularly of their sense of being isolated from the scientific world; but the funny thing about it is that practically all the efforts of the physicists here since 1940 or so have been devoted to war problems of a confidential or secret category." [68]. *Van Vleck became head of the faculty in March 1947 and sent a telegram to Gorter in this capacity:*

"Can you accept summer lectureship harvard june sixteenth to september sixteenth salary twenty five hundred two courses each four hours weekly dynamics relaxation van Vleck." [69]

Such an invitation was nice for Gorter, because he had been stuck in the Netherlands throughout the war.

According to Bloembergen, the intention was to give Gorter two different series of lectures, namely mechanics and a special subject, which could be relaxation phenomena [70]. He remarks that Gorter was famous for his research into paramagnetic[15]

---

[15]Paramagnetism is a form of magnetism whereby certain materials are weakly attracted by an externally applied magnetic field, and form internal, induced magnetic fields in the direction of the applied magnetic field.

relaxation without resonance. Gorter presented the contents of his book in Harvard and the audience, including Bloembergen, read it from A to Z. Gorter studied salts with normal paramagnetism, such as gadolinium, chromium, iron, manganese, nickel, and copper. At very low temperatures, such compounds lose their paramagnetic properties and become either ferromagnetic[16] or the magnetic susceptibility becomes independent of temperature. The higher the temperature, the more internal motion and the less magnetization, as is easy to demonstrate by means of a permanent magnet. If the magnet is heated to a high enough temperature, the magnetism disappears completely. On the other hand, susceptibility increases at lower temperatures. In 1945 it was not yet known how long it would take to reach a balance. Bloembergen found it extremely instructive to study the precursors of resonance [71]. He was fascinated by Gorter and Van Vleck's discussion on paramagnetic relaxation [72].

Do you want to promote in Utrecht, Harvard or Leiden?
In his first letter to Milatz of 27 February 1946, Bloembergen notes that it was not wise to obtain a doctorate in Harvard, since the Dutch doctoral exam was not recognised and he would then be forced to take four English exams with foreign examiners [73]. He consulted with Uhlenbeck and he strongly advised him to take his doctorate in Utrecht. Uhlenbeck went on to say that there would be no objection to this in Harvard. Bloembergen asked Milatz if this was correct and if there were any regulations regarding publications. Purcell agreed with the idea that Bloembergen should receive his doctorate in Utrecht. As thesis, Bloembergen proposed relaxation in NMR and Purcell fully agreed [74].

In a letter to Gorter dated 29 June 1946, when Gorter was still in Amsterdam but already appointed in Leiden, Bloembergen mentioned that, in a letter from Gorter to Purcell et al., he had read that Gorter was not sure whether to start NMR research in Leiden [75]. Bloembergen argued that the new research field opened up so many avenues that he recommended Gorter to do this research in Leiden, and did so at length. In addition, there was considerable interest in NMR in the USA, according to Bloembergen. He referred to the NMR work by Bloch and Hansen in California, at the Nuclear Institute in Chicago, and in MIT, where NMR research had actually started. The letter then gave a description of his experimental setup and the results obtained with it. Bloembergen stated, as in the letter to Milatz mentioned earlier, that he wanted to take his doctorate in the Netherlands. However, although he was not one of Gorter's students, he still wanted to take the doctorate with Gorter in Leiden, because he considered that the subject was more appropriate to Gorter than to Milatz. Gorter answered Bloembergen in a letter of 10 July 1946, saying that he found the proposal interesting. Gorter explained that he had lost some of his fascination for the subject, because he found it more interesting to make a discovery than to improve methods [76]. Moreover, he feared that the relaxation times of solids at low temperatures would be rather long. As a last argument against carrying out NMR research in Leiden, he explained that, because Leiden had been closed in 1941, he could hardly get any staff and also that radio components were difficult to obtain.

---

[16]Ferromagnetism is a form of magnetism whereby spontaneous magnetization occurs without the application of an external magnetic field. The best-known example of this is iron.

Nevertheless, Gorter concludes the letter by saying: *"I will put my nose in it [...] Because after all, the whole area arose from my work on paramagnetic relaxation."* He asked Bloembergen to keep him abreast of developments in the USA, insofar as there were no secrets.

Gorter commented: *"I think it is best if you want to take a Ph.D. with me, but it must be a condition that Prof. Milatz fully agrees with this. People sometimes do not like to see their good students pass to another, and the Leiden-Utrecht relationship is delicate. There is nothing against you possibly telling Prof. Milatz that I made this condition."*

On 11 January 1947, Bloembergen wrote in a letter to Milatz, that everything was going according to plan and that he would probably stay in Harvard for another year [77]. Bloembergen also wrote that he had not yet decided whether he wanted to take the doctorate in Harvard or in the Netherlands. He mentions the subject of his doctoral thesis and noted that, if he wanted to do it in the Netherlands, it would be better to do it with Gorter since the latter had predicted resonance. Bloembergen did indeed follow Gorter's recommendation, saying that he would prefer Bloembergen to do his doctorate in Leiden, but that he made it a condition that Milatz agree. Bloembergen asked Milatz what his position would be on this, also with regard to the idea of doing the doctorate in Harvard.

In March 1947, everything gained momentum. Bloembergen heard nothing from Milatz and he wrote about it to Gorter on 2 March 1947 [78]. He already had a provisional title for his thesis, which would be: *The relaxation time and line width of nuclear magnetic resonance in liquids.* Bloembergen proposed to return to the Netherlands at the end of September 1947 and estimated that the dissertation could be completed in December 1947, at the latest during the Christmas holidays.

On 8 March 1947, Gorter wrote to Milatz:

Dear Milatz,
Last summer, your former employee N. Bloembergen wrote to me about his research on nuclear moments with Purcell and he asked me if I would be willing to act as his supervisor if he wanted to obtain a Ph.D. on this subject. I answered him that I would be very happy to do this, since I had indeed been the first to suggest this method for determining nuclear spins, and the problems are also very similar to those in the area of paramagnetic relaxation, but I made it a condition that you, as his former "boss", would agree to this 100%. Now he writes me that he has written to you about it and that you have not answered in the negative. He thus assumes that you agree, and he proposes to finish his thesis by the end of this year. He also asks me about the layout of the thesis and describes his results, which are really very worthwhile. However, I do not want to supervise him if I am not sure that this will not cause you any bad blood or at least disappointment. I don't care for the idea of trying to steal each other's people, although unfortunately it is now the case that former Leiden students continue to study almost everywhere, except here!
I look forward to hearing what you think about this,
best regards,
yours, CJ Gorter [79]

On 30 March 1947 Bloembergen wrote to Gorter to say that he was certainly interested in the job as a Ph.D. researcher that Gorter had offered him, and stated: *"I think I'd rather come to Leiden for a year from Sept '47–Sept '48 than start '48 hanging around in Stockholm.* [A research position in Stockholm had been offered to Bloembergen, RH.] *I am glad that you wish to take up this research, and I would like to do further research involving very low temperatures."* [80] He proposed to start compiling the dissertation at the beginning of May and to have half of it ready for press in September when he returned home [to the Netherlands, RH].

The second part then had to be ready for discussion. The printer could then deal with the first part in mid-October, and Bloembergen thought that printing would not take more than two months. Gorter told him, however, that it could easily take up to six months and Bloembergen was rather shocked to hear that. In any case, Bloembergen wants *"the whole thing"* to be at the printer's in December 1947. Bloembergen decided to write his thesis in English, because many people were interested in it. In this letter it becomes clear that Gorter wanted to rebuild the Harvard NMR equipment in Leiden, and he asked Bloembergen to bring all kinds of electronics with him from the United States. A radio receiver with *"extra low noise"* was essential. Special pre-amplifiers (the phase sensitive detectors) and tens of meters of shielded cable were needed for high frequency applications. This would probably have been difficult to get in the Netherlands, except perhaps from Philips. A very valuable instrument was the frequency generator, which could generate all frequencies from 16 kHz to 30 MHz, but at a price of $1300.

Bloembergen also wanted to go to Leiden because, in the Netherlands, defending a dissertation in front of a roomful of people was an important event where not only academic colleagues were present, but family and friends could also turn it into something of a social event.

On 31 March 1947 Bloembergen finally receives an answer from Milatz, who had no objection to supervision by Gorter, although he still asked whether Bloembergen might still want to defend his dissertation in Utrecht [81].

This concluded the Utrecht chapter for Bloembergen. He would take his doctorate in Leiden.

In spring Bloembergen gave a lecture at a colloquium in Leiden; Casimir was also there to give a lecture. Hendrik Casimir (1909–2000) was a famous Dutch theoretical physicist from the Philips Physics Laboratory in Eindhoven. He has best known for his research on the two-fluid model of super- conductors (with Gorter) in 1934 and the Casimir effect (with D. Polder) in 1948.

Bloembergen had to get used to Dutch words for things like 'waveguide', where he only knew the correct term in English.

Bloembergen also wrote that, after a few weeks of vacation, he would be back at Harvard on 20 June [82].

Winifred Cadbury (1926–1996)
In 1947, the Friends Service Council (the British Quakers) and the American Friends Service Committee (AFSC, the American Quakers) were awarded the Nobel Peace Prize. They were awarded this prize for their efforts during the Second World War

for Japanese-American citizens who were imprisoned in inhumane concentration camps in the USA, and for their efforts for the poor in the USA and Europe. The prize for was received on behalf of the Quakers in December 1947 by Prof. Henry Joel Cadbury (1883–1974) of the American branch of the famous Cadbury chocolate group (now part of Kraft Foods). Cadbury was chairman of the AFSC, and during the war he visited camps set up for conscientious objectors, where many Quakers, including his two sons, carried out their military service in the form of Civilian Public Service as firefighters, assistants in psychiatric hospitals, or test subjects in medical experiments. He gave speeches there to encourage people [83].

Cadbury was Hollis Professor of Divinity at Harvard. During a festive gathering at Harvard to celebrate the prize, Bloembergen met Kees Boeke, also a Quaker, and he said: "*I am also from Bilthoven*" [84]. Kees Boeke's wife was Betty Cadbury, from the English branch of the family.

Henry Joel Cadbury lived in Cambridge and Bloembergen was invited to his home. There Bloembergen met Winifred, Henry Joel's daughter, who was six years younger than him, and fell in love with her.

In spring 1948 Winifred came to visit Bloembergen's parents, but then she ended the relationship. She liked him, but she didn't like him enough.

Afterwards Bloembergen remained friends with Winifred and her husband Martin Beer. In 1959–1960 Bloembergen, Nico's wife Deli Brink, and their three children went on holiday on the estate of the Cadbury family in the Adirondack Mountains at the invitation of Winifred and her husband. Even later, Bloembergen's son Brink went on a cycling tour through Europe organized by Winifred and Martin Beer [85].

Completion of the NMR study in Harvard and return to the Netherlands

Further career prospects

In summer 1947 Bloembergen presented handwritten copies of various chapters of his thesis to Purcell and Gorter and incorporated the amendments they proposed. In an interview he comments: "*Purcell was so stressed that he said: 'I have to make copies before you leave'.*" [86] Purcell kept a copy and edited it with Pound for a major article in *Physical Review* in 1948 [87]. Purcell wrote the article in better English and was honest enough to mention Bloembergen as first author. Bloembergen said he was not involved in the question of who would be the first author of the article. Pound had suggested that Purcell, as the senior scientist, should be the first author. Purcell thought it was important that Bloembergen be the first to place his scientific flag. Bloembergen himself considered that he had indeed done most of the work, much more than a doctoral candidate, in both the design and construction of the setup and the execution of the experiments, and he was not embarrassed to be mentioned as the first author [88].

In an interview on 20 September 2007, Ramsey commented on this: "*For me, as his best friend, it was unclear which part of the work was from Harvard, which from Leiden, which from Purcell, and which from Pound.*" [89] In an interview on 21 September 2007, Pound made the following remark about the order of the authors: "*I think the fact whose personality is involved is not really useful.*" [90]

Since an article appeared in *Nature* that showed parallels with the one by Bloem-bergen et al., they decided to send a summary of their research as note to *Nature*. This was published in 1947 [91].

Bloembergen was still interested in what was happening in Utrecht. In his letter to Milatz on 27 February 1946, he wrote that George Pegram at Columbia University in New York had all the journals from 1940 ready for Utrecht and asked if he wanted him to pass this on to Van Cittert [92]. He noted that he would speak about Piet van Heerden's crystal counter and asked about Freeman's crystal research. Bloembergen also inquired about the student house in Utrecht where he had been involved and gave the address for something Pieter Endt wanted to buy. In his letter to Milatz on 11 January 1947, Bloembergen noted that it was a *"great pleasure"* for him to hear favorable messages about the University House. He described how Street had registered cosmic radiation with the crystal counter designed by van Heerden, and on 1 July 1947 he passed on another address where Milatz could obtain silicone oil [93].

After completing his Ph.D. in Leiden, Bloembergen had already received invita-tions to work with Rabi in Columbia University and also from Siegbahn in Stockholm in early 1948 [94]. Bloembergen noted that *"Prof. Purcell's group has been rather happy with the experimentation"* and that he could return there too.

On 30 March 1947, Bloembergen wrote to Gorter, that he could not yet make any fixed plans: *"Prof. van Vleck has proposed me as a candidate for a 3-year Harvard fellowship. If that goes ahead, I might be able to return to Harvard early in 1948."* [95]

On 3 May 1947 Bloembergen proposed to do research into relaxation at low temperatures and he hoped that his interest in this research would coincide with Gorter's.

I accept to start in September or October 1947 and in any case for at least one year. [...] On 19 May I will appear in front of the Senior Fellows of Harvard Uni, for a possible fellowship of 3 years, with complete freedom and about $ 3000 per year. Maybe I will still take this opportunity at the end of 1948. [...] I am now practically at the end of the measurements."

He also wrote about the specifications of the frequency generator—which he had mentioned earlier in the letter of 30 March—pointing out that he did not know whether the Philips equipment would meet those requirements. Siegbahn jr., who had also worked in Harvard, had brought such a device from the United States to Sweden, and the order time was about four months. Bloembergen wanted to order such a device in the United States [96].

On 19 May, Bloembergen was due to appear before the Senior Fellows in Harvard for a possible three-year fellowship with complete freedom of research and an annual salary of $3000. He stressed that he might still want to do so at the end of 1948, but that if the work in Leiden was going well and he could continue to work there in that research, he would stay in Leiden. However, he added: *"Of course, it is a pity to miss such an opportunity abroad."* Indeed, he hoped that there would be other possibilities from Leiden to work elsewhere for a short time, for example in the United Kingdom or Switzerland.

Before Bloembergen left for the Netherlands, he visited his brother Evert in Quebec in Canada at the end of August 1947. Evert gave him all kinds of clothes which were in great shortage in post-war Netherlands.

In September he left on the SS *Veendam* of the Holland-America Line from New York to Rotterdam [97]. That was quite different from the Liberty ship on which he had sailed from Rotterdam to Philadelphia in January 1946. The *Veendam* was launched in Belfast in 1923 and damaged by fire on 10 May 1940. On 10 May 1941 it was claimed by the Germans and found its way to Hamburg in May 1945, partially sunk and burned out. After that the ship was repaired, first temporarily in Hamburg, and then permanently in Amsterdam, and it only came back into service again on 30 January 1947 [98].

A retrospective view of Bloembergen's first period in Harvard shows the following.

After Bloch at Stanford and Purcell at Harvard had independently developed nuclear magnetic resonance (NMR), Bloembergen arrived at Harvard at just the right moment, only six weeks after Purcell's discovery of NMR in solids. Bloembergen acquired the necessary know-how of NMR extremely quickly and was soon able to handle the equipment better than Purcell himself.

Bloembergen worked on the relaxation of NMR, which concerns the return of the atom to its ground state following excitation by an electromagnetic field. By describing this relaxation, the theory of NMR was finally elucidated, and Bloembergen played a major role in this.

References

1. Letter from N Bloembergen to JMW Milatz, 27 February 1946
2. Mattson J, Simon M. Pioneers of NMR and Magnetic Resonance in Medicine. The Story of MRI. Dean Books, Jericho, 1996
3. Bloembergen N. (ed) Encounters in Magnetic Resonances. World Scientific, Singapore, 1996
4. Hergé. Tintin in Amerika. Casterman, Brussel, 1947
5. Mattson J, Simon M. Pioneers of NMR and Magnetic Resonance in Medicine. The Story of MRI. Dean Books, Jericho, 1996
6. Bloembergen N and Milatz JMW/ The development of of an a.a. fotoelectric amplifier with a.a. galvanometer. Physica 11:449–464, 1946
7. Interview Rob Herber with N Bloembergen, 7 December 2006
8. Interview Rob Herber with N Bloembergen, 7 December 2006
9. Interview Rob Herber with P Pershan, 18 September 2007
10. Berkel K van. Uhlenbeck, George Eugène (1900–1988) In: *Biografisch Woordenboek van Nederland* (Biographical Dictionary of the Netherlands). www.inghist.nl/Onderzoek/Projecten/BWN/lemmata/bwn6/uhlenbeck
11. Bederson B. Essay: Samuel Abraham Goudsmit (1902–1978) Phys Rev Lett 101.010002
12. Uhlenbeck GE, Goudsmit SA. Ersetzung der Hypothese vom unmechanischen Zwang durch eine Forderung bezüglich des inneren Verhaltens jedes einzelnen Elektrons. Naturwiss. 13:953–954, 1925
13. Casimir HBG. Het toeval van de werkelijkheid. Een halve eeuw natuurkunde (The coincidence of reality. Half a century of physics). Meulenhoff, Amsterdam, 1983

14. Bacher RF. An obituary by Sam's first student and later colleague 1979. Archives of the California Institute of Technology
15. Bor J, Verplancke M. De pijl van de tijd bestaat echt (The arrow of time really exists). Filosofisch Magazine (1997)
16. Interview Rob Herber with M Veltman, 22 November 2010
17. Letter from N. Bloembergen to JMW Milatz, 27 February 1946
18. Asmus A. Edwin C Kemble. Biographical Memoirs. The National Academic Press. At www.nap.edu
19. O'Connor J J and Robertson E F. Julian Seymour Schwinger. MacTutor History of Mathematics. At www-history.mcs.st-andrews.ac.uk
20. Pais A. George Uhlenbeck and the discovery of electron spin, Physics Today 42: 34–40, 1989
21. Anderson PW. John Hasbrouck Van Vleck. Biographical Memoirs. 499–540. National Academy of Sciences, Washington DC, 1987
22. Bleany B. John Hasbrouck Van Vleck. Biogr Mems Fell R Soc 28:627–665, 1982
23. Interview Rob Herber with N Bloembergen 7 December 2006
24. Farmelo G. The Strangest of Men. Faber and Faber, London (2009)
25. Vleck JH Van. The Theory of Electric and Magnetic Susceptibilities. London, Oxford University Press, 1932
26. Letter from N. Bloembergen to JMW Milatz, 27 February 1946
27. Snelders HAM, Gorter, Cornelis Jacobus (1907–1980). In: *Biografisch Woordenboek van Nederland* (Biographical Dictionary of the Netherlands). www.inghist.nl/Onderzoek/Projecten/BWN/lemmata/bwn2/gortercj
28. De Bruyn Ouboter R. C.J. Gorter's life & Science. Www.ilorentz.org/history/gorter/biography.html
29. Gorter CJ. Badluck in Attempts to Make Scientific Discoveries. Physics Today, January 1967:77–81
30. Gorter CJ and Brons F. Physica 4:579,1937
31. Ramsey NF. Early Magnetic Resonance Experiments: Roots and Offshoots. Phys Today 46:41, 1993
32. Rabi II, Zacharias JR, Millman S, Kusch P. A New Method of Measuring Nuclear Magnetic Moment. Phys Rev 53:318, 1938
33. Gorter CJ. Badluck in Attempts to Make Scientific Discoveries. Physics Today, January 1967:77–81
34. Letter from CG Gorter to II Rabi, 5 March 1938
35. Letter from II Rabi to CG Gorter, 17 March 1938
36. Gorter CJ. De magnetische eigenschappen van atomen en ionen (The magnetic properties of atoms and ions). Wolters, Groningen, 1940. Oration
37. Gorter CJ. Paramagnetic Relaxation. Leiden, 1946
38. Mattson J, Simon M. Pioneers of NMR and Magnetic Resonance in Medicine. The Story of MRI. Dean Books, Jericho, 1996
39. Ramsey N. Early History of Magnetic Resonance. Bull Magn Res 7:94–99, 1985
40. Interview Rob Herber with N Ramsey, 20 September 2007
41. Mattson J, Simon M. Pioneers of NMR and Magnetic Resonance in Medicine. The Story of MRI. Dean Books, Jericho, 1996
42. Rogers DCD. Edward Purcell. Profile. The Harvard Crimson, 8 December 1952
43. Waller I. Über die Magnesierung von paramagnetischen Kristallen in Wechselfeldern Zeitschr f Physik 79:370–388
44. Purcell EM, Torrey HC and Pound RV. Resonance Absorption by Nuclear Magnetic Moments in a Solid. Phys Rev 69:37–38, 1946
45. Mattson J, Simon M. Pioneers of NMR and Magnetic Resonance in Medicine. The Story of MRI. Dean Books, Jericho, 1996
46. Bloch F. Principle of Nuclear Induction. Nobel Lecture, 11 December 1952
47. Bloch F. Nuclear Induction 70:460–474, 1946

48. Rigden J. Quantum States and Precession: The Two Discoveries of NMR. Rev Mod Phys 58:433–448, 1986
49. Bloembergen N. (ed) Encounters in Magnetic Resonances. World Scientific, Singapore, 1996
50. Henney K (ed). Index. Volume 28 of MIT Radiation Laboratory Series. McGraw-Hill, New York, 1953
51. Mattson J, Simon M. Pioneers of NMR and Magnetic Resonance in Medicine. The Story of MRI. Dean Books, Jericho, 1996
52. Interview John Rigden with R. Pound, 22 May 2003. Niels Bohr Library & Archives, American Institute of Physics, College Park, MD USA, www.aip.org/history/ohilist/28021_1.html
53. Interview Rob Herber with N Bloembergen, 6 December 2006
54. Letter N Bloembergen to CG Gorter, 29 June 1946
55. Mattson J, Simon M. Pioneers of NMR and Magnetic Resonance in Medicine. The Story of MRI. Dean Books, Jericho, 1996
56. Bloembergen AR. Het gezin van Rie en Auke Bloembergen 1917-1956. Eigen Beheer, Wassenaar, 2003
57. Debye P. Polar Molecules. The Chemical Catalog Company, 1929
58. Purcell EM, Bloembergen N and Pound RV. Resonance absorption by nuclear magnetic moments in a single crystal of CaF_2. Phys Rev 70:988, 1946
59. Purcell EM, Pound RV and Bloembergen N. Nuclear magnetic resonance absorption in hydrogen gas. Phys Rev 70:986–987, 1946
60. Bloembergen AR. Het gezin van Rie en Auke Bloembergen 1917–1956. Eigen Beheer, Wassenaar, 2003
61. Telephone interview Rob Herber with N Bloembergen, 28 April 2010
62. Interview Rob Herber with N Bloembergen 8 December 2006
63. Interview Joan Bromberg J and Paul L Kelly with N Bloembergen, 27 June 1983. Niels Bohr Library & Archives, American Institute of Physics, College Park, MD USA, www.aip.org/history/ohilist/ 4511.html
64. Letter from N Bloembergen to CG Gorter, 2 March 1947
65. Mattson J, Simon M. Pioneers of NMR and Magnetic Resonance in Medicine. The story of MRI. Dean Books, Jericho, 1996
66. Mattson J, Simon M. Pioneers of NMR and Magnetic Resonance in Medicine. The story of MRI. Dean Books, Jericho, 1996
67. Telephone interview Rob Herber with N Bloembergen, 28 April 2010
68. Letter from JH Van Vleck to CG Gorter, 8 August 1945
69. Telegram from JH Van Vleck to CG Gorter, 18 March 1947
70. Letter from N Bloembergen to CG Gorter, 30 March 1947
71. Interview Joan Bromberg J and Paul L Kelly with N Bloembergen, 27 June 1983. Niels Bohr Library & Archives, American Institute of Physics, College Park, MD USA, www.aip.org/history/ohilist/ 4511.html
72. Bloembergen N. (ed) Encounters in Magnetic Resonances. World Scientific, Singapore, 1996
73. Letter from N. Bloembergen to JMW Milatz, 27 February 1946
74. Bloembergen N. (ed) Encounters in Magnetic Resonances. World Scientific, Singapore, 1996
75. Letter from N Bloembergen to CG Gorter, 29 June 1946
76. Letter from CG Gorter to N Bloembergen, 10 July 1946
77. Letter from N Bloembergen to JMW Milatz, 11 January 1947
78. Letter from N Bloembergen to CG Gorter, 2 March 1947
79. Letter from CG Gorter to JMW Milatz, 8 March 1947
80. Letter from N Bloembergen to CG Gorter, 30 March 1947
81. Letter from JMW Milatz to N Bloembergen, 31 March 1947
82. Letter from N Bloembergen to CG Gorter, 30 March 1947
83. www.pabook.libraries.psu.edu
84. Interview Rob Herber with N Bloembergen, 7 December 2006
85. Telephone interview Rob Herber with N Bloembergen, 29 January 2009
86. Interview Rob Herber with N Bloembergen, 7 December 2006

87. Bloembergen N, Purcell EM and Pound RV. Relaxation effects in nuclear magnetic resonance absorption. Phys Rev 73:679–712, 1948
88. Bloembergen N. (ed) Encounters in Magnetic Resonances. World Scientific, Singapore, 1996
89. Interview Rob Herber with N Ramsey, 20 September 2007
90. Interview Rob Herber with R Pound, 21 September 2007
91. Bloembergen N, Purcell EM, and Pound RV. Nuclear magnetic relaxation. Nature 160:475–476, 1947
92. Letter from N. Bloembergen to JMW Milatz, 27 February 1946
93. Letter from N Bloembergen to JMW Milatz, 1 July 1947
94. Letter from N Bloembergen to CG Gorter, 2 March 1947
95. Letter from N Bloembergen to CG Gorter, 30 March 1947
96. Letter from N Bloembergen to CG Gorter, 3 May 1947
97. Bloembergen N. (ed) Encounters in Magnetic Resonances. World Scientific, Singapore, 1996
98. www.arendnet.com/h65.htm

Chapter 8
Doctorate in Leiden: In the Lab of Kamerlingh Onnes (1947–1948)

The response of the United States Consul when Nico Bloembergen describes his profession as *radio-frequency spectroscopist* is: *"How do you spell that?"* [1].

8.1 The Foundation for Fundamental Research On Matter (FOM)

The Schermerhorn/Drees cabinet was the first in the Netherlands after the Second World War, lasting from 24 June 1945 to 3 July 1946. Wim Schermerhorn (1894–1977) had studied civil engineering and became an assistant in geodesy at the Technical University in Delft. In 1926 he became professor of land survey, hydrology, and geodesy. In an interview about Schermerhorn in 1997, Goedkoop (then director of the Netherlands Energy Centre) said: *"He was a pioneer in aerial mapping, especially in the Dutch East Indies. And he became Prime Minister after the war. Yes, he certainly had the idea right away, especially after Hiroshima, that something had to be done with that science."* [2]

According to Jaap Kistemaker (1917–2010, director of AMOLF[1] in Amsterdam in 1988): *"[...] the establishment of FOM is solely related to the situation as it was*

---

[1]An academic institute for fundamental physics with great importance in society, founded by FOM.

Heike Kamerlingh Onnes (1853—Leiden, 1926) was a Dutch physicist, winner of the Nobel Prize for Physics, and professor at Leiden University. "By measurement to knowledge" was the slogan of his laboratory. He was the first to make liquid helium and discovered superconductivity at near absolute zero temperature.

© Springer Nature Switzerland AG 2019
R. Herber, *Nico Bloembergen*, Springer Biographies,
https://doi.org/10.1007/978-3-030-25737-8_8

Fig. 8.1   Jacob Kistemaker. *Source* www.delta.tudelft.nl

known here in the Netherlands, around 1 January 1946, when the Smyth Report[2] was on the table, and there were further rumors of wandering soldiers, and also some physicist in America. And who came here, and who told us about this and everything that had happened. [...] Among others Goudsmit" [3] (Fig. 8.1).

Schermerhorn and others in the cabinet feared the proliferation of nuclear weapons and felt that the Netherlands had fallen behind in physics, especially the study of nuclear energy. In 1945, the Dutch government set up an Advisory Committee on Nuclear Physics (later the Atomic Physics Committee).

According to Lex Boumans, the second director of FOM, the foundation went as follows:

When the Second World War ended in 1945, university research in our country was in a deplorable state. The work had been virtually at a standstill for several years. Developments elsewhere went largely unknown. Some of the equipment had been removed and the rest had not been maintained and was outdated. In addition, there was a large influx of students. The difficulties seemed almost insurmountable. It was in these circumstances, at a time when travelling was still very difficult, that Prof. Dr. H. A. Kramers managed to organize a consultation between leading physicists from Amsterdam, Groningen, Leiden, and Utrecht and contacted the government on their behalf [4].

On 5 April 1946 the Committee for Atomic Physics called for a foundation to be established. This happened on 15 April 1946 and it was called the Foundation for Fundamental Research on Matter.

---

[2]The Smyth Report was named after Henry De Wolf Smyth, who had written up the history of the Manhattan nuclear bomb project in a report. It was made public on 12 August 1945.

Boumans comments:

The founders were three professors, the Minister of Education, Arts, and Sciences, his Secretary-General, and the Prime Minister, represented by his Secretary.

The aim was to catch up as quickly as possible in the field of physics through a joint approach and with the help of government subsidies. This was to be achieved by using the additional resources available to stimulate new research, coordinate programs, and promote exchanges. The government was prepared to provide additional support because, in the context of the industrialization of our country, physics was seen as an essential basic science for technological developments.

The founders were Prof. Dr. G. van der Leeuw (Minister of Education, Arts, and Sciences), Prof. Dr. A. H. Kramers (also on behalf of Prof. Dr. J. M. W. Milatz), Prof. Dr. J. Clay, Dr. H. J. Reinink (Secretary-General of the above ministry), and Dr. H. Bruining, a former pupil of Ehrenfest and secretary to Prime Minister Schermerhorn.

The board was formed by Jacob Clay, professor at the University of Amsterdam, who became chairman of the Board of Directors, Hans Kramers, Anton van Arkel, professor of chemistry at the University of Leiden, J. H. Bannier, the first director of the research organization Pure Scientific Research (ZWO) and former pupil of Minnaert in Utrecht, H. Bruining, HB Dorgelo, professor at Delft University of Technology, Gilles Holst working at Philips Physics Laboratory, Hugo Kruyt, professor of chemistry at Utrecht University, Pim Milatz as treasurer, and Gerard Sizoo, professor at the Vrije Universiteit in Amsterdam.

The relationships between the founders are explained in an interview with Endt, director of the ECN:

There were people like Bakker and Clay, Kramers certainly, Milatz too, yes [who were involved in the foundation of FOM, RH]. *The initiative came from Bakker, Clay, and Kramers. Those were the people who had the ideas and so on. Milatz also participated to some extent.* [5].

Kramers was clearly the most important of the founders at the time, as he was also in charge of day-to-day management.

The universities received little money and the FOM was an important addition to physics research that could be used to finance both the people and the equipment for research. The FOM was a small company, and in the beginning everything was run rather informally, although some had more to say than others.

The following quote from Joan van der Waals, who came to the university from the business community in 1967, illustrates how things were still being done in 1967:

Gorter was definitely the pope you see - and there were something like 40 people in permanent employment at that time who came from physics at the University in Leiden, and 39 of them had done a doctorate there, most of them with Gorter. I was the only one outside and I didn't listen to him. He had never experienced that before. Then came the spring meeting of the FOM on solid-state physics, and you had to put down on paper what you needed. Well, you actually had to make reasonable requests, but Gorter, who was the chairman, had forgotten that, and Poulis brought a piece of paper - that was such a small torn-off piece of paper - it said that he needed 300,000 [about € 140,000] for appliances and I don't remember what. But Wyder and I were new, so we didn't get anything. Well, Kees Le Pair, who was the secretary, then asked for a few things. Gorter was so shocked that he no longer wanted to be president of the working community. [6]

FOM had working groups at the various universities. One of these working groups, the Delft-Leiden Commission for Physical Research on Metals, was chaired by Gorter. On 10 May 1947, the 1947 budget included social security contributions of 10,000 Dutch florins (about 4500 euros) for 'nuclear spin research' [7].

As early as November 1946, Kramers wrote that he had heard of Bloembergen's research on paramagnetic relaxation. At that time Kramers was the Dutch representative in the United Nations Atomic Energy Commission, and when he was in New York, he visited Harvard regularly [8]. On 29 April 1947 Kramers wrote to Gorter: *"Bloembergen really wants to return to Holland. It would be great if he could come to Leiden."* He added: *"My main supporter and assistant here is Pais"* [9].

At the time Kramers was working in Princeton at the Institute for Advanced Study, where Einstein and Pais also worked.

In the 1947 report on research at the Kamerlingh Onnes Laboratory, Gorter wrote the following:

During the course of the year, N. Bloembergen [...] was appointed by the FOM as scientific collaborator to work [...] on the magnetic relaxation of atomic nuclei [...] Dr. Bloembergen accepted his function on 1 September 1947. In collaboration with Purcell and Pound at Harvard University, he had developed methods to measure the magnetic relaxation of atomic nuclei and had succeeded in clarifying the principle of this relaxation in the case of hydrogen nuclei in liquids. Although some tempting offers were made to him from America, he returned to our country at the invitation of the FOM to investigate this relaxation in solids at the Kamerlingh Onnes Laboratory, where a very large temperature range is of course available. [10]

Bloembergen was the second scientific collaborator to join the FOM.

8.2   The Kamerlingh Onnes Laboratory, Leiden

Research and Ph.D. in the cold

When Bloembergen arrived in Rotterdam on the *SS Veendam* in August 1947, his parents had to wait patiently on the quay for hours, while Bloembergen was being questioned by customs. His brother Evert had given him a suitcase full of clothes for their parents.

> A year and a half after the end of the Second World War in Europe, many things were still in short supply in the Netherlands. The distribution of goods and consumables, which began in 1939, would not end until 1952, when coffee finally became freely available again.

Bloembergen was suspected by customs of being a black marketeer and the suitcase was confiscated. Then Bloembergen could embrace his impatient and proud parents. Later, his father Auke came to collect the suitcase after paying a fine [11].

Bloembergen rented a room in the Herenstraat in Leiden, a ten minute bicycle ride from the laboratory [12].

In the United States, Gorter and Bloembergen had already purchased parts for the NMR setup from Radio Shack in Boston. Clay (chairman of the FOM) had a clear opinion on this in a letter to Gorter: *"We will be very happy if you want to make Radio Shack connections because the situation is desperate here."* [13]

As described earlier, the idea was to replicate the Harvard setup in Leiden. There was no need to buy or build a magnet as the Kamerlingh Onnes Laboratory already had such facilities. In Europe, there was only one other NMR setup, in Oxford, so Bloembergen's system in Leiden would be the first on the European continent.

The experiments would be carried out at 'helium temperatures', that is, slightly above the absolute zero of -273 °C. Working conditions were completely different from those in Harvard. The Kamerlingh Onnes Laboratory was a well-oiled semi-industrial laboratory, with a strict division of labor. The equipment was set up by highly skilled instrument makers and technicians in consultation with the scientists. The latter were not expected to work as amateur instrument makers; that would definitely not have been appreciated. By 1882, Kamerlingh Onnes had developed a unique infrastructure through enormous efforts, and researchers like Bloembergen still benefited from this in 1948. The basis of that technical support was provided by the Leiden school of instrument makers. This training organization, which had its tentative beginnings in 1885 and took its final shape around the turn of the century, provided runners, skilled technicians, 'measurement slaves', and glass-blowers, without whom the precision and scale of the Leiden Physics Laboratory would not have been conceivable [14].

By adopting this organization in the laboratory, Bloembergen got beautiful equipment that was manufactured remarkably quickly. The liquefaction of hydrogen and helium was entirely in the hands of the technicians and no mortal scientist was even allowed to touch a thermos containing liquid helium. Liquid helium was only available on Thursdays and Bloembergen had to plan his low temperature experiments well in advance. Helium was expensive, so used helium was recycled.

Bloembergen worked well with staff members Krijn Taconis and Ed van den Handel. He also got an assistant, Nico Poulis (1923–2001), already mentioned above.

The setup was ready for use within a month. Gorter gave Bloembergen complete freedom in the choice, design, and execution of the experiments, because Gorter was the director of the whole laboratory and was busy enough with that; he had no time for any kind of detail. Gorter did not pay great attention to the progress of the research into relaxation phenomena at low temperatures [15].

After the experiments in Harvard with liquids such as water, Leiden now turned to research on the same phenomenon in solids. Atoms in solids move very little compared to those in liquids. Initially, at Gorter's request, Bloembergen tried to repeat Gorter and Broer's inconclusive experiments of 1942 to determine the relaxation time in lithium fluoride crystals. Bloembergen succeeded where Gorter and Broer had failed and was able to determine the relaxation time. For the definition of relaxation, see Chap. 7.

Bloembergen published an NMR study on an antiferromagnetic crystal with Nico Poulis. When Bloembergen left, Poulis took over his NMR research.

Waller had published an article in 1932, in which extremely long relaxation times were predicted for the interaction of nuclear spins with phonons [16],[3] Purcell had been using the same article in 1945, when he expected a relaxation time of a few hours in paraffin. However, paraffin is not a crystal and Purcell was in fact comparing apples to pears.

Bloembergen assumed that the relaxation times in crystals depended on the variation in field strength due to the vibrations of the crystal. Impurities in the crystal play a role here, disrupting the vibrations. Bloembergen tested his assumption 'doping' potassium alum crystals[4] with increasing concentrations of iron, i.e., incorporating more and more iron atoms into the structure, and then measuring the relaxation times at temperatures from 0.9 K to room temperature. The relaxation times appeared to be much shorter than those predicted by Waller, and Bloembergen found a good explanation for his results.

In 1936 Heitler and Teller predicted that the conductive electrons in a crystal would provide relatively short relaxation times [17]. Bloembergen confirmed the theory by measuring the relaxation times of copper powder in paraffin.

Doctorate

In May 1948 Bloembergen obtained his doctorate in the auditorium of the university, before an audience of three hundred people. He was dressed in a tail-coat and came in a carriage. The title of the thesis was, not surprisingly *Nuclear Magnetic Relaxation* [18]. In the foreword Bloembergen mentioned that he was still grateful to the Stedelijk Gymnasium, the school where he had started his studies. He also thanked the people in Utrecht University, Milatz and Rosenfeld, and in Harvard, Purcell, Van Vleck, and Schwinger, and of course Gorter in Leiden.

The dissertation opened with an introduction to the phenomenon of nuclear spin and nuclear spin moments, the principle of resonance, absorption, and dispersion. In the next chapter, he discussed the theory of nuclear magnetic resonance. A chapter on the experimental methods was followed by a comparison between theory and practice. The relaxation times in liquids, gases, and solids were discussed in this chapter. Finally, he discussed relaxation by quadrupole coupling.

During the presentation, a distant relative of Bloembergen's, Adriaan Fokker, curator of Teylers Museum[5] in Haarlem made the remark: *"It is a very nice piece of work, but there is no cow so colorful that there is no spot on it"* [19]. Fokker had discovered that there was a plus in a formula on one page and a minus in the same formula on the next page. Fokker had read the thesis very carefully. Milatz was

---

[3] A phonon is a quantized collective vibration mode of a crystal and is in fact a quasi-particle (Sagara DM. Ordering and low energy excitations in strongly correlated bronzes. Thesis, University of Groningen, 2006).

[4] Potassium alum is the chemical compound $KAl(SO_4)_2.H_2O$.

[5] Teylers Museum in the city of Haarlem was founded in 1778 by Pieter Teyler, a silk merchant, who wrote in his will that a foundation should be formed to promote theology, science, and the arts. Charity and poor relief were also part of the foundation's mission.

invited by Bloembergen and wanted to put on his gown and ask questions. However, Milatz was not allowed to appear in his gown and had to ask questions from his seat among the audience.

The Doctoral Examination Committee decided that Bloembergen should be promoted to doctor in mathematics and physics. This was entered in the graduation records as follows:

Number 32 of 1948

On the twelfth of May. Nicholas Bloembergen, born in Dordrecht, graduated to doctor of mathematics and physics after defending a thesis entitled "Nuclear Magnetic Relaxation".

Public graduation. Supervisor Mr C. Gorter. Doct. Ex. Cum Laude, Utrecht 9-4-43

In the presence of the Rector Magnificus van Oven, Secretary Berg, members of the Faculty of Mathematics and Physics, members of other faculties, Ph.D.s, etc., etc. J. C. Van Oven [20].

To Bloembergen's dismay, he did not graduate *cum laude*. According to him, the reason is not entirely clear, but it may have had to do with the fact that, in a previous case, no *cum laude* had been attributed while the doctoral candidate had clearly deserved it. As a result, it was decided in Leiden not to award any more *cum laude* doctorates [21]. When searching the Ph.D. records, however, it appears that the last *cum laude* doctorate dates from 1937, but it was not from the faculty of mathematics and physics. The last *cum laude* assessment in mathematics and physics was before 1928! It is remarkable, however, that in Bloembergen's case the Ph.D. record does not mention anyone other than the Ph.D. candidate (Bloembergen), supervisor Gorter, Rector Magnificus J. C. van Oven, and Secretary Berg. For previous theses, the entire Doctoral Examination Committee had been mentioned and sometimes also the questioners (opponents). The reason was that Bloembergen's graduation was the last in Van Oven's handwriting; Van Oven was often ill. That could be the explanation. The next person to obtain a Ph.D. in physics in Leiden was Jan Lubbink, who did receive a *cum laude* assessment. Gorter said: *"That is just crazy"* [22] (Fig. 8.3).

Five hundred copies of the thesis were printed. As usual, half went to the physics faculties in Leiden, Harvard, and other universities, as well as to family and friends, while the others were sold quickly by the publisher. Over the next ten years, the thesis would be reprinted in various formats in the Netherlands and Japan. In 1961, it was still popular and was still being published in the series *Frontiers in Physics*, by W. A. Benjamin in New York. It was last published by World Scientific, Singapore, in 1996, under the editorship of Bloembergen, with the title *Encounters in Magnetic Resonances*.

The summer of 1948 also saw the publication of the long 'BPP' paper which Bloembergen wrote with Purcell and Pound in Harvard in 1947 [23]. This paper, for which Bloembergen's thesis formed the basis, has been cited 1506 times from 1948 to the present day and is therefore one of the most frequently cited papers in physics. Commenting on this article in 1977, Bloembergen wrote:

In retrospect, it remains a very basic and seminal paper. It deals with the relaxation times, T1 and T2, introduced by Bloch and Frenkel, in solids, liquids, and gases. The sharp resonances which are based on the concept of motional narrowing are basic to

NMR spectroscopy. The exploitation of this field by the chemists and biochemists, who are more numerous in numbers and more prolific in authoring papers than physicists, is undoubtedly responsible for the high incidence of citation. This does not fully explain, however, why the paper is still so much quoted in the period 1961–1975, 13 to 28 years after its original publication. Many comprehensive books on the subject of magnetic resonance and relaxation now exist, which certainly constitute an improvement on the naive early experimental and theoretical discussions of BPP. Perhaps new workers, confronted with the complexities of modern NMR and its applications, like the account of our early wrestling with some basic problems.... [24]

Bloembergen regularly visited Kramer's home and discussed his research with him. Kramers was then professor of theoretical physics and his room was at the Lorentz Institute, right next to the Kamerlingh Onnes Laboratory where Bloembergen worked. On Wednesday evenings, under the chairmanship of Kramers, the Ehrenfest lectures would be held, with a lively atmosphere. There were regular interruptions of the speakers, a phenomenon not yet common in the 1940s. As Bloembergen had also given a lecture there, he could have put his signature on the wall of the lecture hall. That signature would have appeared next to the names of Albert Einstein, Max Planck, and Felix Bloch, among others. His uncle Adriaan Fokker regularly attended the lectures. Also (Fig. 8.2) present were the students theoretical physics Jan Korringa (1915–2015) and Nico van Kampen (1921–2013). Korringa commented on that period (at the time of the quote in 1982, he was a professor):

They were happy days in mid-twentieth century Leiden. The war was over; the Kamerlingh Onnes Laboratory was under the inspiring leadership of C. G. Corter, and the Lorentz Institute, headed by H. A. Kramers, was once again attracting anyone who was anyone in physics [25].

Nico van Kampen (later professor in Utrecht) felt that Gorter really kept himself at the forefront [26] (Fig. 8.3).

Bloembergen comments on Gorter:

[...] I had great admiration for the way Gorter handled all this. But he was quiet, well-suited to the task, very helpful, and a good leader. But he did not want anyone to be with him. And while all other universities said: we have one chair in experimental physics, we want to have more, in Leiden they had two. In the thirties they had Keesom and De Haas. Before this, there was Kamerling Onnes. What Gorter did, he told Van Itterbeek of the University of Leuven: come here every week for a few days. And Van Itterbeek was nice, but not a big man [27].

Bloembergen becomes a member of the *Christian Huygens* debating society and he regularly took part in the social activities of the society. At weekend, he often stayed with his family in Bilthoven. There he could also play his beloved hockey.

In the summer of 1948, Bloembergen received an invitation to speak at an international conference in Zurich, on the subject of post-war research progress in high-energy physics and atomic physics. Travelling was not as easy as it is today, and Bloch and Purcell could not be present. Waller, Heitler, Pauli, and Weisskopf were there. In the corridors, Bloembergen heard the opinion that experimental physics was ahead of theory. He concluded that the combination of theory and experiment was

Fig. 8.2 Adriaan Fokker.
Source www.gewina-
studium.nl

more than the sum of the two: great progress is only made when theory and practice are united.

After obtaining his doctorate, Bloembergen wanted to finish his work in Leiden. His appointment as an ordinary scientific collaborator was continued until the end of December 1948.

As of 1 June 1948, he received an annual salary of 5500 Dutch florins (about 2500 euros). The Gorter report for the FOM for 1947 and 1948 contains the following information about this period:

During the year 1947 and part of 1948, Dr. N. Bloembergen built up equipment for measuring nuclear magnetic relaxation times. After May 1948, he made a large number of measurements on calcium fluoride, alum, and some alkali halides at temperatures ranging from 1 to 300°K and at frequencies of 20.5 and 9.5 MHz. The relaxation times turned out to be completely at odds with Waller's theory of paramagnetic relaxation. For further details on these measurements and a theoretical

Fig. 8.3 Graduation dinner, 12 May 1948. No women! *Source* 1 Auke Bloembergen Nico Bloembergen; 2 Cor Gorter (supervisor); 3 Auke Bloembergen (Nico's father); 4 Adriaan Fokker (family); 5 Pim Milatz (Utrecht University); 6 Erdo Cornelissen (student friend); 7 Jan Caro (paranymph and friend); 8 W. A. Gugel (uncle); 9 Piet Krediet (year club mate); 10 Wim van Aalderen? (nephew); 11 Bob Lansberg (year club mate); 12 Robert Meindersma (year club mate); 13 A. W. Quint (uncle); 14 Adolf van den Berkhof (paranymph); 15 unknown; 16 Auke Bloembergen (brother); 17 Herbert Bloembergen (brother); 18 B. Engelsman (uncle); 19 unknown; 20 Joop Barents (year club mate); 21 Jaap Nieveen (year club mate); 22 Jan Kluyver? (student friend)

explanation, please refer to a publication that will soon be published in Physica [...] [28].

Rabi measured NMR in isolated molecules in vapor form. Purcell et al. determined NMR in the complex compound paraffin, an alkane with the chemical formula $CH_3(CH_2)23CH_3$, which is a solid at room temperature. Bloembergen set himself the task, not only to measure relaxation times, but also to explain them. One could say that Bloembergen used NMR as a method to explain the properties of liquids and solids. He himself referred to the research on magnetic relaxation and magnetic susceptibility prior to the NMR study. He recalls that, already in 1939, Zavoisky in the Soviet Union published the first positive indications of NMR in condensed matter (liquids and solids).

It has already been mentioned that some of the work, as laid out in the BPP publication, was often cited. The dissertation itself also attracted attention. Erwin Hahn (1921–2016) says on this subject:

I built a Pound-Purcell-Bloembergen apparatus for my thesis, and I was in fact inspired by Bloembergen's thesis, in which he outlined the theory of what was called mutations in the nuclear magnetic moment, a calculation which had in fact been made by Schwinger [29].

Back to Harvard

In the spring of 1948, Bloembergen was informed by Van Vleck from Harvard that he had been chosen as Junior Fellow in the Society of Fellows. Although Bloembergen had had positive experiences in Leiden, life in the post-war Netherlands was not easy for him. He also found scientific life in Harvard extremely exciting. These factors persuaded him to return to Harvard. He wrote to Van Vleck to say that he accepted the appointment, but regarding the completion of the work in Leiden, he did not want to start on 1 July 1948, but on 1 January 1949. Van Vleck agreed.

Leiden was not happy with this, even though Gorter could not offer Bloembergen any prospect of an appointment. Jaap Kistemaker, an FOM employee, said about this period: *"There were perhaps 5 or 10 of us at the most who could do something, and then you've got to row with the oars you've got. The most important thing was whether someone was loyal. We did not need deserters. That was the point. These are very important points when choosing people."* At the request of interviewer Klomp: *"But how do you mean…yes, do you mean from a FOM perspective?"* Kistemaker was clear: *"Yes, I mean, lazy people who can get a good job somewhere in America tomorrow and then say, "Well, I'll just go there". You can't work with those people."* [30]

Gorter, too, was not at all happy that Bloembergen was leaving for America. Bloembergen explained that he was going to America for three years, but that after that everything was still open. At a meeting of the FOM board in October 1948, Gorter says: *"Bloembergen has been offered a fellowship in America for 3 years and it is to be feared that they will try to keep this very capable young man there. How should FOM view this case?"* There was agreement on the FOM board that Bloembergen should be retained if they could find any way to do that, but they expected little result. The chairman [Kramers, RH] and Gorter agreed to consult with Bloembergen and perhaps propose the possibility of 'leave without objection'.

At the farewell party in December 1948, Gorter expressed the hope that Bloembergen would eventually return to the Netherlands.

Bloembergen celebrated New Year's Eve 1948 at home with his family in Bilthoven. His brother Evert was also there; he had just returned to the Netherlands from a two-year stay in Quebec [31].

However, Bloembergen's departure for Harvard had to be postponed because he had problems obtaining his visa. He wrote about this to Harvard and James Fisk, professor at Harvard and Senior Fellow in the Society of Fellows, approached the State Department to plead on his behalf. Bloembergen was informed via the American consulate in Rotterdam that he had to provide further information about his visa application at the consulate. The consul had a list of instructions from the *State Department* and questioned Bloembergen about it. The consul asked what kind of physicist he was and he answered *"radio-frequency spectroscopist"*. The consul smiled and said: *"How do you spell that?"* and continued *"What is the title of your Ph.D. thesis?"* to which Bloembergen replied: *"Nuclear Magnetic Relaxation"*. Bloembergen then had to explain that he was not really a nuclear physicist. The consul also asked Bloembergen which party he had voted for in the elections. Bloembergen explains that he had never voted, because there had been no elections during the occupation. In the first post-war year that elections were held in the Netherlands, Bloembergen was in Cambridge. The consul asks: *"Had you been here, how would you have voted?"* and Bloembergen answered: *"Your guess is as good as mine. Probably somewhere in the center."*

According to Bloembergen, these questions pointed to the burgeoning McCarthyism in American politics (we shall say more about this later).

Based on the answers to the questions, Bloembergen's visa could not be refused and he left for the United States in January 1949. Just like on his first outward journey, he left Rotterdam on a freighter [32]. This Victory ship was also a type of cargo ship designed during the war, and was the successor to the Liberty ship, which Bloembergen had used to leave Rotterdam three years earlier. The first Victory ship was built in February 1944 and it was not only more robust, but thanks to its more powerful steam turbine, also faster than its predecessor.

The Cold War and science

After the end of the Second World War in September 1945, an uncomfortable relationship had arisen between the United States and the Soviet Union. The economy of the United States had to switch from a war economy to a peace economy, there were massive strikes, and there was a shortage of housing and consumer goods. Moreover, inflation was high. On top of that, the power of the atom had come on the scene, casting a shadow over the newly acquired peace. This combination of international tension, economic instability in the US, and fear of the atomic bomb created a perfect breeding ground for conservative nationalism.

A number of nuclear physicists who had been involved in the development of the atomic bomb were appalled at the terrible consequences of its use. These physicists were committed to civil control of the atomic weapon. They set up the Atomic Energy Commission to take stock of nuclear research outside the United States and to set up an international control system for atomic energy. In addition, the National Science Foundation was established to support scientific research in the United States. In politics, and in particular, in the House Un-American Activities Committee of the House of Representatives (founded in 1938) and the Federal Bureau of Investigation, there was increasing resistance to what was seen as the naive attitude of nuclear physicists to Soviet communism. During the war, any projects that had something to do with defense were only accessible after a one-off investigation called 'clearance'. From then on, if there was any doubt over someone's reliability, a new investigation had to be carried out each time. In addition, the number of projects requiring clearance was extended to include, for example, the Brookhaven National Laboratory, established in 1947, where fundamental research was carried on matter out using an accelerator. Several researchers were refused membership because suspected of sympathies with the communist party or organizations associated with it, on the grounds of their antecedents. According to Wang, the atmosphere could be described as follows: *Fear, Suspicion and the Surveillance State* [33].

The House Un-American Activities Committee should not be confused with Joe McCarthy, whom Bloembergen spoke about earlier. McCarthy was chairman of the Standing Subcommittee on Investigations of the Government Operations Committee of the Senate and did not begin his activities until 1953.

8.3 Deli Brink—A Family of Ministers

A sailing camp with consequences
In the summer of 1948, Bloembergen went on a sailing camp with the *Christian Huygens* debating society. Sailing camps used to be places where boys and girls could meet each other outside the spying eyes of their parents. Another participant was Clara Brink, who had known Bloembergen since 1944. Clara introduced Bloembergen to a member of her family, Deli Brink. The two immediately got on well with each other (Fig. 8.4).

The Brink family [34]
The Brink family, like the Bloembergen family, came from Friesland. Hendrik Gerard (Hein) (1875–1944) was born in Ravestein and died in Oegstgeest. He was a preacher. He was married in 1900 to Huberta Deliana (Deli) Scholte, who was born in 1872

Fig. 8.4 Deli Brink. *Source* Nico Bloembergen

in Zegveld and died in 1949 in Nijmegen. Deli Scholte was the sister of the husband of his sister Dorothea Brink. Hein and Deli Scholte were the grandparents of Deli Brink.

Two children were born from Hein's first marriage. The eldest son was Rinse, who was born on 9 February 1902 in Heerhugowaard and died on 27 November 1980 in The Hague. Rinse was married in 1926 to Nantje Jantien Annechien (Nanny) Meijer, born on 2 May 1898 in Groningen. The youngest son of Hein and Deli, Antoni, was born on 12 October 1907 in Ruinerwold and died on 11 August 1987 in Delfzijl.

Rinse and Nanny had two children, Huberta Deliana (Deli), born on 1 October 1928 in Solo (Indonesia) and Antoni Hendrik (Anton), who was born on 16 July 1930 in Solo and died on 18 February 1997 in Harris (USA).

Hein Brink divorced Deli Scholte and remarried in 1917 with his housekeeper, Hendrikje Smilde, who was born in Ruinerwold in 1893 and died in Baarn in 1982. He was severely criticized for the divorce in the pastor's family and his remarriage with the housekeeper made things worse. The family rejected Hein and his second wife.

Hendrikje was eighteen years younger than Hein and of simple origins. She had only had a primary school education. Hein Brink and Hendrikje Smilde had two children, Grietje, born in Rolde on 18 April 1919, and Clara, also born in Rolde on 20 June 1921. Clara died on 30 September 2009 in Corvallis (USA) This is the Clara mentioned in Chap. 5. Clara was a kind of aunt for Deli.

Grietje worked as an assistant to the first female professor, Johanna Westerdijk, at the Central Bureau for Fungal Cultivation in Baarn, now the Westerdijk Fungal Biodiversity Institute in Utrecht. Victor Jacob Koningsberger had used his influence to get this job for her. When Grietje married Oswald de Bruin (a chemist), she was dismissed as a civil servant, as required by the legislation in force at that time [35].

Clara studied chemistry in Utrecht (she was one year younger than Bloembergen). Due to the closure of the university, she was forced to stop her studies. However, she was able to take the exams at the professors' homes, just as Bloembergen had done. She graduated in 1946. Clara obtained her Ph.D. in Leiden with van Anton van Arkel. She then became an assistant to Van Arkel and studied X-ray crystallography with Caroline McGillavry in Amsterdam. Clara obtained her doctorate in 1950 [36,37].

Deli's parents

Deli's mother Nanny had a diploma as a tailor, while her father Rinse was a biologist. The latter obtained his doctorate in Groningen in 1925 with a thesis entitled: *Contributions to a systematic study of hydroids*. After his doctorate, he became a teacher at the gymnasium in Gorinchem. After that he received a scholarship from an oil company to go to the Dutch East Indies. Then Rinse and Nanny went to Solo in Surakarta for three months, where Rinse worked at the Soil Chemical Composition and Crop Testing Station. The children Deli and Anton were born here, Deli on 1 October 1928 and Anton on 16 July 1930. Later the family went to Paseroean, to the Sugar Industry Testing Station in Java. From 1926 to 1934, the director of the test station was Jacob Christian Koningsberger (1867–1951), the father of Victor Jacob

Koningsberger (1895–1966), whom we met in Chap. 4 as professor of biology at the University of Utrecht. Later the family moved to Malang (near Semarang).

War

On 7 December 1941 Japan attacked Siam (Thailand), Malaysia (Malacca), the Philippines, and Pear Harbor in Hawaii. This made the Second World War a fact in Asia. The Japanese were coming ever closer to the Dutch East Indies: Singapore fell, then Hong Kong (both British at the time) in December. Bali and Timor were taken in February 1942 and Java in March. On 8 March 1942 the Dutch East Indies capitulated. The Dutch were locked up in camps, the men separated from the women.

In a letter to the family in the Netherlands, Rinse gave an overview of what had happened to him in telegram style. After being held briefly as a prisoner of war in March 1942, he was able to stay in his house or in a small house behind it until 31 May 1944. After that he and his family were interned in Semarang until 13 September 1944, when he and his son Anton were taken to another camp in Semarang, and after that he had to go alone to a labor camp (until August 1945) [38].

On 6 August 1945, the first atomic bomb was dropped on the Japanese city of Hiroshima. As a result, 78,000 people lost their lives immediately. Radioactive radiation subsequently killed thousands more people, resulting in a total of 140 000 victims. Because Japan did not surrender after the first bomb was dropped, a second bomb was dropped on the city port of Nagasaki on 9 August. There were approximately 65,000 victims there too. On 15 August, the Emperor Hirohito gave a historical speech on the radio (the Japanese had never before experienced the emperor speaking directly to them), in which it was reported in archaic Japanese that Potsdam's declaration (only unconditional surrender would be accepted by the Allies) would be accepted. On 2 September the surrender was officially signed on board the American battleship *Missouri.*

Deli gave the following report about the last day of the occupation in a kind of diary:
WHAT DATE IS IT TODAY?
[Deli had asked someone earlier what day it was RH]:
"What? Who cares? Did you see the plane?"
"Plane?" I did not understand.
There were clusters of women talking excitedly.
There were more shouts and an unusual sound right above the roof. I saw a small plane with a red, white, and blue circle on it. English? American? Dutch? It flew low, we could see two white men in the transparent domed cockpit. They smiled and made V-signs with their hands. A stillness came over the crowd, as if hypnotized. My right hand rose involuntarily, following the plane's circling movements over the camp. Then a woman waved a tiny Dutch flag, made with a scrap of paper and crayons. Another started singing the "Wilhelmus van Nassouwe" the Dutch national anthem [...]
Long after the plane had disappeared, I still stood with my hand ridiculously pointing towards the sky, tears streaming over my face. And, all I could think was: What date is it today? THIS IS A HISTORICAL MOMENT.

Liberation, the reality

> Unfortunately, the liberation was not as the Dutch in the Dutch East Indies had imagined. It was quite different for the Indonesians. After the capitulation of Japan, a political vacuum arose in the Dutch East Indies. On 17 August 1945 Indonesian activists declared the independence of Indonesia. Under political pressure from the leaders Soekarno, Hatta, and others, the Japanese allowed them to set up an Indonesian army. That army was not given weapons, but was trained by Japanese instructors. In addition, the *Banteng* groups founded a kind of guerrilla force. When the Netherlands wanted to restore its authority over the archipelago, violence broke out in a number of places, resulting in several thousand deaths, mainly among Indo-Europeans and Chinese. In large parts of Sumatra, the power of the indigenous people was violently crushed. Historically, this is known as the Bersiap period, which lasted from 1945 to 1947.
>
> Dutch citizens (totoks) were still in Japanese internment camps, designated by the Japanese as protected areas, so that relatively few of them had been killed. In total, about 50,000 Dutch civilians were interned in camps in Indonesia. For many the liberation only took place about a year later when they were 'exchanged' or liberated by British-Indian troops (especially the Ghurka's). Presumably, most of the victims (hundreds) fell in Surabaya, just before the direct war between the English army and the newly formed Republic broke out. [39, 40].

Rinse wrote about the period after the liberation in a letter:

25 September 1945. Deli was not skinny, but amaemic [presumably anemic, suffering from anemia, RH] *and very weak. Yet they were already better than they had been (because there had been more food for the past month) [...] but it was not even a month ago* [hence, October 1945, RH] *that the bullets were whistling over our heads. We, too, had no freedom: we were still locked up in an internment camp; the same dirty camp in which the Japanese had tormented us for so long [...] The reconstruction of the country is not yet an issue; the absence of the Allies (who first appeared only a few weeks ago) has meant that anarchy, racial hatred, and banditry have been able to spread unhindered. After the war the family was reunited in the camp.*

Deli

Deli went to primary school in Semarang before the Second World War. After that she spent another four months at the HBS, a former Dutch High School for the 12–18-year age group, in Malang. After the war the family finally left the internment camp in Semarang to a house in Batavia. On 21 November 1945, Deli received another general expression of sympathy from Queen Wilhelmina [41].

In 1946 the family went to the Netherlands with a troop transport ship. Deli and Anton stayed in rooms in the Netherlands; the parents returned to Java in 1947. Deli completed a bridging HBS for three years and passed her final exam in 1948 [42].

Deli and Nico in the Netherlands

After her exam, Deli studied medicine at the University of Amsterdam. Clara (Klaartje) Brink was busy with her chemistry dissertation and was a member of the Christiaan Huygens debating society. There were only a few girls in the society. Every year the group organized cheap with holidays. Deli had very little money: she went into town every day to eat a hot meal for one guilder. She also went to see Hendrikje, Klaartje's mother in Amsterdam and was always welcome there. Klaartje told Deli: *"Our group is going on holiday, and it'll be very cheap. We are going to rent a barn. Do you want to come along?"* Deli really wanted to go along: she had already heard that it would be a group of seventeen men and three girls and that they would go to Friesland with a few sailing boats. Of the three girls, two were already engaged to another member of the society. So, Deli would find herself among a large number of unmarried men. Bloembergen, like two others, had just received his doctorate. According to Deli, he was different from the other men; he had been in the United States for a while and that was noticeable. Deli came from the Dutch East Indies and had a different background to the other girls. They felt attracted to each other. After the second day Deli thought to herself: *"Well, that's someone for me."* So the relationship was clear for Deli. They met regularly throughout the summer of 1948 and went sailing together in the Bloembergen's sailing boat (Fig. 8.5).

They went to see Nico's parents in Bilthoven. Deli was not found to be the right girl by Nico's mother. First, Deli did not go to church, and second, she had not studied (Deli commented: *"How can you have studied now if you haven't had a school for four years and are only 19 years old?"*). Furthermore, Deli was not nicely dressed and she was wearing lipstick!

Fig. 8.5 Nico and Deli in 1948 on a sailing camp during their first meeting. *Source* Nico Bloembergen

Nico did not really know what to do about this, especially since he would soon leave for the United States again. Despite his 28 years, he still listened carefully to his mother and said to Deli: *"You are far too young to get married."* Deli's answer: *"Well, then you shouldn't have started. But anyway, okay, let's call it off"* [43].

However, things would turn out differently to what Bloembergen thought. Chapter 9 will describe how the relationship between Deli and Nico would get going again.

References

1. N. (ed) Encounters in Magnetic Resonances. World Scientific, Singapore, 1996
2. Interview HA Klomp with JA Goedkoop, 11 February 1997. In: FOM and Dutch physics after the Second World War. An 'oral history'. Stichting FOM, Nieuwegein, 1998
3. Interview HA Klomp with J Kistemaker, 26 February 1997. In: FOM and Dutch physics after the Second World War. An 'oral history'. Stichting FOM, Nieuwegein, 1998
4. Boumans AA. Fifty years of the FOM. In: FOM and Dutch physics after the Second World War. An 'oral history'. Stichting FOM, Nieuwegein, 1998
5. Interview HA Klomp with PM Endt, 3 February 1997. In: FOM and Dutch physics after the Second World War. An 'oral history'. Stichting FOM, Nieuwegein, 1998
6. Interview HA Klomp with JH van der Waals, 10 March 1997. In: FOM and Dutch physics after the Second World War. An 'oral history'. Stichting FOM, Nieuwegein, 1998
7. Minutes of the Board of Directors of the FOM, 10 May 1947
8. Letter from HA Kramers to CG Gorter, 25 November 1946
9. Letter from HA Kramers to CG Gorter, 29 April 1947
10. Gorter CG. Report of the research for the Foundation F.O.M. performed in the Kamerlingh Onnes Laboratory in Leiden in 1947
11. Mattson J, Simon M. Pioneers of NMR and Magnetic Resonance in Medicine. The Story of MRI. Dean Books, Jericho, 1996
12. Interview Rob Herber with N Bloembergen, 6 December 2006
13. Letter from J Clay to CG Gorter in Boston, 25 June 1947
14. Delft D van. Heike Kamerlingh Onnes. Een biografie. Een man van het absolute nulpunt (A biography. The man of absolute zero). Bert Bakker, Amsterdam, 2005
15. Bloembergen N. (ed) Encounters in Magnetic Resonances. World Scientific, Singapore, 1996
16. Waller I. Über die Magnetisierung von paramagnetischen Kristallen in Wechselfeldern. Zeit Phys 79:370-388, 1932
17. Heitler W and Teller E. Time Effects in the Magnetic Cooling Method. Proc Roy Soc 155:629-639, 1936
18. Bloembergen N. Nuclear Magnetic Relaxation. Nijhoff, den Haag, 1948. Thesis
19. Interview Rob Herber with N Bloembergen, 7 December 2006
20. Catalogus Candidatorum qui gradum adepti sunt. Promotieboek 20-1-1928 t/m 10-7-1953 (list of Ph.D.s). Universiteit Leiden
21. Interview Rob Herber with N Bloembergen, 7 December 2006
22. Interview Rob Herber with N Bloembergen, 7 December 2006
23. Bloembergen N, Purcell EM, and Pound RV. Relaxation Effects in Nuclear Magnetic Resonance Absorption. Phys Rev 73:679–712, 1948
24. Bloembergen N, Purcell EM, Pound RV. Citation Classics: Relaxation Effects in Nuclear Magnetic Resonance Absorption. Cit Clas 18:7, 1977
25. Korringa J. Nuclear magnetic relaxation and resonance line shift in metals. Physica 16:601–610, 1950
26. Interview Rob Herber with NG van Kampen, 18 February 2009

27. Interview Rob Herber with N Bloembergen, 7 December 2006
28. Gorter CG. Werkgroep Leiden (Leiden workgroup) 1947–1948 Stichting voor Fundamenteel Onderzoek der Materie (Foundation for Fundamental Research on Matter)
29. Interview Joan Bromberg with EL Hahn, 21 August 1986. Niels Bohr Library & Archives, American Institute of Physics, College Park, MD USA, www.aip.org/history/ohilist/4652.html
30. Interview HA Klomp with J Kistemaker, 26 February 199. In: FOM and Dutch physics after the Second World War. An 'oral history'. Stichting FOM, Nieuwegein, 1998
31. Bloembergen AR. Het gezin van Rie en Auke Bloembergen 1917-1956. Eigen Beheer, Wassenaar, 2003
32. Bloembergen N. (ed) Encounters in Magnetic Resonances. World Scientific, Singapore, 1996
33. Wang J. American Science in the Age of Anxiety: Scientists, Communism and the Cold War. The University of North Carolina, Chapel Hill, 1999
34. www.tjhbrink.eu
35. Interview Rob Herber with O de Bruin, 30 November 2009
36. Singleton MF. Biographical Sketch of Clara Shoemaker, 1998. osulibrary.orst.edu
37. Letter from R Brink to his family in The Netherlands, 9–11 November 1945
38. Letter from R Brink to his family in The Netherlands, 9–11 November 1945
39. www.wikipedia.nl
40. www.bersiapkampen.nl
41. Letter Wilhelmina to former forced laborers in the women's camps, 21 December 1945
42. Interview Rob Herber with D Bloembergen-Brink, 11 November 2009
43. Interview Rob Herber with D Bloembergen-Brink, 7 December 2006

Chapter 9
A Permanent Position in Harvard. Marriage

In 1952 Bloembergen receives an invitation to attend a congress in Japan. He travels by military aircraft. During the Korean war, however, this is far from simple. It takes Bloembergen almost a week to get from Boston.

Instead of being organized chronologically like the previous chapters, this chapter is organized in terms of research area. The next two chapters go even further with NMR, but other issues are also discussed.

9.1 Bloembergen as a Junior Fellow in the Harvard Society of Fellows (1949–1951)

The Society of Fellows is an institution of Harvard University. There are about 24 Junior Fellows and 9 Senior Fellows. To be eligible as a Junior Fellow, you must be between 20 and 30 years old and have been chosen by *Seniors from among the recent graduates of American universities for their promise of excellent contributions to knowledge and thought.*

The Junior Fellow is appointed for a period of three to six years and then *given board and lodging, and a stipend and set at liberty to pursue any intellectual adventures that they find interesting and important. All courses, seminars, and laboratories in Harvard are open to them. Their expenses for necessary travel and equipment are met. They must fulfil no academic requirements but a negative one: during their terms, they may not be candidates for a degree.*

The Society is a kind of training school for talent, but that does not mean that an appointment will follow afterwards, nor that the Junior Fellow is bound exclusively to Harvard. Approximately one third of Junior Fellows receive a permanent appointment at the end. In 1981, 5 of the 10 Nobel Prizewinners had been Junior Fellows at Harvard [1]. Many well-known scientists have made important contributions to the

© Springer Nature Switzerland AG 2019
R. Herber, *Nico Bloembergen*, Springer Biographies,
https://doi.org/10.1007/978-3-030-25737-8_9

university and society following a Junior Fellowship, such as physicists John Bardeen (double Nobel Prizewinner), Harvey Brooks (who did much to incorporate science into public policy), Robert V. Pound, Edward Witten (string theorist and the most cited living physicist), Noam Chomsky (linguist), Peter Elias (information theory), Daniel Elsberg (military analyst who revealed the Pentagon Papers on the Vietnam War), Donald Griffin (biologist), and Arthur M. Schlesinger (historian and Pulitzer prizewinner) [2].

NMR Work from Leiden

The money for Bloembergen's research was no problem. The Joint Services Electronic Program was adopted in 1946. This can be seen as a continuation of the scientific efforts of the United States during the Second World War, such as the Manhattan Project and the MIT Radio Laboratory. The program was founded by the U.S. Army, the U.S. Navy, and the U.S. Air Force. In Harvard, E. L. Chaffee led the program and NMR was accepted as a research area (Fig. 9.1).

Bloembergen moved into Eliot House, a student house at the university, so that he could be close to his work, as in his student period at Cambridge. He would work in the Vanserg building, a wooden construction from the First World War.

Fig. 9.1   Emory Chaffee. *Source* www.siarchives. si.edu

In an experiment with copper powder in paraffin, Bloembergen checked the relaxation times predicted by Heitler and Teller for metals in 1936 [3]. The experimental values fitted well with the theory and Bloembergen stated that the mechanisms for NMR of most solids, liquids, and gases were by then well understood [4].

An important breakthrough in Leiden in 1948 was the measurement for a copper sulphate crystal that made Bloembergen had grown himself. At room temperature the resonance was clearly measurable, but at very low temperatures 10 peaks became visible. These corresponded to the 10 positions the proton can take in the crystal. This was the first measurement of a so-called 'chemical shift', which reveals the dependence of NMR frequencies on the electron environment of the molecule. The neighbors of each atom in the crystal influence each other. Today, chemical shifts form the basis for NMR spectroscopy, a technique that can elucidate the chemical structure of a molecule by means of NMR. Bloembergen described the theory after his return to Harvard. He sent the manuscript to Gorter on 14 June 1949 and proposes to send it in as a joint publication by Leiden and Harvard. He presented it to the *Journal of Chemical Physics* [5], a clear indication that he was aware of the significance of the research for chemistry.

Bloembergen's research went unnoticed. Knight's research on metals, which led to the same result and was published later in the same year, was the first to be recognized. Since 1949, Ramsey had published a series of articles in which the chemical shift was explained. Ramsey only heard of Bloembergen's experiments in 1993. The reason was that Bloembergen had intended to present the research at the annual congress of the *American Physical Society* on 26–29 January 1949, but visa and transport problems had prevented this (see Chap. 8). However, a summary of the study was included in *Physical Reviews* [6].

However, at Gorter's request, the research would be published in *Physica*, the official journal of the Kamerling Onnes Laboratorium and the leading Dutch journal of physics. In a letter to Bloembergen, Gorter explained: *According to the old agreement, all Leiden results are published in "Physica". The text is then immediately used for the Communications. So I would appreciate it very much if your article could follow this usual path. Of course, it would also be possible to publish the measurements here and the theory in America; however, that would not be very elegant* [7]. Gorter reiterated this in a subsequent letter: *As I said, I would like to offer the piece for Physica and include it in the Leiden Communications, and in whichever case the non-theoretical part* [8]. In 1950, it was published in *Physica* [9].

Although this journal is in English, it is not read by Americans. Publication in *Physical Review* would have been preferable.

Bloembergen later stated in a meeting: *I claimed priority of all chemical shifts in NMR. It was way before Walter Knight found the temperature independent shift in metals* [10].

Bloembergen still worked closely with Gorter in Harvard and with Poulis, Bloembergen's successor in Leiden. Bloembergen was curious about the Leiden results for lithium. According to him, this was one of the most important elements, together with sodium [11]. Gorter also suggested to Slater, who was one of the organizers of a conference at MIT, that Bloembergen could give a lecture on nuclear spin at low temperatures [12]. That conference was in September 1950.

Microwaves. Health Problems

Bloembergen worked in the ERL Laboratory. This was part of Harvard's Radio Research Laboratory, where anti-radar equipment was developed during the First World War. Since 1946 the Harvard Joint Services Electronics Program had been established here under the direction of E. L. Chaffee. Bloembergen was viewed with some mistrust: his colleagues, mostly engineers and physicists, working on applications of microwaves, though of Bloembergen as a foreigner who did not fit well in the group. There was a culture of practical jokes and sometimes coarse language. Despite the prejudices, Bloembergen adapted quickly and his colleagues were relieved about this [13]. In a letter to Gorter, he commented: *Very cozy.* The ERL Lab had access to microwave equipment and had a better workshop than the Lyman Lab, where Bloembergen had worked previously [14]. He shifted his attention from paramagnetic resonance (which he had studied up until then) to ferromagnetic resonance. For this purpose, a vibrating cavity was made from nickel. He also shifted his focus from low-temperature radio-frequency spectroscopy to high-temperature microwave spectroscopy.

With Edward Purcell and Brebis Bleany, Bloembergen regularly had discussions about 'physics', as he himself put it [15]. He probably didn't consider his colleagues at the ERL Lab to be real physicists! The British physicist Bleany (1915–2006) had come from Oxford and was Research Fellow at Harvard and MIT in 1949. During the Second World War he had done a lot of research on radar and then on the applications of microwave techniques in the investigation of condensed matter [16].

In the Netherlands, Bloembergen had an attack of biliary colic before he left for Cambridge again, and in Cambridge he had another. In the hospital it was decided that the gallbladder should be removed. A week before admission, Bloembergen had another pain in his abdomen and told the surgeon, that he thought it was appendicitis. The surgeon promised Bloembergen that he would look. After the operation, the surgeon told Bloembergen that there had indeed been an acute appendicitis and that both the appendix and the gallbladder had been removed. Bloembergen wrote to Gorter: *After 10 days, out of hospital again, thanks to American surgery* [17]. Bloembergen had a large scar, as used to happen for such operations, and later concluded ironically in an interview: "*The appendix is a clear example of intelligent design*" [18]. Fortunately, the Society of Fellows paid for the hospital costs.

On 26 June Bloembergen went for a month to Ann Arbor (University of Michigan) to attend the Physics Summer School [19]. He flew from Boston to Detroit, the first time he had taken a plane, and there he attended lectures by Luis Alvarez, Richard Feynman, and Frederick Seitz. Bloembergen himself ran a working group on magnetic resonance [20].

Working conditions were not optimal in the ERL Laboratory. In summer 1949, the temperature under the roof rose to 35 °C!

Deli and Nico in Cambridge

Deli started studying medicine in September 1948. For nine months, she had heard nothing from Nico and then decided to look for him, in spite of Nico's decision to end the relationship.

Travelling in 1949 was no easy task for someone without money. Vassar College, a foundation from Poughkeepsie in New York, offered students from other countries the opportunity to come to the United States in a kind of exchange program. Deli enrolled, together with 14 other students, and paid 150 guilders ($150) so that she could travel for free on a boat carrying many emigrants from the Netherlands to Canada. On such ships there were often many empty cabins. In the summer of 1949, she left for Canada. Bloembergen still knew nothing about it. Deli stayed with several families in the United States. As soon as she was in the US, she wrote to tell Bloembergen that she was in the country. The letter took a long time to reach its destination and in the meantime Deli had a quiet and happy time with the people she stayed with. When Bloembergen finally received the letter, he invited her to Cambridge [21]. Bloembergen was still at the summer school in Ann Arbor when Deli was traveling from one Vassar address to another. When the school was over, she arrived in Boston after a two-and-a-half day bus trip from Minneapolis (!) [22].

In an interview in 2006, Deli explained: *"He said he would pay to let me sleep in the International Student House, which cost 1 dollar a night. He paid for the food, because I didn't have a red cent left. The first day I called and he said: 'Do you need a ring' and I said 'yes'. After that we started looking at rings in Boston. Then I sent a telegram to Holland and another to Indonesia. My father later said: "in the telegram you had already put 'engaged' and we felt you should have asked first." But he was nevertheless pleased. My mother had already said that it was okay. We had signed the telegram to Nico's parents with Deli and Nico as one word: 'Delinico', then you only paid for one word. But they made a mistake at the post office, they had put an 'm' for the 'i' of Deli and dropped the 'n' of Nico, so it read 'Delmico'. Nico's father was in London that day and Nico's mother opened the telegram and at first didn't understand what was going on; but when she saw it came from Cambridge, USA, she immediately realized. They couldn't imagine how I came to be America. Nico's mother then did a 180 degree turn. She wrote us a very nice letter to the boat I had taken to return to Holland on the first or second of September. It began: 'Dear youngest daughter'".*

Deli left Minneapolis two days earlier than planned in the summer of 1949 and wrote to her last Vassar address in New York that she would come two days later. She stayed with Nico in Cambridge for four days. When the people at the Vassar address in New York found out why Deli would arrive late, they also invited Nico and the two stayed for two days in New York.

Deli and Nico agreed that they would marry in the Netherlands in June 1950. Nico arranged to go to the Netherlands for three months, during which he could do research in Leiden.

Nico waved Deli out of New York on 2 September 1949, and wouldn't see her again until May 1950 [23].

NMR Collaboration with Leiden

In early 1950, Bloembergen visited New York, Chicago, and both East Lansing and Ann Arbor in Michigan, giving a number of lectures on nuclear magnetic resonance. In March he would go to Oak Ridge in Tennessee.

In Chicago in the summer of 1949, Bloembergen had a conversation with Willard Stout about manganese fluoride. Bloembergen was interested in the resonance of ^{19}F and Stout had the largest 'single crystal' of manganese fluoride (MnF_3) that had ever been made. The crystal weighed 2 g and Bloembergen had calculated that 2 g was enough to show the fine structure at a temperature between 1 and 20 K. He wanted to do measurements with Poulis in the Kamerlingh Onnes Laboratory in June 1950 [24].

Gorter invited Bloembergen to attend a congress on spectroscopy in Amsterdam from 18 to 23 September 1950 [25]. Bloembergen answers that he would be leaving Rotterdam for the United States on 8 September and could not attend the congress. In the same letter he mentioned that he had received the manganese fluoride crystal and that it was a very fine specimen. He also asked Poulis to adjust the diameter of the tube in which he wanted to make the measurement to the diameter of the crystal. He said that the work had to be done quickly because Stout wanted the crystal back in September. Bloembergen did wonder, however, whether he would still get a visa, following the recent Fuchs affair [26].

The Cold War Intensifies

As already mentioned in Chap. 8, the House Un-American Activities Committee had been set up to work against what was seen at the time (1948) as the naive attitude of atomic physicists towards the Soviet Union.

After the end of the Second World War, the countries of Central and Eastern Europe were brought under the influence of the Soviet Union: Poland, Romania, Bulgaria, Albania, and finally Czechoslovakia in February 1948. Parliamentary democracy was replaced there by a communist dictatorship following a coup d'état. This produced a series of reactions, such as the establishment of the North Atlantic Treaty Organization (NATO) in April 1949. In Berlin, a crisis broke out in the summer of 1948 because the Soviets blocked access for the Western allies. The latter reacted by setting up an airlift from West Germany. In May 1949, an agreement was reached on the entrance to Berlin and the airlift was terminated. In June 1950, South Korea was invaded by communist North Korea. Soon this became a regional war, involving the United Nations, the United States, China, and the Soviet Union. Another cause of increasing tension was the atomic bomb. In 1946, the Soviets had succeeded in building a working nuclear reactor. A larger reactor followed in 1948. On 29 August 1949, the Soviet Union exploded its first atomic bomb. It is uncertain whether much information about nuclear reactions and the atom bomb really was obtained through espionage in the United States and the United Kingdom, but politicians like Richard Nixon were convinced that this was the case.

Klaus Fuchs (1911–1988), Julius Rosenberg (1918–1953), and Ethel Rosenberg (Née Greenglass) (1915–1953)

Fuchs was a theoretical physicist from Germany and was involved in the Manhattan project which developed the atomic bomb during the Second World War. Before the war he had been a member of the communist party in Germany and had fled his homeland after riots between the communist activists and the Nazis. After France, he ended up in the United Kingdom, where he obtained his doctorate in Bristol in 1937. In 1939 he obtained his second title (Doctor of Sciences) with Max Born in Edinburgh. After the outbreak of the Second World War he was interned as a German, first on the Isle of Man and later in Quebec in Canada. Born intervened in his favor and Fuchs was released in early 1941. He then started working on the British nuclear bomb program. When Germany invaded the Soviet Union on 22 June, Fuchs started to spy for the Soviets. In 1942 he became a British citizen and in 1943 went to Columbia University in New York, where he worked on the Manhattan project. In 1946, Fuchs was arrested in the United Kingdom and sentenced to 14 years in prison [27].

In June 1950 Julius Rosenberg was also arrested on suspicion of espionage for the Soviet Union. Rosenberg was trained as an electrical engineer and joined the Army Service Corps in 1940 to work on the development of radar. Rosenberg was an important member of the Communist Party. Ethel Greenglass was an actress and singer and also a member of the communist party, where she met Rosenberg. In 1943 they cancelled their membership of the party in order to spy for the Soviet Union. In June 1950, Julius was arrested and in August 1950, so was Ethel. A trial followed in 1951, after which they were executed in June, although they insisted that they were innocent [28].

The Cold War at Its Height

All the previous cases, but especially the participation of the United States in the Korean war and the espionage cases, persuaded the American government that they had to stop international communism. The political climate in the United States hardened. The scientific world was also affected: physicists became dependent on the 'national security state'. Progressive and left-wing scientists were no longer given grants or asked to serve as experts.

Since January 1947, Ramsey had made particular efforts to help John P. and M. Hildred Blewett, who wanted to work at Brookhaven National Laboratory. Brookhaven was an institute where fundamental research was carried out on nuclear physics, some for defense purposes. The Blewetts wanted to do fundamental research and had already been hired by the physics department run by Ramsey, when the Federal Bureau of Investigation (FBI) started to make difficulties. The Blewetts were not members of the communist party and had no communist sympathies, but had progressive left-wing ideas. Rabi heard about the problems and offered John Blewett a place at Columbia University. Seven months later, after the intervention of numerous organizations, 'interrogations' by the FBI and dark committees, the Blewetts were finally allowed to conduct non-civil research on 14 August [29].

There are many other examples of scientists that the US federal government excluded or tried to exclude in ways that were definitely not legal. In the McCarthy era, after 1953, Harvard physics professor Wendell H. Furry was suspected of communist sympathies. Purcell and Ramsey defended him and Nathan M. Pusey, president of Harvard, supported Furrell and refused to dismiss him. In 1954 Oppenheimer was suspected; he received support from Rabi and Ramsey [30].

And Bloembergen? He got on with his research and encountered no difficulties.

Broader insights?

At the beginning of 1950, Bloembergen moved to the cyclotron laboratory: he wanted to work his way into nuclear physics in the coming year [31]. Gorter congratulated him on this choice [32]. Bloembergen contacted his friend Pieter van Heerden from Utrecht and they decided to investigate the range of dispersion of a proton beam in condensed matter with energies between 35 and 120 MeV (5.6×10^{-12} to 10.2×10^{-12} J) [33]. This research went back to work that Bloembergen and van Heerden had done in Utrecht 10 years before.

Bloembergen became friends with another Junior Fellow, Karl Strauch (1922–2000). He came from Giessen in Germany and had emigrated with his family from hostile Nazi Germany via Paris to California in the mid 1930s. Strauch later (in 1957) became professor at Harvard and then a colleague of Bloembergen.

A Short Stay in Leiden. Marriage in Amsterdam

In May 1950 Bloembergen returned to the Netherlands for a period of three months. Deli stayed in a room in Amsterdam until May, but usually spent the weekends with Nico's parents in Bilthoven. Deli got along well with Bloembergen's father Auke: according to Deli they had a similar character, but Deli didn't get on very well with Rie [34]. Deli also met Bloembergen's sisters, Diet and To. Nico moved in with his parents in Bilthoven. Deli's parents came over from Indonesia and because Deli's landlady in Amsterdam still had one room left, her parents moved in with her. In those days it was customary to get married from the bride's house. Deli's parents did not have a house in the Netherlands, so Nico's mother suggested that they get married from the house in Van Ostadelaan in Bilthoven. However, that would have offended the honour of Deli's parents. The parents suggested that they go to the Park Hotel on the Stadhouderskade in Amsterdam the day before and get married from the hotel. On 26 June 1950, they went from the hotel to the town hall on the Oude Zijds Voorburgwal and from there to the United Baptist Church on the Singel. This was a rather discreet place, in the sense that it was not recognizable as a church from the public road. The church wedding took place at the insistence of Bloembergen's mother Rie. The bridesmaid was Clara Brink, who had introduced Nico and Deli to each other [35]. The honeymoon goes to Texel, Grouw (where they went sailing), and Paris. While they were in the Netherlands, Nico and Deli lived in Deli's room in Amsterdam [36] (Figs. 9.2 and 9.3).

In Leiden, Bloembergen worked with Poulis on antiferromagnetism in manganese (II) fluoride (MnF_2) [37].

Back in Harvard. Wasted Time. America Forever

Deli and Nico moved into a terraced house in Cambridge. Bloembergen went by

Fig. 9.2 Wedding photo, Amsterdam, 12 June 1950. From left to right: Nanny Brink Meijer (Deli's mother), Auke Bloembergen (Nico's father), Rie Bloembergen-Quint (Nico's mother), Rinse Brink (Deli's father). *Source* Nico Bloembergen

bike to work, in the typically Dutch way [38]. Life in Cambridge was not easy for the newlyweds. They had no family, no friends, and no money [39].

After this Bloembergen worked with others on short investigations. With W. C. Dickinson of MIT, he published a short paper on the measurement of gyromagnetic ratios in liquids [40].

Bloembergen worked with Walter Kohn on the so-called Knight shift [41], due to which the NMR frequency of an atom in the metallic state is higher than the NMR frequency of the atom in a non-metallic compound in the same external field [42]. Kohn received the 1988 (shared) Nobel Prize for Chemistry *for his development of the density-functional theory.*

Bloembergen worked briefly on the detection of radioactivity using a scintillation counter.[1] He did this with Maurice S. Raben, a medic at the New England Center Hospital in Boston. Raben had suggested that an assay[2] for the determination of beta radioactivity (emission of electrons) could be performed by dissolving the sample in

---

[1]Detection device for ionizing radiation by recording the flashes (scintillations) produced by the radiation in certain materials, called scintillators. A scintillator is a material that emits light flashes (fluorescence) upon exposure to radiation. To detect gamma radiation, NaI (T1) single crystals are particularly suitable; for beta radiation, anthracene or diphenyloxazole dissolved in toluene are situable (www.nrg.eu.).

[2]An assay is a method in molecular biology to demonstrate the activity of a substance in an organic sample. The method described here is known as a radioimmunoassay (RIA). In 1977, Rosalyn Yalow received the shared Nobel Prize for Medicine for the development of an RIA method for insulin.

Fig. 9.3 Wedding photo Nico and Deli. *Source* Nico Bloembergen

a fluorescent solution. A short publication by the two of them appeared in *Science* [43].

Bloembergen had to teach for the first time. With Clayton Swanson he gave a lecture series on a special topic: low-temperature physics. Bloembergen dealt with conduction and magnetism. In an interview, he described the period in which he worked on nuclear physics as follows:

"I later tried to be a general physicist. In 1950–1951 I worked on the Harvard cyclotron. Just to have an idea of the general development, to master all the physics, but it was a dead end, in hindsight. A waste of time. I should have stayed with magnetic resonance, then I would have gained two precious years. I wrote one more article about the range and distribution of protons with Pieter van Heerden [44]. This was a historically interesting article, but from the point of view of my scientific career, a wasted of time [...] I really didn't enjoy working in a large group." [45]

Bloembergen's family and friends thought that, when his appointment with the Society of Fellows was over, he and Deli would return to the Netherlands. Bloembergen believed, however, that the academic climate in the United States was better than in the Netherlands and he wanted to continue his career in the United States. Deli also preferred the United States: she had never felt at home in the Netherlands given the strict cultural norms of the time. Furthermore, in the United States, she found the same kind of freedom she was used to in the Dutch East Indies before the Second World War [46].

In the summer of 1950, Gorter tried to persuade Bloembergen to come and work at the Institute for Nuclear Research (IKO) in Amsterdam, which was founded in 1946. Gorter wrote the following about this in a letter to Bakker of the IKO[3]:

If he gets his visa, Bloembergen is supposed to return to America on 8 September. So you will have the opportunity to speak to him here before, if your own journey is not delayed. I suggested he should ask you if there would be a place for him at the I.K.O. in the future, and in principle he was not opposed to that. He will have finished in Harvard about a year from now and already has offers from Chicago and Berkeley. He is the man who will choose the best he can get [47].

In the summer of 1951 Bloembergen ended his research on nuclear physics.

9.2   In Permanent Employment (Tenure, from 1951)

Associate Professor

When his Junior Fellow appointment ended, in 1951, Bloembergen was offered a position as Assistant Professor of Physics at the University of Chicago. However, this

---

[3]The Institute for Nuclear Physics (IKO) was set up a few months after the FOM in 1945 to carry out nuclear physics research in the Netherlands and was located in the Amsterdam Watergraafsmeer, near the FOM institute of which Kistemaker was director. The successor to the IKO is the National Institute for Subatomic Physics (Nikhef), a collaboration of FOM, the University of Amsterdam, Radboud University Nijmegen and Utrecht University.

was not a permanent job. Harvard also offered Bloembergen a position as Assistant Professor, but not in physics, because Van Vleck, Ramsey, Purcell, and Pound were already in these positions. What could Harvard offer?

In 1945 Vannevar Bush (Vice President of MIT) had written a report on the future of scientific research in the United States, commissioned by President Roosevelt [48] (See also Fig. 6.2). This report stated, among other things, that *Publicly and privately supported colleges and universities and the endowed research institutes must furnish both the new scientific knowledge and the trained research workers.* James Bryant Conant was president of Harvard from 1933 to 1953 and a fellow member of the Vannevar Bush Committee. Conant, a chemist, 'translated' the Bush report into physics at Harvard. He stated that the Physics Department needed all its resources for high-energy physics and nuclear physics, so there was no place for solid state physics, the field in which Bloembergen was the only one working at Harvard at the time (see Chap. 11). According to Bush's report, the money from the Gordon McKay legacy could be used for mechanical engineering; but what was that in 1950? Conant stated: solid-state physics will become the basis of mechanical engineering. Harvard, however, did not want to spend the money on the kind of research that large institutions like MIT was already investing in. It was also stated that solid-state physics was an important area of research [49].

Gordon McKay (1821–1903) earned a fortune by supplying machines for the shoe industry. McKay lived in Arrow Street in Cambridge and got to know Nathaniel Shaler, a well-known geologist. Shaler advised him to invest in a gold mine in Montana that would yield McKay 10% of his capital. Because of his friendship with Shaler and the expectation that broadly trained engineers would be needed, McKay left a significant part of his capital to Harvard, and not to MIT.

This capital could not be paid out until all the interested parties mentioned in McKay's will had died. In addition to his last wife Minnie and his two sons from that marriage, 13 women were mentioned in the will. His mother-in-law and her sister were heirs, and so were others who were not family members.

In 1949 the last heir died and the money was finally released. The amount Harvard received was more than everything received by MIT at the time [50]!

Bloembergen was able to get a place in the new Applied Physics Department, in which he would do solid-state physics. That research was funded by the Gordon McKay legacy. However, Bloembergen felt that, after almost four years at Harvard, he would rather go to Chicago, because then he would get an appointment in physics and Chicago was offering to pay 30% more than Harvard. Purcell told Bloembergen that it would be better if he stayed in Harvard. Bloembergen went to Van Vleck and Harvey Brooks, who had just become Associate Professor of Applied Physics. Van Vleck asked Bloembergen whether he would stay in Harvard if he got a permanent position as Associate Professor of Applied Physics. Bloembergen accepted the offer, got an even higher salary, and could just keep doing the research he was already doing as a Fellow [51]. However, the decision on this lay with Harvard President Conant and was not to be taken until May 1951. With the help of Van Vleck, Bloembergen wrote to Chicago's dean Cyril Smith to explain the situation to him and ask Smith to postpone the deadline for Bloembergen's response to the invitation to become

Assistant Professor. In the summer of 1951, the case was closed and Bloembergen obtained a permanent position or tenure [52]. Permanent employment was important for the young family, because it ensured a fixed income [53].

Bloembergen wrote elatedly to Gorter:

You will probably be pleased that another of your PhD students has obtained a professorship. I have been appointed Associate Professor of Applied Physics and am therefore a permanent member of Harvard. The Applied Physics Department will be completely reorganized. Van Vleck has been appointed dean of the Applied Science School and a new laboratory is being built just behind the Lyman Laboratory, where special attention will be paid to solid state physics. The whole venture is backed by the $6,500,000 Gordon McKay endowment [54].

In 2006 Bloembergen commented on the appointment in an interview: *"So that was when I learnt that, at Harvard, if you want to move forward, you need to get offers from the outside."* [55]

In the summer, teachers in the USA were not paid, because they were only paid for actual teaching and that stopped in summer, so in order to obtain some income, Bloembergen gave elementary courses on physics at the Summer School in the summer of 1951 [56].

As Bloembergen mentioned in a letter, the new Applied Physics Department received generous resources for research in solid-state physics [57].

The cooperation with Purcell terminated, because Bloembergen was now working in another department. Bloembergen wrote about his period of collaboration with Purcell in the following terms:

My close collaboration with Ed Purcell had ended, but I shall never forget his helpfulness and kindness during my years 1946 and 1947 as his first graduate student. He invited me several times to his house at 5 Wright Street in Cambridge, Massachusetts, which was less than a minimum walk from his office in the Lyman laboratory of physics. His wife Beth was also very hospitable and I remember playing with their two young sons, Dennis and Frank. My contacts during the following decades obviously became more sporadic, but were always enjoyable [58].

9.3   Family Life in Cambridge and Lexington. Death of His Parents

Deli found everything in Cambridge equally adventurous: marriage, furnishing their house, and just living in the United States. She didn't have it easy in Cambridge at first. Her English was rather poor and she knew no one. On 27 April 1951, their daughter Antonia is born. Nico and Deli couldn't afford a taxi, so they had to walk to the nearby hospital (Fig. 9.4). They come to the main entrance, but there they were referred to another building: *"Take your car and go..."* They didn't have a car, so Deli was transported by Nico in a wheelchair. They had nothing to pay the doctor in the hospital. Although their health insurance did cover medical costs, the cover for

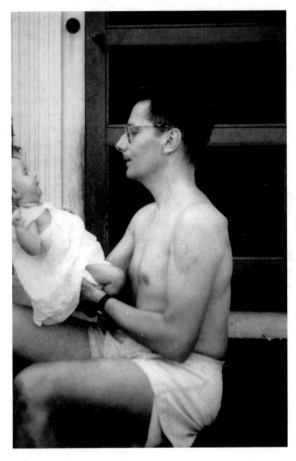

Fig. 9.4 Bloembergen with daughter Antonia in Cambridge, Massachusetts, June 1951. *Source* Nico Bloembergen

a birth only started one year after the birth. Nico wanted to borrow money for the birth from his father, but that required a special permit, with a note from the doctor and a note from the hospital.

In 1952 the family visited Holland, and of course Nico's parents in Bilthoven [59].

On 28 January 1953, their son Brink Auke was born, and in 1955, their daughter Juliana. All three children were born in Cambridge. Deli and Nico never went out, because they couldn't afford it. Nico persuaded Deli to go to the tea parties at Harvard to meet the wives of the other professors. At one point, the Chancellor of the Federal Republic of Germany, Konrad Adenauer, came to Harvard with his daughter. Deli comments: *"[...] and then those stupid Americans thought that German and Dutch must be more or less the same. I was invited to sit next to her* [the daughter of Adenauer, RH] *and I was supposed to translate from German to English and from English to German. My English was bad and my German was even worse."* Conant, the president of Harvard, noted that Deli was a regular visitor to the monthly tea party and asked if she would like to join a committee for new arrivals. Deli accepted the offer and was very flattered. She sat on the committee for a year and had a monthly lunch at the home of Harvard's president, which she enjoyed enormously.

Deli made her own clothes and hats. Because they didn't have a car, she would go shopping with the stroller. Deli would go by bike to see the doctor, and 30 years later he was still talking about it: *'The Dutch lady that came on a bike.'* A woodworking course was another of the fun things Deli did, together with a neighbor. Bloembergen then looked after the children. He was always at home by six o'clock for dinner [60]. Deli also did a tiling course in adult education, which cost almost nothing [61].

When Bloembergen didn't go by bike, he would take the bus. At some point he was waiting for the bus when Gerald Holton, a Harvard colleague, gave him a lift. Holton asked Bloembergen, why he didn't have a car and Bloembergen answered that he wanted to save first and only then buy the car. Holton thought this a typical example of Dutch thrift [62].

In 1953 a friend from student days, Maarten Bouman, came to visit him [63]. And in 1953, the Bloembergen's parents came to visit for a few days. Officially, it was a business trip in which his father Auke was accompanied by his mother Rie, but Auke was actually already out of the business. It was more like a farewell gift from the company, and a farewell trip, and they also visited New York for a week [64].

With financial support from Nico's father, Nico and Deli were able to buy a piece of land to build a house in Lexington. This is a town of 30,000 inhabitants, actually more of a large commuter village which had striking similarities with Bilthoven, and which also had about the same number of inhabitants. The proximity to Boston meant that many Lexington residents worked in Cambridge or Boston. Lexington was famous because on 19 April 19 1775, the first shot of the American War of Independence was fired here; a fact that is recalled by a statue of the *Minuteman* in the center.

The family came to live in the Stonewall Road close to the Five Fields neighborhood in 1955. The environment was dominated by young families and the homes

Fig. 9.5 Children's playground at Stonewall Road, Lexington, Massachusetts, near the house of the Bloembergen family. *Source* Rob Herber, 2009

were built by the *The Architects Collaborative* under the direction of Walter Gropius.[4] The neighborhood had the same English landscape architecture as Bilthoven: curved lanes bordered by detached houses with lawns, shrubs, and trees (Fig. 9.5). The neighborhood had a swimming pool, a shared tennis court, a playground, and a small ice pond for skating. The children in Lexington went to state schools (not to expensive private ones). Deli taught them to play music and Bloembergen took them with him for walks, swimming, or skiing on weekends [65]. In this way, he thus learnt to ski when he was forty (in 1960). In the first few years, the Bloembergens had no money to go on holiday, but the environment was so beautiful that it was hardly a problem to stay at home. In particular, they would spend their weekends outdoors. They got along well with their neighbors. Everyone nearby was in a similar situation to the Bloembergens: *Unconventional, and nobody wanted a typical American average house. We had the same views on raising children* [66].

On 28 June 1955, Bloembergen's father Auke died in a car crash near his home in Bilthoven. All the children except Nico came to their parents' home and stayed together until the cremation. According to his brother Auke Reitze, Nico had said

---

[4]Walter Gropius (1883–1969) was a German architect who founded the Bauhaus, an academy for architecture and applied art in Weimar, in 1919. After the Nazis seized power, Gropius left for the United Kingdom in 1934, and the United States in 1937, where he became a professor in Harvard. In 1938, Marcel Breuer came to Harvard and founded an architectural firm with him. In 1946 Gropius founded The Architects Collaborative with seven others (www.walter-gropius.com). The office built homes in *Five Fields* from 1951 to 1959 (en.wikipedia.org).

that he wouldn't come across for the dead. In the family chronicle, Auke Reitze said about his father: *So suddenly there came an end to this energetic doer, who had lived intensely and vibrantly and had meant so much his family and the people around him.* Nico wrote: *It suited him to go out wham, boom, without suffering* [67].

In the summer of 1956, it became clear that the health of his mother Rie was not so good. His brother Evert wrote to Nico that it would be good if he could come to Bilthoven. Auke Reitze comments: *For the living, he wanted to come and so he arrived in mid-September for a week's visit. Nico wrote to his mother in advance that he would tell her 'everything'. That applied to the solid-state laser (predecessor of the laser) that Nico had invented that summer.* (see Chap. 11, RH). *That week he spent a lot of time sitting at his mother's bedside, which she enjoyed enormously. When it was time to say goodbye: See you next year on the quay in New York (she was planning to visit Deli and Nico in America). Nico understood that this was the last time he would see her. It was as if his mother had been holding on until Nico's arrival. For when he had gone, she didn't remain for long in the land of the living.*

Rie died on 24 October 1956 [68].

9.4 Career Until the End of the 1950s

Bloembergen was one of the few permanent employees who could remain in Harvard. Most had a five-year contract, after which they would have to leave.

In 1952 Bloembergen was placed at number two on the list of the appointment committee for a professor of experimental physics at the University of Amsterdam (UvA). Bloembergen knew nothing about it, but he was only on the list so that the Amsterdam City Council (which appointed the professors of the UvA until 1961) would have something to choose between. Nico's father was pleased, because he imagined Nico coming back to the Netherlands. When Bloembergen was approached about the list by the people in Amsterdam, he said he was not interested. Bloembergen says that the nomination committee should take someone like Pieter Endt from Utrecht, but not him. However, the committee wanted someone who would not do it. Number one on the committee's list was Gerhart Rathenau (1911–1989), who works at the Philips Physics Laboratory [69]. Rathenau was a cousin of the German Foreign Minister Walther Rathenau, who was murdered by nationalists in 1922 because of his Jewish origins. Gorter thought that Bloembergen should be number one on the nomination, but the rest of the committee disagreed [70]. Gorter kept good contacts with the Bloembergens, and in October 1952 he visited Deli's parents in Surabaya on their invitation, during a trip to Japan [71].

In 1952 Purcell received the Nobel Prize with Bloch *for their development of new methods for nuclear magnetic precision measurements and discoveries in connection therewith*. According to Bloembergen, Purcell received the prize partly on the basis of the BPP publication [72]: *And he* [Purcell, RH] *said: Don't you feel you've been wronged? I told him: Well, I owe my tenure at Harvard to it. I got a permanent appointment in 1951 on the basis of this work, and I find that enough reward."* [73]

In 1955 Bloembergen received an offer to become Rathenau's successor in Amsterdam. Gorter had worked hard to achieve this [74]. Casimir stated in a letter that if Bloembergen wants to take over the management of the Philips team in Amsterdam, the salary could be supplemented. Casimir was an important man in Dutch physics after the Second World War. Initially, Kramers and Gorter had dominated, but later Casimir and De Boer became the most important [75].

Bloembergen wrote back to Gorter: *Although we are having a great time here in America and especially at Harvard, we have very seriously considered using this unique opportunity to return to the Netherlands. However, after discussions with Van Vleck and others, I have decided to stay here. I really appreciate that you thought of me, but please do not make further offers* [76]. An interview throws a different light on the matter. Bloembergen went to Van Vleck and said he had an offer to come to Amsterdam: *I don't really know what to do. It is a choice either to become a big fish in a small pond, or to stay here like a small fish in a big pond. Well, then I got promotion and some extra money from Harvard."* [77] In 1956 Bloembergen thanked Gorter in a letter for his efforts to appoint him as correspondent of the Royal Academy of Sciences. Bloembergen hoped to attend a few meetings in 1957 [78]. However, whenever he had his first 'sabbatical leave', Bloembergen would ask about the possibilities for spending a semester or academic year in Amsterdam or Leiden, perhaps in 1956, but probably in 1957–1958 [79]. The promotion in question was his appointment in 1957 as Gordon McKay Professor of Applied Physics [80]. Gorter's reaction to this was to accuse Bloembergen of being a deserter. Bloembergen left the accusation for what it was and a while later they became good friends again [81].

Gordon McKay Professor

The Harvard newspaper described the appointment as follows:

Two scientists who have pioneered in basic developments for microwave communications will become Gordon McKay Professors of Applied Physics at Harvard University on July 1.

Professor John H. Van Vleck, Dean of Engineering and Applied Physics, announced promotion of Nicolaas Bloembergen, an authority on magnetic resonance, and C. Lester Hogan, whose field is solid-state electronics.

Dr. Bloembergen had led in the development of solid-state "masers" which give promise to vastly increase efficiency in reception by radar and radio telescopes, and may facilitate the wireless transmission of telephone messages and television beyond the horizon. Solid-state masers are microwave amplifiers of a new type. It is possible for them to operate at the temperature of liquid helium, so as to have signal-to-noise characteristics as much as 100 times better than conventional vacuum-type amplifiers [82].

Another proposal came from the Netherlands, in which Bloembergen could become director of a new institute that was about to be founded. He would then oversee the construction and interior design. The last thing Bloembergen wanted, however, was to have to deal with all kinds of red tape associated with running such an institute. Later, a senior official of the Ministry of Education even asked him under what conditions he (Bloembergen) would agree to return to the Netherlands [83]. What pleased him the best by far was Harvard, where he was responsible only for his own research [84].

In 1958, Nico and Deli were naturalized without any problems [85].

Bloembergen became a member of the National Academy of Sciences in Washington in 1960 and was one of the youngest members [86]. Deli talked about Nico's early membership and the reaction of the wives of other academics to this nomination: *"And then a woman comes over to me and says: 'Hello dear, are you here with your father?' I reply: 'No I am here with my husband.' 'Your husband? Aren't you very young?' 'Yeah, that's OK. He is young, too.'"* [87]

9.5   Research on Crystals and Ferromagnetism

In the autumn of 1951, Bloembergen was to give a series of lectures on solid matter physics. During his time in the Society of Fellows, it seems as if he tried out all kinds of different things. Not only did he try his hand as a general physicist by immersing himself in nuclear physics, but even within the field that held his attention, NMR, he investigated all sorts of things without discovering much connection between the results. This was different once he had his permanent appointment in his pocket. Bloembergen then had educational obligations, which meant he could spend less time on research, although on the other hand he then had employees to whom he could 'outsource' parts of the research. His first PhD student was Ted Rowland and he did the lion's share of the laboratory work; Bloembergen would gradually do fewer and fewer measurements himself. In the spring of 1952, he restarted his research on magnetic resonance [88]. This now included mainly magnetic resonance in crystal lattices, paying particular attention to lattice defects. These disturbances are often caused by contaminants, such as 'holes' in the lattice or by dislocations between the atoms. The first research was carried out on alloys.[5]

In the summer of 1952, Bloembergen received an invitation to give a lecture at the first physics conference in Japan since the Second World War. This symbolised the reintegration of Japan into the scientific world after the Second World War. Bloembergen accepted the invitation on the condition that he could get transport to and from Japan. To his surprise, Purcell did not receive an invitation. Bloembergen

---

[5] In metallurgy, an alloy is a solid mixture of a metal with one or more other elements (usually another metal). Examples are bronze (copper and tin), brass (copper and zinc), cast iron (iron with a high carbon content), and steel (iron, carbon, and many other components, such as nickel, chromium, manganese, and vanadium).

suspected that the conference organizers felt he was the most important man, because he was mentioned as the first author of the BPP article. When the Nobel Prize was awarded to Purcell in October 1952, the Japanese discovered the situation and invited Purcell. However, the latter was far too busy and rejected the invitation. Luckily for Bloembergen, his invitation was not withdrawn and he was able to go to Japan, although that was not so easy in 1952. He received what looked almost like a set of marching orders from the Military Air Transport Service (MATS): Bloembergen's research was being partly financed by the Office for Naval Research. Bloembergen took the train from Boston to Washington Union and from there by bus to Patuxant River Air Force Base near Washington. From there he took a 12-h flight in a Lockhead Constellation to a base north of San Francisco Bay. After that, he had to go to Moffit Air Base, south of the bay, at midnight and was put up in the Bachelor Office Quarters. The next morning he was told that no flights over the Pacific Ocean from that base would be able to leave San Francisco and he had to go back to the other side of the bay. Because of the Korean war, military cargo had priority and Bloembergen was unable to leave for another 72 h. He and fellow physicist, Keith Bruecher, decided to make a virtue of necessity and spend the next three days in San Francisco. It was Labor Day weekend when Bloembergen saw a palm tree for the first time in his life. Finally, he got a place in a small Douglas DC 4. It couldn't fly right across the ocean in one go, but had to land in Wake Island to refuel. Then it went on to Honolulu and finally to Haneda near Tokyo. After 40 h in a noisy plane, Bloembergen had finally reached his destination. He had been on his way from Boston for about a week! He arrived just in time for the congress, which started the next day. Bloembergen played billiards with Louis Néel from Grenoble and discussed antiferromagnetic resonance with him.[6] Bloembergen gave a lecture at a meeting chaired by Van Vleck. He noticed that, when the Japanese heard his first name, it would bring a smile to their faces. It turned out that his name sounded just like the name of the national park, Nikko. There is a Japanese saying that you shouldn't call something beautiful (kekko) if you haven't seen Nikko. Bloembergen was enthusiastic about the meetings with numerous international colleagues. He was very surprised by some of the things he saw. For example, in Japan in 1952 there were still very few cars around. Moreover, most women were still dressed in kimonos and sometimes carried large loads on their backs, respectfully walking a few steps behind their husbands. After Kyoto, Bloembergen went to a special conference on magnetism in Osaka, chaired by Takeo Nagamiya. He gave a lecture about the latest NMR results and received the compliment from Gorter, also present, that his lecture was the only original one. The next day Bloembergen was supposed to give a colloquium at the University of Kobe, but that night a hurricane raged through Kobe. Bloembergen was nevertheless picked up by a university car and managed to reach the main entrance of the institute via the deserted streets, while metal rooftops flying through the air caused beautiful fireworks if they hit the power lines. However, the

---

[6]Antiferromagnetism, like ferromagnetism, is an expression of orderly magnetism. It occurs when the temperature goes below a certain value, the Néel temperature. Examples of antiferromagnetic substances are hematite, chromium, and alloys such as iron manganese and nickel (II) oxide.

main entrance was inaccessible due to deep pools of water and he was led into the institute through the back entrance. He gave his lecture and was brought back to his hotel in Osaka. On the way Bloembergen saw many collapsed railway bridges. Train services were suspended indefinitely and 5000 stranded passengers spent the night at the station.

Bloembergen's hosts managed to put him on a plane to Tokyo. There he had to wait a few days in officers' quarters until MATS was able to provide a plane to take him back to the United States. The return trip was in a DC4 via the Aleut and Alaska to Seattle-Tacoma. Bloembergen then went to the physics department in Berkeley, where he met a number of colleagues who were also investigating magnetic resonance. He gave a lecture on nuclear magnetic resonance in metals that elicited the remark *"that old nonsense"* from Edward Teller [89].

Bloembergen was back in Harvard before the start of the autumn lectures [90].

In 1953, articles by Bloch and Purcell appeared in Science, revealing for the first time the principle of NMR outside the discipline [91, 92].

Ted Rowland was a student looking for a doctoral supervisor. His previous supervisor, John Hobstetter, went emeritus in 1950 and Rowland didn't know how he would graduate. In 1951, he joined Bloembergen, who was just an Associate Professor. Three years earlier, in 1948, Bloembergen had conducted research in Leiden into the resonance of copper (and in particular, isotopes[7] ^{63}Cu and ^{65}Cu) (see earlier). He had also investigated brass (a copper and zinc alloy), and found certain inexplicable phenomena. This had haunted Bloembergen until he met Rowland, who was one of the few people interested in research into alloys [93]. Rowland described his contacts with Bloembergen thus:

When Nico described his observations, I was impressed by the man as much as by the problem. He was tough minded, told one what he knew, and had very little patience with talk in terms of talk. I was to find that he never bothered trying to explain to me anything he thought I was incapable of understanding. If I did not understand X then I probably could not, and discussion was a waste of time [...] Nico's early advice and comment was relatively sparse; when I asked what to read to get started, his answer was, "Read about noise." After a week or so of noise, I decided that our tastes must differ and that the connection between alloys and noise was not as clear to me as it was to him. He loaned me a copy of his thesis [94].

Rowland appropriated the instrumentation and investigated a number of alloys. It turned out that there was quadrupole broadening and that was the reason why Bloembergen had not been successful with his experiments in 1948. Rowland's experience with metals, combined with Bloembergen's experience in NMR, led to a number of interesting results that would be published in a new magazine on metallurgy.

---

[7]A chemical element is determined by the number of protons in the nucleus. One proton in the nucleus is hydrogen, two is helium, and so on. In addition, there may be different numbers of neutrons in the nucleus. Examples are deuterium (hydrogen with a proton and a neutron) and the above-mentioned copper isotopes. Copper has 29 protons; there is an isotope with 34 neutrons (^{63}Cu) and an isotope with 36 neutrons (^{65}Cu). As can be seen, the notation indicates the total number of protons and neutrons.

In 1953 Bloembergen and Rowland investigated the spin exchange between two isotopes of thallium, ^{203}Tl and ^{205}Tl. Enriched thallium was needed to do this research, which had to be obtained from the Los Alamos National Laboratory for atomic physics. They thus obtained practically the entire world supply of enriched thallium from 1953. Rowland succeeded in reducing thallium oxide to elemental thallium without much loss [95]. While Rowland was working on his experiments, Ruderman and Kittel developed a theory in Berkeley according to which the connection between the spins was made via the free electrons in the metal. In their paper, published in 1954, the introduction quoted the first paper by Bloembergen and Rowland, as well as the broadening of the silver line in another article [96].

Yosida in Berkeley investigated copper-manganese alloys for the polarization of the aforementioned conduction electrons. He cited the research by Ruderman and Kittel and the second investigation by Bloembergen and Rowland on the abnormal line broadening [97].

The theory later became known as the RKKY interaction. Kittel spoke about the broadening of thallium and its explanation at a congress in 1954, to which Bloembergen added:

"Charlie, we fully recognize the priority of your theory, but we claim priority for the pseudo-dipolar interaction in conductors, for both types of interactions in insulators, and for the systematic experimental proof of all these effects by variation of the isotopic abundances." We shook hands and have remained on very friendly terms both before and after [98].

Peter Sorokin wanted to be supervised by Bloembergen in 1954. Sorokin describes what he had done prior to working with Bloembergen:

"I had planned to go into theoretical solid-state physics. In my second year of graduate work, another student, Don Weinberg, and I signed up for a reading course on nuclear magnetic resonance given by Bloembergen, which we thought would be an easy course. At the time, Bloembergen wasn't a very polished lecturer, and we let everything go over our heads, but we weren't too concerned. Then he announced that he wanted a term paper from each of us as proof that we took the course, and we wrote the papers and handed it to him. We got unsatisfactory grades, so we went to Prof. Bloembergen, who said 'These papers don't say anything about what I was teaching'
We couldn't get unsatisfactory grades, so I spent part of the summer to understand NMR, then wrote up a paper which Bloembergen accepted. By that time, I felt I had invested so much time in the subject, which actually seemed interesting, that I might as well sign up and do a thesis with him. So did Don." [99]

Bloembergen suggested that he conduct research into caesium halides (halides include fluorides, chlorides, bromides, and iodides). At that time, Bloembergen and Rowland had just found the above interaction in thallium and Bloembergen thought it would be interesting to study these effects in a single crystal with heavy isotopes. Sorokin spent a lot of time on theory and speculation. Bloembergen felt that Sorokin was not getting very far and suggested that he should first experiment and then build on the theory. Sorokin, however, wanted to continue with his calculations. Bloembergen rather bluntly noted that Sorokin would not become a brilliant theorist and that he should concentrate on research into caesium and halogen nuclei. Bloembergen: *I may have offended his ego, but he did get interesting and unanticipated results. I became very interested in the interpretation of these double resonance relaxation phenomena which are described in this paper* [100]. Among other things, Sorokin determined the chemical shift [101].

George Benedek, a research fellow, and T. Kushida, a senior research fellow from Japan, did research into the influence of high pressures on the magnetic resonance of the ^{63}Cu nucleus in copper (I) oxide and the ^{35}Cl nucleus in potassium chlorate (a substance used in match heads) and in *para*-dichlorobenzene (used in mothballs, formerly in WC cubes) [102]. Both substances are solids at room temperature. Bloembergen was only concerned with the dependence of the quadrupole resonance frequency on pressure and temperature [103]. Copper (I) oxide is one of the most studied compounds in the history of semiconductor physics. Benedek continued with this kind of research and wrote a book about it in 1963 [104].

In 1951, Bloembergen resumed the research on ferromagnetic resonance that he had done in his first year as a fellow in 1949. Two former students of E. L. Chaffee were interested in doing postdoc research on ferromagnetic resonance. Shyh Wang was Bloembergen's first postdoc, the Richard Damon was the second. Bloembergen drew up a manuscript on the research into ferromagnetic resonances and after Van Vleck had made some corrections [105] and Purcell had approved it, it was published in 1952 [106]. Bloembergen was sure that it was reviewed for *Physical Review* by Hans Bethe. The latter later noted that the work had been well done, but that the simplest way had not been chosen [107]. Bloembergen stated that, although there were no particularly striking results, he now understood ferromagnetism better and had also learnt something about microwave technology.

A study which did not receive much attention, but later proved to be of importance in Bloembergen's attempts to establish the principles of the maser (see Chap. 11), was the study published with Pound in 1954 on the damping of radiation by a magnetic field at magnetic resonance [108].

In a study carried out by Bloembergen with the students Sidney Shapiro, Peter Pershan, and the postdoc J. O. Artman, they investigated the interaction of nuclear and electron spin resonance in an atom [109].

In a study done by Bloembergen with Tom Morgan, a professor at Texas University who was in Harvard for a sabbatical in 1960, they investigated the relation between NMR in solutions and electron magnetic resonance [110]. The article attracted the interest of chemists, biologists, and medical researchers and was widely quoted by them.

Between 1960 and 1965, Bloembergen slowly brought the NMR research to an end. He had spent almost 20 years on NMR research since 1946, and his attention now shifted entirely to non-linear optics.

During his 20 years of NMR research, Bloembergen exposed the foundations of the nuclear magnetic resonance and, together with others, erected an impressive construction, which provided a platform for further physical research, especially in solid-state physics. There were also applications outside physics, in chemistry, biology, and medicine (MRI), exploiting the chemical shift. The following chapter will discuss these topics further.

Contacts with Gorter

Bloembergen still had contacts with Gorter. In 1957 he took his first sabbatical with a Guggenheim Fellowship. He left on the *S. S. Maasdam*, a regular passenger ship of the Holland America Line, and arrived in Rotterdam on 15 June 1957. He remained in the Netherlands from June to September and then visited Gorter. He also went to Amsterdam, Delft, and Utrecht, and visited Philips in Eindhoven [111]. Gorter wished to discuss several problems with him and wrote: *and several young people are also very happy that you come* [112]. From October to December 1957, Bloembergen was in Paris, where Kastler and Abragam had invited him. In January 1958, he and his family returned to the United States [113]. In February 1958, Gorter invited Bloembergen to the Kamerling Onnes conference: *where you should not be missing* [114]. In 1958, Gorter briefly visited Harvard and also visited the Bloembergen family in Lexington [115]. In 1959, Gorter recommended George Seidel, who had studied under P. H. Keesom and worked for a year in Leiden with a grant from the National Science Foundation: *I definitely recommend you take him if you can get him for your lab* [116]. Seidel did indeed come to work as a postdoc under Bloembergen. In September 1959, Gorter asked Bloembergen to participate in a series of books [117], and when Bloembergen agreed, wrote in October: *I am very happy that you want to participate in the third part of "Progress in low temperature physics"* [118]. In January 1962 Gorter thanked Bloembergen for sending a copy of his dissertation and expressed his expectation to see Bloembergen in Jerusalem during that year [119]. This was the last letter from Gorter to Bloembergen.

It is striking how Gorter's and Bloembergen's roles gradually changed. As a PhD student in 1946, Bloembergen was still the student and Gorter the teacher, and following Bloembergen's graduation, this would remain the case for some time. However, from 1951, they were both professors and therefore colleagues. In the letters from Bloembergen to Gorter, the address changed from *Most learned* in 1946 to *Dear Professor Gorter* in 1947, *Dear Professor Gorter* in June 1949, and *Dear Gorter* from November 1949. Gorter increasingly asked Bloembergen for advice on NMR research in Leiden, and Bloembergen would comment on the results. Gorter also asked Bloembergen to mediate in order to obtain crystals for research from the United States, for example. Gorter, on the other hand, would comment on Bloembergen's NMR and maser research, but only on the results of the research, and not on its design.

References

1. Interview Rob Herber with N Bloembergen, 6 December 2006
2. Brinton, C. (Ed.). (1959). *The Society of Fellows*. Cambridge, MA: Harvard
3. Heitler, W., & Teller, E. (1936). Time Effects in the Magnetic Cooling Method-I. *Proc Roy Soc, 155*, 629–639
4. Bloembergen, N. (1949). Nuclear Magnetic Resonance in Copper. *Physica, 15*, 588–592
5. Letter from N Bloembergen to CG Gorter, 14 June 1949
6. Bloembergen N. Fine Structure of the Magnetic Resonance Line of Protons in CuSO4.5H2O. Phys Rev 75:1326, 1949
7. Letter from CG Gorter to N Bloembergen, 22 June 1949
8. Letter from CG Gorter to N Bloembergen, 5 July 1949
9. Bloembergen N. Fine Structure of the Magnetic Resonance Line of Protons in $CuSO_4.5H_2O$. Physica 16:95–112, 1949
10. J. H. Van Vleck: Quantum theory and magnetism (audio tape). 22 March 2011, American Physical Society, Dallas, Niels Bohr Library & Archives, American Institute of Physics, College Park, MD USA
11. Letter from N Bloembergen to CG Gorter, 29 November 1949
12. Letter from CG Gorter to N Bloembergen, 22 June 1949
13. Bloembergen, N. (Ed.). (1996). *Encounters in Magnetic Resonances*. Singapore: World Scientific
14. Letter from N Bloembergen to CG Gorter, 14 June 1949
15. Letter from N Bloembergen to CG Gorter, 14 June 1949
16. www.independent.co.uk
17. Letter from N Bloembergen to CG Gorter, 14 June 1949
18. Interview Rob Herber with N Bloembergen, 7 December 2006
19. Letter from N Bloembergen to CG Gorter, 14 June 1949
20. Bloembergen, N. (Ed.). (1996). *Encounters in Magnetic Resonances*. Singapore: World Scientific
21. Interview Rob Herber with D Bloembergen-Brink, 7 December 2006
22. Bloembergen, N. (Ed.). (1996). *Encounters in Magnetic Resonances*. Singapore: World Scientific
23. Interview Rob Herber with D Bloembergen-Brink, 7 December 2006
24. Letter from N Bloembergen to CG Gorter, 16 February 1950
25. Letter from CG Gorter to N Bloembergen, 22 February 1950
26. Letter from N Bloembergen to CG Gorter, 8 March 1950
27. www.worldlingo.com/ma/enwiki/nl/Klaus_Fuchs
28. www.atomicarchive.com/Bios/Rosenberg.shtml
29. Wang, J. (1999). *American Science in the Age of Anxiety: Scientists*. Chapel Hill: Communism and the Cold War. The University of North Carolina
30. Mattson, J., & Simon, M. (1996). *Pioneers of NMR and Magnetic Resonance in Medicine*. Jericho: The Story of MRI. Dean Books
31. Letter from N Bloembergen to CG Gorter, 16 February 1950
32. Letter from CG Gorter to N Bloembergen, 22 February 1950
33. Bloembergen, N., & van Heerden, J. (1951). The range and straggling of protons between 35 and 120 MeV. *Physical Review, 83*, 561–566
34. Interview Rob Herber with D Bloembergen-Brink, 6 July 2010
35. Interview Rob Herber with D Bloembergen - Brink, 7 December 2006
36. Interview Rob Herber with D Bloembergen-Brink, 6 July 2010
37. Bloembergen, N., & Poulis, N. J. (1950). On the Nuclear Magnetic Resonance in an Antiferromagnetic Crystal. *Physica, 16*, 915–919
38. Interview Rob Herber with D Bloembergen-Brink, 7 December 2006
39. Interview Rob Herber with D Bloembergen-Brink, 7 December 2006

40. Bloembergen, N., & Dickinson, W. C. (1950). On the Shift of the Nuclear Magnetic Resonance in Paramagnetic Solutions. *Physical Review, 79,* 179–180
41. Kohn W. Autobiography. Nobelprize.org
42. Kohn, W., & Bloembergen, N. (1950). Remarks on the Nuclear Resonance Shift in Metallic Lithium. *Physical Review, 80,* 913
43. Raben, M. S., & Bloembergen, N. (1951). Determination of radioactivity by solution in a liquid scintillator. *Science, 114,* 363–364
44. Bloembergen, N., & van Heerden, P. J. (1951). On the range and straggling of protons between 35 and 120 MeV. *Physical Review, 83,* 561–564
45. Interview Rob Herber with N Bloembergen, 6 December 2006
46. Bloembergen, N. (Ed.). (1996). *Encounters in Magnetic Resonances*. Singapore: World Scientific
47. Letter from CG Gorter to CJ Bakker, 27 June 1950
48. Bush V. Science – The Endless Frontier. A Report to the President by Vannevar Bush, Director of the Office of Scientific Research and Development. United States Government Printing Office, Washington (1945)
49. Interview Joan Bromberg and Paul L Kelley van Nicolaas Bloembergen, 27 June 1983. Niels Bohr Library & Archives, American Institute of Physics, College Park, MD USA, www.aip.org/history/ohilist/4511.html
50. Lewis HR. Gordon McKay. Brief life of an inventor with a lasting Harvard legacy: 1821–1903
51. Interview Rob Herber with N Bloembergen, 7 December 2006
52. Bloembergen, N. (Ed.). (1996). *Encounters in Magnetic Resonances*. Singapore: World Scientific
53. Interview Rob Herber with N Bloembergen, 7 December 2006
54. Letter from N Bloembergen to CG Gorter, 29 June 1951
55. Interview Rob Herber with N Bloembergen, 7 December 2006
56. Letter from N Bloembergen to CG Gorter, 29 June 1951
57. Bloembergen, N. (Ed.). (1996). *Encounters in Magnetic Resonances*. Singapore: World Scientific
58. Bloembergen N. Purcell and NMR. American Physical Society meeting, February 29, 2012, Boston, MA
59. Bloembergen, A. R. (2003). *Het gezin van Rie en Auke Bloembergen 1917-1956*. Wassenaar: Eigen Beheer
60. Interview Rob Herber with J Bloembergen, 30 May 2009
61. Interview Rob Herber with D Bloembergen-Brink, 7 December 2006
62. Interview Rob Herber with G Holton, 21 September 2007
63. Interview Rob Herber with M Bouman, 8 March 2007
64. Bloembergen, A. R. (2003). *Het gezin van Rie en Auke Bloembergen 1917-1956*. Wassenaar: Eigen Beheer
65. Bloembergen, N. (Ed.). (1996). *Encounters in Magnetic Resonances*. Singapore: World Scientific
66. Interview Frank Elstner met N Bloembergen. Die Stillen Stars. Nobelpreisträger privat gesehen. Heute: Prof. Nicolaas Bloembergen. ZDF-documentaire (1987)
67. Bloembergen, A. R. (2003). *Het gezin van Rie en Auke Bloembergen 1917-1956*. Wassenaar: Eigen Beheer
68. Bloembergen, A. R. (2003). *Het gezin van Rie en Auke Bloembergen 1917-1956*. Wassenaar: Eigen Beheer
69. Interview Rob Herber with N Bloembergen, 7 December 2006
70. Letter from J de Boer to CG Gorter, 16 February 1953
71. Letter from CG Gorter to N Bloembergen, 6 January 1954
72. Bloembergen, N., Purcell, E. M., & Pound, R. V. (1948). Relaxation Effects in Nuclear Magnetic Resonance Absorption. *Physical Review, 73,* 679–712
73. Interview Rob Herber with N Bloembergen, 7 December 2006
74. Letter from N Bloembergen to CG Gorter, 14 March 1955

75. Interview Rob Herber with MJG Veltman, 22 November 2010
76. Letter from N Bloembergen to CG Gorter, 11 April 1955
77. Interview Rob Herber with N Bloembergen, 7 December 2006
78. Letter from N Bloembergen to CG Gorter, 12 May 1956
79. Letter from N Bloembergen to CG Gorter, 14 March 1955
80. Bloembergen, N. (Ed.). (1996). *Encounters in Magnetic Resonances*. Singapore: World Scientific
81. Interview Rob Herber with N Bloembergen, 7 December 2006
82. Morning Papers of Friday, 12 April 1957. Harvard University. University News Office
83. E-mail from N Bloembergen 19 July 2010
84. Bloembergen, N. (Ed.). (1996). *Encounters in Magnetic Resonances*. Singapore: World Scientific
85. Bloembergen, N. (Ed.). (1996). *Encounters in Magnetic Resonances*. Singapore: World Scientific
86. Interview Rob Herber with N Bloembergen, 7 December 2006
87. Interview Rob Herber with D Bloembergen-Brink, 7 December 2006
88. Letter from N Bloembergen to CG Gorter, 19 December 1951**THIS FOOTNOTE IS NOT IN THE PDF**
89. Bloembergen, N. (Ed.). (1996). *Encounters in Magnetic Resonances*. Singapore: World Scientific
90. Bloembergen, N. (Ed.). (1996). *Encounters in Magnetic Resonances*. Singapore: World Scientific
91. Bloch, F. (1953). The Principle of Nuclear Induction. *Science, 118,* 425–430
92. Purcell, E. M. (1953). Research in Nuclear Magnetism. *Science, 118,* 431–436
93. Bloembergen, N. (Ed.). (1996). *Encounters in Magnetic Resonances*. Singapore: World Scientific
94. Rowland TJ. Nuclear Magnetic Resonance in Metals: A Selective Review of the Beginning. In: Levenson MD, Mazur E, Pershan PS, Shen YR (eds) Resonances. A volume in honor of Nicolaas Bloembergen. World Scientific, Singapore, 1990
95. Bloembergen, N., & Rowland, T. J. (1955). Nuclear Spin Exchange in solids: ^{203}Tl and ^{205}Tl Magnetic Resonance in Thallium and Thallium Oxide. *Physical Review, 97,* 1679–1698
96. Ruderman, M. A., & Kittel, C. (1954). Indirect Exchange Coupling of Nuclear Magnetic Moments by Conduction Electrons. *Physical Review, 96,* 99–102
97. Yosida, K. (1957). *Physical Review, 106,* 893–898
98. Interview Rob Herber with N Bloembergen, 7 December 2006
99. Hecht J. Laser Pioneer Interviews: Peter Sorokin. An IBM Fellow Talks About the Second Laser and the Dye Laser. Laser and Applications. March 1985, 53–57
100. Bloembergen, N. (Ed.). (1996). *Encounters in Magnetic Resonances*. Singapore: World Scientific
101. Bloembergen, N., & Sorokin, P. (1958). Nuclear Magnetic Resonance in the Cesium Halides. *Physical Review, 110,* 865–875
102. Kushida, T., Bebedek, G. B., & Bloembergen, N. (1956). Dependence of the Pure Quadrupole Resonance Frequency on Pressure and Temperature. *Physical Review, 104,* 1364–1377
103. Bloembergen, N. (Ed.). (1996). *Encounters in Magnetic Resonances*. Singapore: World Scientific
104. Benedek, G. B. (1963). *Resonance at High Pressure*. New York: Interscience
105. Bloembergen, N. (Ed.). (1996). *Encounters in Magnetic Resonances*. Singapore: World Scientific
106. Bloembergen, N., & Damon, R. W. (1952). Relaxation Effects in Ferromagnetic Resonance. *Physical Review, 85,* 699
107. Interview Rob Herber with N Bloembergen, 6 December 2006
108. Bloembergen, N., & Pound, R. V. (1954). Radiation Damping in Magnetic Resonance Experiments. *Physical Review, 95,* 8–12

109. Bloembergen, N. (1959). Shapiro, Pershan PS, Artman JO. *Cross-Relaxation in Spin Systems. Phys Rev, 114,* 445–459
110. Bloembergen N, Morgan LO. Proton Relaxation Times in Paramagnetic Solutions. Effects of Electron Spin Relaxation. J Chem Phys 34:842–850, 1961
111. Statement of Plans for Research of N. Bloembergen. John Simon Guggenheim Memorial Foundation
112. Letter from CG Gorter to N Bloembergen, 1 May 1957
113. Letter from N Bloembergen to CG Gorter, **12 May 1957**
114. Letter from CG Gorter to N Bloembergen, 7 February 1958
115. Letter from CG Gorter to N Bloembergen, 14 September 1959
116. Letter from CG Gorter to N Bloembergen, 18 February 1959
117. Letter from CG Gorter to N Bloembergen, 14 September 1959
118. Letter from CG Gorter to N Bloembergen, 14 October 1959
119. Letter from CG Gorter to N Bloembergen, 19 January 1960

Chapter 10
From Atoms to Humans: Chemical Shift and Magnetic Resonance Imaging

We measured the relaxation time of protons in water, in aqueous solutions, the influence of viscosity and temperature. Those data are now the bases in which the MRI pictures can be taken. Interview with Bloembergen in 2004. [1]

10.1 Chemical Shift

In his NMR experiments in 1939, Ramsey found six different peaks instead of the expected single peak for molecular hydrogen (H_2) and deuterium (D_2) [2]. It turned out that, in the hydrogen molecule, each hydrogen atom influenced the other (see Chap. 7). Ten years later, Knight found a shift in the resonance in metals [3]. In that same year, Bloembergen also published his discovery of the shift in copper sulphate [4]. From 1949, Ramsey published a number of articles explaining the chemical shift (for the definition of chemical shift, see Chap. 9), such as in *Magnetic Shielding of Nuclei in Molecules.* [5] However, not everyone was convinced of the applications of NMR at the time. In an interview, Bloembergen remarked: "*Purcell did not even see its importance for determining chemical structure. He said in 1950 that it was only academic trickery. which had no practical value*" [6]. Independently of Knight, Proctor and Yu also encountered the chemical shift in 1950 [7].

Erwin Hahn (1921–2016) came across the NMR publications by Bloch and the authors of BPP in Illinois in 1948 and immediately became very interested. He set up a replica of Bloembergen's apparatus in Harvard from the publications, but learned even more from a visit to Purcell and Pound, where he could see the working equipment himself. Gutowsky, who was hired at the Illinois Chemistry Department at the same time that Hahn visited Harvard, was curious to see what Hahn would do with the equipment. Hahn found inspiration in the BPP publication and Bloembergen's dissertation. The latter briefly described 'wobbles'. In 1948 Jacobsohn and

© Springer Nature Switzerland AG 2019
R. Herber, *Nico Bloembergen*, Springer Biographies,
https://doi.org/10.1007/978-3-030-25737-8_10

Wangsness gave a theoretical explanation for these wobbles [8], but neither group realized that it was free precession.[1] This precession could be explained by the fact that the resonance goes from fast to slow and back again. Another phenomenon described in Bloembergen's dissertation was *nutation*. This was the reaction of the spin to a rotating or oscillating magnetic field. The spins then go from the excited energy level to the ground level and back again. Hahn decided to study this effect for his thesis and published the first results of his research in 1948 [9]. In 1949, after his doctorate and a postdoc place in Illinois, Hahn made the discovery of his life: the spin echo. Unlike the Harvard group, Hahn worked with pulses. One day he turned up the power of the pulse and considerably shortened its duration, whereupon he got a strange, symmetrical V-shaped signal on his oscilloscope,[2] quite different from the so far pulse-shaped signals. He first thought it was an artifact, a glitch, an unwanted signal that occurs because there is a fault in the circuit. Luckily, having manipulated the signal in every possible way, he was able to show that it was an echo. The size of the signal appeared to depend on the intensity of the applied pulse and on the pulse width [10].

Hahn's publication has become one of the most cited publications in physics, with 5850 citations [11]. High-resolution pulsed NMR of liquids and pulsed NMR of solids would from the time of Hahn's publication dominate the applications of chemical shift in physics, chemistry, and other discussed below. The development of two-dimensional and three-dimensional NMR, e.g., to study the spatial structures of proteins, is also largely due to Hahn's research.

From 1950 onwards, articles began to appear on the subject of 'chemical shift', with research carried out not only by physicists, but also by chemists, such as the article by Gutowsky (1919–2000) and Hoffman of the Noyes Chemical Laboratory, University of Illinois, Urbana, in which it was stated:

Several instances have been reported recently of the dependence of nuclear magnetic resonance on the nature of the chemical compound containing the nuclei. This "chemical effect" or magnetic shielding has been discussed theoretically but the complexity of the calculations permits their application only to the simplest molecules. We are presently making an experimental survey of the wide variety of existing polyatomic fluorine compounds to determine the influence of structural factors [12].

Gutowsky worked in Harvard with Purcell, Pake (see Chap. 9), and Kistiakowsky to determine the structure of crystals. This was in fact the breakthrough in chemistry. The complicated calculations for simple molecules were replaced by simple pictures

---

[1]Precession is a motion typical of a rotating object like a top under the influence of an external force. The axis of rotation of the top wobbles. In the case of nutation, this wobble gains an extra boost of power, creating extra wobble in the precession.

[2]An oscilloscope is the measuring instrument that succeeded the galvanometer. The latter used mechanical registration with a roll of paper and pen. The galvanometer was an electromechanical instrument, but the oscilloscope is completely electronic without moving parts. The oscilloscope contains a cathode ray tube (CRT). In the CRT, electrons are excited in the cathode by means of thermal emission (as in a television tube) and then accelerated by the anode toward the display. That display has a luminescent layer that illuminates when the electrons hit it. Today, almost only digital oscilloscopes are made with built-in memories.

for complex molecules, which could be used to identify them. In 1952 NMR was first used to study a chemical reaction [13]. In 1953, Powles and Gutowsky published a study of the methyl group in organic compounds in a chemistry journal [14]. This marked the next important step: from the application of NMR to physical chemistry (which is close to physics) to its application to structural analysis in organic chemistry. In the previous publications, Bloembergen's results, as recorded in the BPP publication, were still mentioned as the basis. Gutowsky was the first to coin the term *Nuclear Magnetic Resonance*, which has been widely used since then [15].

From 1953, there was a flood of publications on NMR, initially almost exclusively by physicists, but gradually also by chemists. Shoollery of Varian Associates gave for the first time an overview of the (commercial) equipment that could be used and the applications, such as the structural determination of organic compounds (Fig. 10.1). Here, too, the BPP article is mentioned again [16]. Varian Associates was established in California in 1948 by the brothers Russell and Sigurd Varian, and the company's first activity was NMR research. In 1946, with the help of Russell Varian, Bloch and Hansen in Stanford had registered a patent with the title: *Method and Means for Chemical Analysis by Chemical Induction* (Fig. 10.2). They were not thinking of chemistry at all, but they thought it would be useful for a few physicists. The Varian brothers received the exclusive license to exploit the patent and that was the beginning of the company. To begin with, they developed a magnetometer to measure magnetic fields. In 1950, Varian developed the first NMR spectrometer, but it took until 1953 before the first device was marketed (Fig. 10.3). It was ordered by Humble Oil from Texas for no less than $30,000. DuPont Experimental Station and Shell Development followed. In 1954, it became apparent for the first time that NMR was superior to

Fig. 10.1 One of the first pieces of commercial NMR apparatus, 1962. In the foreground the built-in magnet. *Source* Graphic Services, College Station, Texas A&M University, Texas

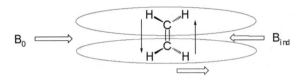

Fig. 10.2 Example of a chemical shift. Induced magnetic field of alkenes in external magnetic fields. Ellipses are the field lines. Anisotropic induced magnetic field effects are the result of a local induced magnetic field experienced by a nucleus resulting from circulating electrons that can either be paramagnetic when it is parallel to the applied field or diamagnetic when it is opposed to it. It is observed in alkenes where the double bond is oriented perpendicular to the external field with pi electrons likewise circulating at right angles. The induced magnetic field lines are parallel to the external field at the location of the alkene protons which therefore shift downfield to a 4.5–7.5 ppm range. The three-dimensional space where there is a diamagnetic shift is called the shielding zone with a cone-like shape aligned with the external field. *Source* V8rik at the English language Wikipedia, CC BY-SA 3.0, commons.wikimedia.org/w/index.php?curid=17218884

Fig. 10.3 Russell (left) and Sigurd Varian. *Source* sempervirens.org

the infrared spectrometry used so far in organic chemistry as a method for structural determination. In 1956 Russell Varian filed a patent application for *Gyromagnetic Resonance Methods and Apparatus.* This was an invention to increase stability over a certain time using Fourier analysis.[3] The invention was far ahead of its time, because the Fourier analysis could not yet be carried out. Russell Varian died in 1959, seven years before the patent was finally registered [17].

After introducing an electromagnet instead of the permanent magnet used so far, stabilizing the magnetic field and increasing the field strength to 1.4 T (for comparison, Bloch used a field of 0.2 T), the NMR technique finally broke through in 1961 [18]. More and more chemistry laboratories used the technique for structural analysis: the present author remembers that the Laboratory for Organic Chemistry of the University of Amsterdam had one of the first NMR devices in the Netherlands for structural determination in 1964.

The development of NMR techniques in chemistry moved ahead very quickly. In 1955, the first ^{14}N (nitrogen) [19] and ^{17}O (oxygen) [20] NMR chemical shifts were reported, followed by ^{29}Si (silicon) [21]. In 1956 the first application to the nutrients in potatoes and apples was published [22], and in the same year an early medical application to red blood cells [23]. In 1957 followed the ^{13}C (carbon) shift [24] and the application of NMR to polymers by Wilson and Pake [25]. In 1958, an internal standard, tetramethyl silane, was proposed for the structural determination [26]. This standard is still used.

From 1957, Richard Ernst (born 1933), a chemist at the ETH in Zurich, improved the equipment for high-resolution NMR. The technique was still in its infancy at the time. An important point was the improvement in sensitivity and Ernst was one of those who devised the methodological refinement known as Fourier transformation NMR (FT-NMR) [27]. In 1963 Ernst joined Varian Associates in Palo Alto, California. Wesley Anderson also worked on Fourier transformation at Varian. However, the scientific world did not want it. The manuscript was submitted twice to the *Journal of Chemical Physics* and twice refused, at which point it was presented to the *Review of Scientific Instruments* in 1966 and finally published. Anderson showed that the analysis time could be reduced by a factor of 100 [28]. Although Varian had the patent for FT-NMR, the company did not want it, so the German company Bruker Analytische Messtechnik came up with the first FT-NMR spectrometer in 1969. Another technique to obtain a better signal-to-noise ratio was time-averaging.[4] These two techniques resulted in an enormous improvement in the signal-to-noise ratio, which either allowed much smaller samples to be taken or significantly reduced the analysis time. By 1968, Ernst had returned to the ETH, but he remained a consultant at Varian.

---

[3]In a Fourier transform, a transient signal, i.e., a signal like a pulse that fluctuates in time, is transformed (converted) to a frequency representation. One then obtains a frequency distribution of the signal and can clearly distinguish noise frequencies. The latter (lower and/or higher than the signal frequencies) are then removed and the reverse transformation reverts to the time domain where the pulse will be clearly visible because virtually devoid of noise.

[4]Time-Averaging is a technique from astronomy, where the average noise may approach zero after repeated measurements (for example, 40 repeats).

In 1992, Ernst was awarded the Nobel Prize for chemistry for his contributions to the development of the methodology of high-resolution nuclear magnetic resonance (NMR) spectroscopy [29].

After the application of NMR to physical and organic chemistry, the next area to come under the spotlight was biochemistry, around 1964. One of the first publications appeared in 1965, on the binding of penicillin to albumin in serum [30]. This was followed by many publications on structural determinations.

In 1971 Ernst went to a congress attended by Jean Jeener of the Free University of Brussels, who had been a postdoc under Bloembergen in 1964 [31]. Jeener pointed out that, if a Fourier transformation is performed after two consecutive pulses, a two-dimensional spectrum can be produced.

Ernst developed this idea from 1974 onwards, and two-dimensional FT-NMR, followed by three- and multidimensional techniques, made it possible to analyze increasingly complex compounds.

Wüthrich did NMR research on hemoproteins in the 1970s,[5] and thus determined the three-dimensional structure of these compounds in solutions [32]. In 2002 he was awarded the shared Nobel Prize for Chemistry. Reaction kinetics was already under investigation with NMR in 1961 [33]. An early example of an application to cell biology came in 1967, with the leakage of sodium from muscles [34].

Related research fields such as biophysics, physical biochemistry, structural biology, molecular reactions, pharmacology, physiology, drug applications also followed. The number of applications of NMR research into chemical shift was so great that simply listing all the research directions was not feasible.

In almost all publications on the (application) of the chemical shift, the pioneers of NMR were mentioned, including Purcell, Bloch, Bloembergen, Ramsey, and also Gorter. The BPP publication was most frequently cited, making it clear that the application of the NMR chemical shift in chemistry followed directly from the discovery of NMR. It also showed that the work of Bloembergen, Purcell, and Pound played a key role in the applications of the chemical shift.

10.2 Magnetic Resonance Imaging (MRI)

Raymond Damadian (born 1936), a physician at the State University of New York Downstate Medical Center, worked on kidney examination. He began to take an interest in NMR when he attended Purcell's lectures on NMR around 1964. Damadian thought that NMR might be a useful instrument for the study of sodium in cells he was working on at the time [35]. He referred to earlier research by others using NMR on cell water, in which the precession of the protons differed from that in

---

[5]Hemoproteins are metal proteins which bind to the heme (iron group). An example is hemoglobin (the protein found in vertebrate animals), in which iron is bound to four porphyrins. This whole molecule is bound to the protein globulin. Hemoglobin is responsible for oxygen transport in the body.

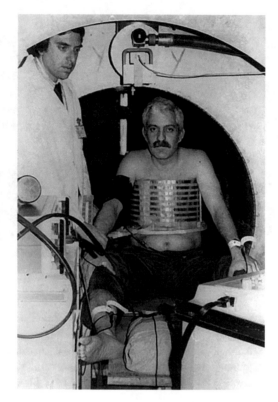

Fig. 10.4 Raymond Damadian demonstrates his MRI apparatus in 1977. *Source* www.inc.com

pure water. Furthermore, Damadian had been able to detect renal disorders (tumors) using his own NMR research (Fig. 10.4). He showed that the resonance values in rat tumors deviated from those in normal tissue [36]. Until then, X-rays had been the only technique that could detect tumors, and everyone tried to reproduce Damadian's findings. In fact, he was not the first to come up with the idea of examining biological samples with NMR, but he was the first to come up with the idea of examining abnormal tissues, in this case cancerous tissues. Damadian was also the first outside physics, chemistry, or biology to dare to use NMR on biological samples. Although Damadian suggested that the technique could be used to diagnose cancer, it later became apparent that the differences found between healthy tissue and cancer tissue were too small to serve as a reliable diagnostic tool. In 1974, he patented his method with the title *Apparatus and Method for Detecting Cancer in Tissue* [37]. Although Damadian had only done experiments with rat tissue, he also claimed with foresight the application for humans in the patent. He stated that a human could be surrounded by a large coil which would in turn be surrounded by an electromagnet. This patent would prove to be of great commercial significance. Later Damadian realized that his relationships between the relaxation times T_1 and T_2 in cell fluids were the same

Fig. 10.5   Paul Lauterbur. *Source* www.acs.org

as those found by Bloembergen twenty years earlier in a mixture of glycerol and water [38].

In an interview in 2004, Bloembergen remarked: "We measured the relaxation time of protons in water and in aqueous solutions, and the influence of viscosity and temperature. Those data are now the bases on which the MRI pictures can be taken" [39].

The chemist Paul C. Lauterbur (1929–2007) notes that, in 1971, he had seen Sayan's experiments, which confirmed Damadian's results:

Even normal tissues differed markedly among themselves in NMR relaxation times, and I wondered whether there might be some way to noninvasively map out such quantities within the body [40].

Lauterbur was attached to New York State University and had just become director of NMR Specialities, a company with some financial difficulties (Fig. 10.5). As an expert on NMR shifts, he was asked if there were some way to save the company. Saryan was a graduate student from Johns Hopkins University, who repeated Damadian's experiments at NMR Specialities. Lauterbur came up with the idea of using magnetic gradients to translate a one-dimensional FT-NMR scan of inhomogeneities in tissues into a two-dimensional map.

In 1972 there was a conference in Bombay (Mumbai), where Lauterbur presented his results. Bloembergen was there and found the method very slow. He asked why there was no pulsing. Waldo Hinshaw from the University of Nottingham heard that and took the idea home [41]. In 1973 the first publication on what would later become Magnetic Resonance Imaging (MRI) appeared in *Nature*, authored by Lauterbur [42]. This contained the following claim:

A possible application of considerable interest at this time would be to the in vivo study of malignant tumours, which have been shown to give proton nuclear magnetic resonance signals with much longer water spin-lattice relaxation times than those in the corresponding normal tissues.

Lauterbur had made the first two-dimensional picture of two 1 mm diameter tubes containing water. Both methods, Damadian's and Lauterbur's, were used side by side during the first years of MRI. In 1974, Lauterbur investigated the first biological samples with NMR: the water distribution in the branch of a pine tree, the oil distribution in a pecan nut, the water distribution in a living shellfish, and the water distribution in the lungs of a living mouse [43].

Ernst realized in the mid-1970s that the two-dimensional spectroscopic technique described in the *Chemical Shift* paper could be used for MRI [44]. It was easier to make a two-dimensional picture with this technique than with Lauterbur's one-dimensional technique.

Hinshow developed Bloembergen's idea of pulsing to devise a faster method, and a colleague of his, Peter Mansfield, designed a new MRI method that made lengthy computer calculations superfluous: selective radio-frequency irradiation [45]. This allowed a small part of the sample to be depicted and then combined into a larger image. Three-dimensional MRI now became possible [46].

Due to the use of ever larger magnets, cooling of the magnets with liquid nitrogen or helium (superconducting magnets), and the application of ever better measuring methods involving computers, the opening between the magnetic poles could be made even larger. As a result, ever larger objects could be scanned. Instead of electromagnets, permanent magnets were then brought back, whence cooling was no longer necessary and an open system could be designed.

In 1976 Mansfield produced the first MRI scan of a part of the human body: a finger [47]. In 1977 Damadian produced, among other things, an MRI scan of the heart [48].

A fairly recent development is magnetic resonance microscopy (MRM), in which small organisms can be scanned at microscopic level by means of high-resolution MRI. Even the metabolism of bacteria can be determined with MRM [49]. The MRM scan of the pistil of a geranium flower is also a nice example of this technique [50].

Functional MRI (fMRI) came on the scene in 1990. Here it is the function of the brain that is determined and not the structure, as with conventional MRI. This includes, for example, medical research into Alzheimer's disease, research into fear and aggression, and research into the function of certain parts of the brain in, for example, speech. In *Nature*, in 2008, there was a critical article about this application, explaining the limitations of the technique, such as the fact that only snapshots are made [51].

In 2003, the Nobel Prize for Medicine was awarded to Lauterbur and Mansfield *for their discoveries concerning magnetic resonance imaging.* Damadian did not get a prize and reacted furiously. In Damadian's hometown, the local newspaper *The Sydney Morning Herald* carried the headline: *The man who did not win.* Damadian thought he should have been awarded the Nobel Prize in the first place and Lauterbur in the second place. The latter, in turn, felt that he himself should have been the only one awarded the prize. Damadian called the Nobel Committee, but they stuck to their position. Bloembergen was also approached by the Nobel Committee with the question whether Damadian should not have been considered for the prize in physics. According to Bloembergen, however, Damadian did not have the necessary qualities to qualify for the Nobel Prize in Physics [52].

Damadian had already had a controversy with Lauterbur, because Lauterbur hadn't cited him in his first article, but did quote Saryan (who had merely repeated Damadian's work). He responded with full-page advertisements in *The New York Times (NYT)* and *Washington Post* which cost $200,000 and $290,000 respectively. A first series of ads in the *NYT* called for support for Damadian with a coupon that could be sent in, and carried the headline: *The Shameful Wrong That Must Be Righted*, with a picture of a Nobel Prize medal that had been turned upside down. The preamble of the slip to be sent in was: *Dear Members of the Nobel Committee: The TRUTH must have a place. I/We believe this year's Nobel Prize for Physiology or Medicine should include Dr. Raymond Damadian* [53]. The series ended on 2 December 2002, when it became clear that the Nobel Committee would not allow itself to be influenced by a newspaper advertisement which read: *This is the great voyage of scientific discovery that gave the world the MRI. It will be ignored on the shameful night of December 10th* [2002, when it became known who received the Nobel Prize] The bottom line was: *The Nobel Prize will make itself irrelevant to the true history of the MRI. It will therefore lose its credibility as an award for scientific achievement* [54]. The advertisements were drafted by *The Friends of Raymond Damadian* and published by FONAR, Damadian's own company. Damadian had earned a lot with his patents and his company, but this did not help Damadian, who remarked that, now that he saw how easily the Nobel Prize could be manipulated, he had lost all respect for the prize and was no longer even sure if he wanted it [55]. It was particularly ironic that the book mentioned earlier on NMR and MRI [56] was specially released to support Damadian's candidacy for the Nobel Prize for Medicine. Bloembergen himself supported Damadian's candidacy, while Lauterbur declined the offer to contribute to the book [57].

Finally, a quote from Pound, which makes it clear that the people who originally worked on NMR only had research on atomic nuclei in mind

NMR […], initially conceived as a way to reveal properties of atomic nuclei, has become a major tool for research in materials science, chemistry, and even medicine, where magnetic resonance imaging (MRI) is now an essential tool. [58]

References

1. Interview Marika Griehsel with N Bloembergen, June 2004. 54th Meeting of Nobel Laureates in Lindau, Germany. Nobel Web, 2004
2. Kellogg JMB, Rabi II, Ramsey NF, Zacharias JR. An Electrical Quadrupole Moment of the Deuteron. Phys Rev 55:318–319, 1939
3. Knight WD. Phys Rev 76:1259–1260, 1949
4. Bloembergen N. Fine Structure of the Magnetic Resonance Line of Protons in CuSO4.5H2O. Phys Rev 75:1326, 1949
5. Ramsey N. Phys Rev 78:699–703, 1950
6. Interview Rob Herber with N Bloembergen, 7 December 2006
7. Proctor WG, Yu FC. The Dependence of a Nuclear Magnetic Resonance Frequency upon Chemical Compound. Phys Rev 77:717, 1950
8. Jacobsohn B, Wangsness RK. Shapes of Nuclear Induction Signals. Phys Rev 73:942–946, 1948
9. Hahn EL. Nuclear Induction Due to Free Larmor Precession. Phys Rev 77:297, 1950
10. Mattson J, Simon M. Pioneers of NMR and Magnetic Resonance in Medicine. The Story of MRI. Dean Books, Jericho, 1996
11. Hahn HC. Spin Echoes. Phys Rev 80:580–594, 1950
12. Gutowsky HS, Hoffman CJ. Phys Rev 80:110–111, 1950
13. Hickmott TW, Selwood PW. The Use of Nuclear Induction in the Kinetic Study of the Reaction $Eu^{+3} + \varepsilon\, Eu^{+2}$. J Chem Phys 20:1339-1341, 1952
14. Powles JG, Gutowsky HS. Proton Magnetic Resonance of the CH_3 Group. II. Solid Solutions of t-Butyl Chloride in Carbon Tetrachloride J Chem Phys 21:1704–1709, 1953
15. Gutowsky HS. Nuclear Magnetic Resonance. Ann Rev Phys Chem 5:333–356, 1954
16. Shoolery JN. Nuclear Magnetic Resonance Spectroscopy. Anal Chem 26:1400–1403, 195
17. Mattson J, Simon M. Pioneers of NMR and Magnetic Resonance in Medicine. The Story of MRI. Dean Books, Jericho, 1996
18. Shoolery JN. NMR Spectroscopy in the Beginning. Anal Chem. 65: 731A–741A, 1993
19. Holder BE, Klein MP. Chemical Shifts of Nitrogen. J Chem Phys 23:1956, 1955
20. Weaver HE, Tolbert BM, La Force RC. Observation of Chemical Shifts of ^{17}O nuclei in Various Chemical Environments. J Chem Phys 23:1956–1957, 1955
21. Holzman GR, Lauterbuur PC, Anderson JH, Koth W. Nuclear Magnetic Resonance Shifts of ^{29}Si in Various Materials. J Chem Phys 25:172, 1956
22. Shaw TM, Elsken RH. Moisture Determination, Determination of Water by Nuclear Magnetic Absorption in Potato and Apple Tissue. J Agric Food Chem 4:192–164, 1956
23. Ordeblad E, Bhar BN, Lindström G. Proton magnetic resonance of human red blood cells in heavy water exchange experiments. Arch Biochem Biophys 63:221–225, 1956
24. Lauterbur PC. C^{13} Nuclear Magnetic Resonance Spectra J Chem Phys 26:217–218, 1957
25. Wilson CW, Pake GE. Nuclear Magnetic Relaxation in Polytetrafluoethylene and Polyethylene J Chem Phys 27:115–122
26. Van Dyke Thiers G. Reliable Proton Nuclear Resonance Shielding Values by "Internal Referencing" with Tetramethyl-silane. J Phys Chem 62:1151–1152, 1958
27. Mattson J, Simon M. Pioneers of NMR and Magnetic Resonance in Medicine. The Story of MRI. Dean Books, Jericho, 1996
28. Ernst R, Anderson WA. Application of Fourier Transform Spectroscopy to Magnetic Resonance. 37:93–102, 1966
29. Mattson J, Simon M. Pioneers of NMR and Magnetic Resonance in Medicine. The Story of MRI. Dean Books, Jericho, 1996
30. Fischer JJ, Jardetzky O. Nuclear Magnetic Relaxation Study of Intermolecular Complexes. The Mechanism of Penicillin Binding to Serum Albumin. J Am Chem Soc 87:3237–3244, 1965
31. Walsh WM, Jeener J and Bloembergen N. Temperature-Dependent Crystal Field and Hyperfine R Interactions. Phys Rev 139: A 1338, 1965

32. Wüthrich K. Biophys J 32:549–560, 1980
33. Myers OE, Sheppard JC. J Phys Chem 83:4769–4670, 1961
34. Cope FW. Bull Math Biophys 29:691–704, 1967
35. Mattson J, Simon M. Pioneers of NMR and Magnetic Resonance in Medicine. The Story of MRI. Dean Books, Jericho, 1996
36. Damadian R. Tumor Detection by Nuclear Magnetic Resonance. Science 171:1151–1152, 1971
37. Damadian RV. Apparatus and Method for Detecting Cancer in Tissue. US Patent 3.789.832, 1972
38. Mattson J, Simon M. Pioneers of NMR and Magnetic Resonance in Medicine. The Story of MRI. Dean Books, Jericho, 1996
39. Interview Marika Griehsel with N Bloembergen, June 2004. 54th Meeting of Nobel Laureates in Lindau, Germany. Nobel Web, 2004
40. Lauterbur PC. Cancer Detection by Nuclear Magnetic Resonance Zeugmatographic Imaging. Cancer 57:1899–1094, 1986
41. Mattson J, Simon M. Pioneers of NMR and Magnetic Resonance in Medicine. The Story of MRI. Dean Books, Jericho, 1996
42. Mattson J, Simon M. Pioneers of NMR and Magnetic Resonance in Medicine. The Story of MRI. Dean Books, Jericho, 1996
43. Lauterbur PC. Pure Appl Chem 40:149–157, 1974
44. Mattson J, Simon M. Pioneers of NMR and Magnetic Resonance in Medicine. The Story of MRI. Dean Books, Jericho, 1996
45. Mattson J, Simon M. Pioneers of NMR and Magnetic Resonance in Medicine. The Story of MRI. Dean Books, Jericho, 1996
46. Garroway AN, Grannell PK, and Mansfield P. Image Formation in NMR by a Selective Irradiative Process. J Phys C 7: 457–462, 1974
47. Mansfield P, Maudsley AA. Medical Imaging by NMR. Brit J Radiol 50:188–194, 1976
48. Damadian R, Goldsmith M and Minkoff L. NMR in Cancer: XVI. FONAR Image of the Live Human Body. Physiol Chem Phys. 9:97–100, 1977
49. Aguayo JB. Gamscik MP, and Dick JD. High Resolution Deuterium NMR Studies of Bacterial Metabolism. J Biol Chem 263:19552–19557, 1988
50. Lee SC, Kim K, Lee S, Han Yi j Kim SW, Ha KS and Cheong C. One Micrometer Resolution NMR microscopy. J Magn Res. 150:207–213, 2001
51. Logothetis NK. What we can do and what we cannot do with fMRI. Nature. 453:869–878, 2008
52. Interview Rob Herber with N Bloembergen, 27 April 2010
53. Advertisement in the New York Times of 9 October 2002
54. Advertisement in the New York Times of 2 December 2002
55. The man who did not win. Sydney Herald, 17 October 2003
56. Logothetis NK. What we can do and what we cannot do with fMRI. Nature. 453:869–878, 2008
57. Interview Rob Herber with N Bloembergen, 27 April 2010
58. Pound RV. Edward Mills Purcell. Books.nap.edu/html/biomems/epurcell.html

Part IV
Masers and Lasers

Chapter 11
The Birth of the Maser—Oil
and the Milky Way: Applications
of the Maser

> Bloembergen has made a great discovery: he writes to his sick mother that he
> will tell her everything when he comes home [1].

During the time that Bloembergen was working on NMR, many others were doing
research on microwaves for radar. The radar technology was continuously improved,
but there were still problems with the signal-to-noise ratio. The technical question
of how to obtain lower noise led to a fundamental physical principle: stimulated
emission.

11.1 The Birth of the Maser—Einstein and Microwaves

Stimulated emission—From theory to practice with nuclear magnetic resonance
As early as 1916 Einstein suggested that a (gas) molecule could change to a lower
energy level under the influence of incident radiation if the energy difference corresponded to the energy of the radiation. He called that *Zustandsänderungen durch
Einstrahlung* [2], now known as stimulated or induced emission.

Spontaneous emission occurs, for example, when a pan is heated on the gas. We
often see yellow radiation, coming from sodium atoms (derived from table salt, NaCl)
that emit when heated. The same colour can be seen in the well-known sodium lamps
used for street lighting.

Stimulated emission can also occur in atoms, and this phenomenon has been
identified for the whole electromagnetic spectrum, from radio frequencies through
ultraviolet light to X-rays. In 1924 Tolman stated that, if more molecules of a higher
level move to a lower level than the other way around, the primary radiation will be
amplified by 'negative absorption' [3]. In that same year, Kramers stated that both
negative absorption and negative dispersion (e.g., frequency-dependent refraction of
radiation in a glass prism) could occur [4].

© Springer Nature Switzerland AG 2019   227
R. Herber, *Nico Bloembergen*, Springer Biographies,
https://doi.org/10.1007/978-3-030-25737-8_11

Ladenburg came close to stimulated emission from gas discharges in 1933 [5] but saw only anomalous dispersion.[1]

Stimulated emission, amplification, oscillation: who was first?

Valentin Fabrikant (1907–1991)
In 1939, the Soviet physicist Valentin Fabrikant published a remarkable thesis *The emission mechanism of a gas discharge*. One chapter was entitled: *On experimental evidence for the existence of negative absorption* [6]. However, the dissertation was written in Russian and was not noticed outside the Soviet Union.

Fig. 11.1 Valentin Fabrikant. *Source* ru.wikipedia.org, Biberman L.M.—Personal photo archive

---

[1]In normal dispersion (color shift, for example by dispersion in a prism), the refractive index decreases with increasing wavelength. In anomalous dispersion, the refractive index increases with increasing wavelength.

Brossel and Kastler suggested the double resonance method in 1949. This method combines optical resonance with magnetic resonance and they introduced the concept of an 'optical pump' as a mechanism for achieving a higher energy level [7, 8].

Fabrikant, whose work had been at a standstill during the Second World War, resumed his research in 1945 and this led to a patent, which in fact laid down the principle for both the maser and the laser: *Method for amplification of electromagnetic radiation (ultraviolet, visible, infrared and radio wavebands).* He and his group subsequently conducted research on the optical pumping of cesium gas, which was published earlier than the research by Townes and Schawlow (see below) [9]. These studies were also published in Russian and were not noticed in the West! (Fig. 11.1)

Fabrikant had already used the terms *negative absorption* and *amplification* in his thesis in 1939. He had filed a patent in the Soviet Union in 1951, but it was not published until 1959. Townes found that Fabrikant had published *a rather obscure thesis*, presumably describing it this way because it was published in Russian. Townes wrote about the patent: *Fabrikant was definitely working on relevant concepts as early as 1939, but unfortunately he did not get very far and no one picked up his work as being particularly interesting* [10]. Fabrikant himself admitted that he *"did not pay attention to the practical value of this idea"*. in reference to stimulated emission [11]. Nevertheless, it can be concluded that Fabrikant was indeed the first to successfully extract results from these theoretical explorations into stimulated emission.

Willis Lamb (1913–2008)
In 1947 Willis Lamb Jr. and Retherford experimented with the hydrogen atom to achieve a transition from the lowest energy level to a higher energy level with radio

Fig. 11.2   Willis Lamb Jr. *Source* www.mediatheque.lindau-nobel.org

frequencies (1 GHz). Furthermore, Lamb foresaw that there would be population inversion in which the energy difference between the low and high levels corresponded to a radio frequency photon [12]. Later they discovered that hydrogen has a metastable level [13] (Fig. 11.2).

Purcell, Pound, and Ramsey

The NMR researchers started to realize something, partly as a result of Lamb's experiments. As explained in earlier chapters, in NMR the spins (rotation) of the protons in a magnetic field are reversed by an electromagnetic pulse. The protons of a lower energy level are 'lifted' to a higher energy level by the pulse. In 1950 Purcell, Pound, and Ramsey experimented with the relaxation times of a lithium fluoride (LiF) crystal in NMR. They found that the relaxation times were long enough to cause spin inversion [14]. First they constructed a magnetic field in which the spins of the nuclei pointed in the direction of the magnetic field. The atomic nuclei would thus have been in the state of lowest possible energy, known as the ground state. They then turned the magnetic field over so that the spins would now point in the opposite direction to the magnetic field and would therefore have been at a higher energy level, i.e., in an excited state. The system would then slowly return to equilibrium [15]. They called the excited state a state with a negative temperature[2] and this was in fact the first observation of population inversion [16]. In population inversion, there are more atoms in the excited state (higher energy levels) than in the ground state: this is a condition for negative absorption and therefore for the operation of masers and lasers. Ramsey wrote in a letter to Dixon: *As we were the first scientists reporting such work, we had to choose our own terminology for describing and chose, for historical reasons, to do so largely in terms of negative absolute temperatures, but we clearly realized and also reported that we were observing amplification by stimulated emission of radiation, as evidenced by my review paper* [17]. In the relevant review article of 1952, it was stated that: *In a negative temperature state the high energy levels are occupied more fully than the low, and the system has the characteristic that, when radiation is applied to it, stimulated emission exceeds absorption* [18].

Bloembergen was not involved in this research by Purcell, Pound, and Ramsey [19]. But he knew what was going on [20].

In his book about the Purcell group and others, Townes wrote: *A number of scientists had inverted nuclear-spin populations, using radio-wave excitation, though nowhere near enough to provide a net amplification, and had studied the results* [21]. Ramsey disputed this in an interview and stated that the Purcell group had indeed demonstrated amplification by stimulated emission, albeit at radio frequencies. According to Ramsey, Townes initially refused to admit this, but after a year of

---

[2]In normal matter, negative kelvin temperatures are not possible, whence 0 K (−273 °C) is the lowest possible temperature. There are systems (not ordinary matter) in which negative kelvin temperatures are possible. Those systems can reach a temperature of, for example, −400 °C. The Purcell and Pound system loses internal energy and was therefore described by the authors as having a negative temperature.

correspondence and discussions between Ramsey and Townes, the latter acknowledged that the Purcell group had achieved reinforcement by stimulated emission [22].

In their 2003 publication, Meyers and Dixon say the following about patents for masers and lasers: *The negative temperature concept [...] has been introduced by Edward Purcell, Robert Pound, and Norman Ramsey as an elegant, quantifiable method of describing population inversion* [23].

Ramsey wrote the following in a letter to Dixon in 2006: *Our work can correctly be described by the statement: E. M. Purcell, R. V. Pound and N. F. Ramsey produced, correctly analyzed and published the first observations of amplifications of electromagnetic radiation by stimulated emission of radiation using atoms with a deliberately modified energy distribution to make stimulated emission exceed absorption. Although the net system amplification was not mentioned, there was in fact a net amplification of the input radio frequency power.*

Purcell, Pound, and Ramsey were thus the first to achieve both stimulated emission and amplification experimentally.

Microwaves

The development of the radar (see Chap. 6) provided many insights into microwave technology. Together with the previously discovered radio technology, physicists were given two new tools to perform resonance research.

In addition to nuclear magnetic resonance, where research was conducted with radio frequencies, electron spin resonance (ESR) also came on the scene. ESR uses microwaves with a higher frequency and hence greater energy content than radio waves.

Charles Townes (1915–2015) [24, 25]

Rabi was working at Columbia University in 1947. The Pentagon offered him $500,000 a year for further defense research, an astronomically high salary. Rabi thought this was too much and the amount was reduced, but it was certainly enough to pay the $6000 salary of a new Associate Professor. This would be Charles Townes.

During the Second World War, Townes had worked at the Bell Labs on 1.25 cm radar that could be used for defense purposes (with a frequency of 24 GHz, while a kitchen microwave has a frequency of 2.7 GHz). By 1933 Cleeton and Williams had designed an oscillator suitable for generating wavelengths between 1 and 3 cm [26]. In 1934 the same authors had used a wavelength of 1.1 cm to determine the absorption spectrum of the ammonia molecule (NH_3) [27] (Fig. 11.3).

The previously mentioned wavelength of 1.25 cm was not suitable for defense purposes, because water vapor would absorb the radar radiation after just a few hundred meters (see also the same problems encountered by Purcell at the Rad Lab in Chap. 7). In his book, Townes comments: *I really woke up when I saw a memorandum written by John Van Vleck of Harvard University [...] He pointed out that there should be an absorption by water molecules at about this wavelength [...]* [28] Townes realized that fundamental research could be done on these systems after the war. He had to remain at the Bell Labs for another half a year after the war to

Fig. 11.3 Charles Townes (left) and James Gordon with their second maser. *Source* www.gizmode. com.au

finish his work there, and then published a study on the ammonia spectrum [29]. This research would prove to be of great importance for research on stimulated emission.

In 1948 Townes went to Columbia University to fill the vacancy created when Norman Ramsey moved to Harvard. The work Townes did on Columbia was a continuation of the work he had done at the Bell Labs, dealing with the microwave spectra of molecules. Townes' interest shifted to wavelengths of a few millimeters (with a frequency up to 300 GHz), almost into the infrared. Microwave radiation is generated by klystrons (see Radar 'Radio Detection And Ranging' in Chap. 6) and vibrating cavities. At wavelengths of a few millimetres, the vibrating cavity becomes very small, on the order of several millimeters. An additional problem is that a lot of heat is produced, and this also has to be removed.

On 26 April 1951, the breakthrough came. There had been a meeting the previous day, at which a number of people had been discussing the problem of stimulated emission, without success. Townes was sitting in a park early in the next morning when he found the solution. He noted his ideas on the back of an envelope and read them to his collaborator and later brother-in-law Ard Schawlow (1921–1999). Townes' idea was that energy could be absorbed by a molecule. That energy could then make the molecule resonate and thus produce millimeter waves. However, this was a non-equilibrium state. The path from thought experiment to working experiment was far from simple. Townes realized that a few decades earlier Rabi had worked with atomic and molecular beams and had separated excited particles from particles with lower energy by deflecting them with a magnetic or electric field. Rabi had thus produced a beam of excited particles. Townes also referred to the previously mentioned description by Purcell and Ramsey of the negative temperatures of inverted populations. At first he was not interested in stimulated emission, because he did not think it was advanced research. That changed when he saw stimulated emission as a microwave source.

Calculations showed that it should be possible to pass a beam of ammonia molecules through a gap, deflect the molecules with low energy using a magnetic or electric field and resonate the rest in a resonant cavity.

Further calculations suggested changing the wavelength from 0.5 mm (600 GHz, the rotation transition of ammonia) to 1.25 cm (24 GHz, the vibration transition). The latter frequency was a lot easier to work with. Two members of Townes' group, Zeiger and Gordon, spent a year building an experimental setup. In 1953 Zeiger left Columbia, but Gordon continued the research. At the end of 1953, Gordon and Townes finally saw the light of day, stating that: *An experimental device, which can be used as a very high resolution microwave spectrometer, a microwave amplifier, or a very stable oscillator, has been built and operated* [30]. In 1955, they published a comprehensive paper, in which the term 'maser' was used for the first time for this kind of setup. Maser is an acronym for 'Microwave Amplification by Stimulated Emission of Radiation' [31].

Before that publication, however, Townes was to have an unpleasant surprise. In April 1955 he attended a Faraday Society congress in Cambridge (United Kingdom). Townes had wanted to give a lecture on the maser, but the conference organizers decided that he should talk rather about magnetic effects in molecules. Townes did that, but then he listened bewildered as the Soviet physicist Alexander Prochorov gave a talk describing the theory of ammonia masers! Prochorov had read the summary Townes had submitted for his lecture and realized that the work lacked a good theory. This was now provided by Prochorov. Townes, however, hardly knew anything about Prochorov's work, because at the time it was never certain that Soviet scientists would show up, let alone know what they were going to talk about. After Prochorov's presentation at the congress, Townes got up and said: *"Well, that is very interesting, and we have one of these working."* Then he described his experiments. Townes cited Prochorov and his student Basow in his first paper on the maser. Prochorov had also been working for years on the development of an experimental setup, just like Townes [32].

This collision between the two groups was a foretaste of the battle for the maser and laser. Ramsey notes: *"I was a little disappointed, that he* [Townes, RH] *did not refer to our paper."* [33]

It is clear that Townes and Gordon built the first working maser, while Prochorov and Basow claimed that they were the first to develop the theory.

But the principle that Townes and Schawlow applied was Bloembergen's: *"Historically, it's quite clear that the pumping scheme came first with the microwave masers and then Townes and Schawlow immediately saw the opportunity to use the same scheme at optical frequencies."* [34]

Further developments of the maser—Bloembergen enters the arena

Basow and Prochorov drew up another theory that went beyond the description of the ammonia laser. Townes used magnetic fields to separate excited ammonia molecules from ones that were not excited. However, this was a difficult matter. Basow and

Prochorov suggested a maser with three levels instead of two (ground state and excited state). The idea was that, for example, from the ground state the molecule would first reach the highest excited level. This level should then be metastable, i.e., the molecule should remain in this state for a while before falling back to a lower energy level. That lower level would of course still be above the ground state. Stimulated emission would occur when the molecule fell from the metastable to the intermediate level under the influence of incident radiation. How exactly this should be done, however, remained vague.

Initially, Townes' publications went almost unnoticed, but gradually there was more and more interest in maser research. The fact that the ammonia laser was a laboratory setup that was difficult to use outside of that environment encouraged researchers to look for a maser that could be used outside the laboratory. Townes himself felt that, now that he had demonstrated the principle of the maser in a working setup, he had to continue with other research and that others should concern themselves with applications.

Bloembergen initially had little interest in masers. Such research was carried out using molecular beams and Bloembergen was busy with his NMR research on solid matter (see Chap. 9). He did not see any application of the solid maser. He knew of the negative temperatures introduced by the Purcell-Pound-Ramsey group and he knew that it was a transient[3] phenomenon: after a while the normal situation would return. It was also clear to him that a transient phenomenon was inherent in a system with two energy levels [35].

On 17 May 1956 Bloembergen attended a lecture by Jim Gordon from Townes' group in Columbia, and he knew that this group was working with molecular beams. Everyone was talking about magnetic resonance and masers [36]. Woody Strandberg from MIT gave a lecture about a new amplifier he had developed. Strandberg called the amplifier the *Versitron*. According to him, this 'quantum mechanical' amplifier was especially useful for TV and radio frequencies, but also for radar and microwave applications in communications [37]. The participants in the colloquium discussed the inversion of magnetism in paramagnetic salts. Bloembergen said later in an interview [38]: *That Versitron, that's just paramagnetic resonance; it was already known for nuclear spins that you can flip them over that way. Carr and Purcell published something about that in 1953, that it could be done with a 180° pulse.* [39] *In 1946, Bloch had already demonstrated 'spin inversion' as he called it. In any case, in my opinion it was no longer scientifically important.*

Bloembergen asked Strandberg: "*Now, why would you work on getting inversion in condensed matter, because it's no good for a frequency standard?*" To which Strandberg replied: "*Well, it may be useful as a low noise microwave receiver amplifier*" [40]. Bloembergen's interest was now aroused and he pointed out that such an amplifier must be 'continuous wave', so with a signal that is continuously present (in contrast to the pulses mentioned earlier). The most important thing is that there must be a permanent inversion [41]. At the time, Bloembergen was also working

---

[3]A transient event is a short-lived burst of energy in a system caused by a sudden change of state. Wikipedia.

as an advisor to the Lincoln Laboratory, which dealt with radar, communications, and computers for the Ministry of Defence. He understood the importance of a low noise amplifier for radar [42]. On 12 June, he writes in his notebook: *Got an idea. Overhauser effect may be used to obtain negative temperatures under certain conditions* (Fig. 11.4). Underneath, he drew three equidistant horizontal stripes and noted: *Equal splitting won't do.* On 13 June he made another drawing with three stripes: two close to each other and the third just below representing the ground level. Between the upper and lower levels, he drew an arrow and wrote: *Saturation here. This will do it!* [43].

Basow and Prochorov had previously published a theory about three levels in a journal published in the Soviet Union, but that was for molecular beams. This theory had never been tested experimentally. Furthermore, they did not use relaxation, a necessary condition when continuous wave inversion is used. Later, Bloembergen claimed that he was not aware of Basow and Prochorov's publication; he only learnt about it when the people at the Bell Labs who were working on his patent application pointed it out to him. In an interview, Basow commented: *We were not very much surprised by the appearance of Bloembergen and Scovil's work (1956 and 1957) on amplification in paramagnetic crystals. As is known, our own proposals made together with A.M. Prokhorov on the use of three-level systems for the production of inversion by radiation pumping dated to 1955* [44].

Prochorov, however, admitted that: *Bloembergen had the right idea, to do it in solids* [45].

In Bloembergen's mechanism, the first point was that, in a solid substance, an atom has many energy levels. Earlier experiments had shown that, if a lot of energy is 'pumped' into one transition of such a system, over-saturation and population inversion can occur between two energy levels. An important factor is how quickly the atoms fall from a higher energy level to a lower level. Bloembergen went on to describe specific materials that could meet the conditions [46]. Hecht says in his book: *It was a brilliant step forward that drew on his earlier nuclear resonance research* [47].

Discussing Bloembergen's three-level mechanism, Townes made the following remarks:

It was perhaps this considerable experience [with NMR, RH] that led him to envisage a clever system [...] Thus was born the three-level electron-spin maser. In it, microwave energy first boosts electrons from their lowest level to the third level up, bypassing a second or intermediate level. Thus energized, the material can amplify a somewhat lower frequency of input radiation, one corresponding to the energy released as the electrons fall to the intermediate level. This meant that one could pump the maser medium to its excited state with energy of a different wavelength than the maser's output.

For such a maser to operate, its electrons have to be in a crystal with a particular type of asymmetry. As I was not very familiar with paramagnetic materials at the time, I had not even realized that this arrangement of energy levels was possible in such materials. Bloembergen was better versed than I was in this field, and he envisioned the three-level paramagnetic maser [48].

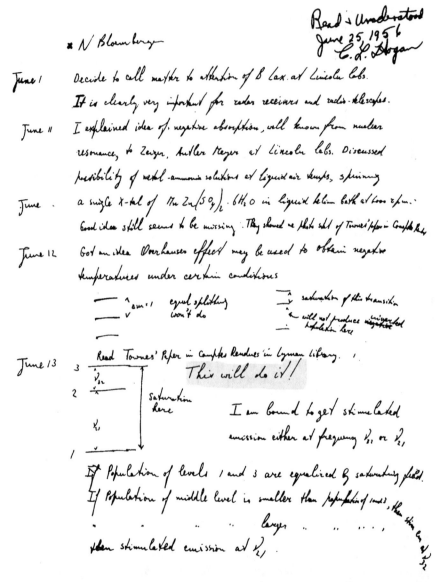

Fig. 11.4 Page from Bloembergen's notebook, June 1956. Herein he describes the three-stage principle for a maser and laser. Saturation occurs between levels 1 and 3 due to the applied field. If the population of level 2 is smaller than levels 1 and 3, stimulated emission takes place between 3 and 2. If the population of 2 is greater than 1 and 3, stimulated emission occurs between 2 and 1. *Source* Nico Bloembergen

Bloembergen himself was well aware that he had gold in his hands. In the summer of 1957, his brother Evert wrote, saying that it would be good if he could come to Bilthoven, owing to the poor health of their mother. Bloembergen wrote to his mother to say that he was coming and that he would tell her 'everything'. He was referring to the three-level maser [49].

Bloembergen tried to confirm the theory with experiment. In Harvard, astronomers were working with a radio telescope. They were particularly interested in the interstellar hydrogen wavelength of 21 cm (1.4 GHz) discovered in Harvard, and they needed to be able to make as sensitive a measurement as possible. However, the 21 cm wavelength was not in fact a good choice and the experiments were not going well [50].

Derrick Scovil of Bell Labs heard of Bloembergen's theory and was better equipped at the industrial laboratory than Bloembergen in Harvard (Fig. 11.5). In 1957, Scovil succeeded in making the first working solid state laser, using gadolinium ethyl sulphate (a material that was also suggested by Bloembergen) [51].

In 1958, Bloembergen's group also published a paper on experiments with a 21 cm maser [52]. The solid substance used was a single crystal of potassium hexacyanocobalt, $K_3Co(CN)_6$, a complex salt, doped[4] with 0.5% potassium hexacyanochromate (III), $K_3Cr(CN)_6$, as active ingredient. The maser was cooled to a temperature of 2 K (-271 °C) and pumped at a frequency of 8 GHz. Autler of MIT used Bloembergen's maser as a basis for later research in radio astronomy [53].

Fig. 11.5 Nico Bloembergen with his maser in 1958. *Source* Nico Bloembergen

---

[4]Doping makes a non-conductor into a semiconductor like silicon by adding small quantities (impurities) of substances such as phosphorus and boron. Wikipedia.

Fig. 11.6 The energy diagram for Cr (III) in synthetic ruby for a maser and laser. Notice the differences in the vertical axis between the microwave laser and the optical laser. v = frequency. 1, 2, 3, 4, $\bar{E}$, $2\bar{A}$, $^4F_2$, and $^4F_1$ are energy levels. *Source* Bloembergen N. Nonlinear Optics and Spectroscopy, Reimpressions de les prix Nobel en 1981. The Nobel Foundation (1982)

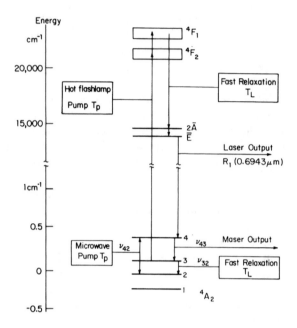

In 1958 Kikuchi found the ideal material for masers (and lasers): ruby [54]. His group at Michigan university used the ruby laser[5] and Bloembergen's three-level principle. The pump frequency used was 24.2 GHz and the generated frequency 9.2 GHz.

How did Bloembergen later view the extension of his research from NMR to the maser? (Fig. 11.6)

In a 1984 interview, he threw some light on this: *In nuclear resonance, I had always been interested in saturation phenomena, T-1's and T-2's. I studied those same problems in microwave magnetic resonance and also ferromagnetic resonance. So the maser came along with the very interesting application of saturation in magnetic resonance* [55].

11.2   From Oil to the Milky Way: Applications of Masers

In spontaneous emission by ordinary kitchen salt, monochromatic light is produced, that is, light of one wavelength, or one color, namely yellow. With the radiation emitted when emissions are stimulated, something very special is going on. The radiation emitted by masers is not only of one wavelength, but also coherent. One can imagine this as follows: in the case of spontaneous emissions, it is like a crowd

---

[5]Ruby is a red gem with the mineralogical name corundum and the chemical formula Al_2O_3 (aluminum oxide). Natural ruby is expensive, but much cheaper synthetic ruby is used in physics. The red color is caused by "doping" of the mineral with chrome.

walking down the street at the same speed, but not necessarily in step with each other. With coherent radiation, it is more like a group of marching soldiers, all of whom have exactly the same speed and all of whom are also in step with each other. The power of such a group is known from what happens on bridges: soldiers must interrupt their marching rhythm on a bridge to avoid the risk of resonance. Coherent radiation is therefore much more powerful than non-coherent radiation.

Bloembergen succeeded in developing a principle for the maser that made it possible to use it outside the laboratory. The three-level mechanism with a continuous pump frequency appeared to be directly usable for different materials and frequencies. On 15 October 1956 Bloembergen applied for a patent for the three-level principle [56]. The patent was not granted until 20 October 1959.

Townes was the first to apply for a patent on the ammonia laser, in May 1955 [57]. He initially called it the *Microwave Oscillator*. The patent was granted in November 1957, but in the meantime it had become clear to Townes that the restriction to microwaves and molecular beams would unnecessarily impede the application of the patent. On 5 December 1956, hence well after Bloembergen's work had become known (Townes had already received a pre-print of the Bloembergen patent in October 1956), Townes submitted an amendment. This was possible under US law at the time. Among other things, the intention had been to add the solid-state maser and the three-level principle, but in the end, a reference to *solid paramagnetic material* and possible higher energy levels was included (as used by Bloembergen), but not the three-level solid-state maser. Townes' final patent was finally granted on 24 March 1959. It will not surprise the reader that numerous passages in the two patents are identical. Myers and Dixon: *It would appear that the authors of '922* [Townes' patent, RH] *benefited significantly from the teachings of '654* [Bloembergen's patent, RH]. *After all, Bell Laboratories attorneys drafted both patents* [58]. It is significant that in Townes' patent there is no reference whatsoever to Bloembergen's.

The American patent legislation was subsequently amended and has since been in line with European legislation, which, once a proposal for patenting has been submitted, can no longer be supplemented, let alone amended. The patent battle would become even fiercer with the laser (see Chap. 12).

The three-level solid-state ruby maser was now the standard setup for maser research. Bloembergen himself was still working on the cyanate laser, in which he and his group were studying the relaxation effects [59].

Deli and Nico Bloembergen told an anecdote about this. At a meeting of the Institute of Radio Engineering in New York in 1959, Bloembergen and Townes both received an award [60]. Townes asked Deli at a congress where she came from and when he heard that she was from Indonesia and missed school for four years as a result, Deli thought Townes seemed worried. Frances Townes, Townes' wife, thought Deli was wearing a very nice dress (and indeed, Deli had made it herself from French silk). She told Deli that she wanted to show her something in her hotel room. Deli went with her and Frances showed her a necklace with a pendant containing a ruby and said: "*Charlie made it for me, he polished the stone, and made the silver thing around it.*" Deli told her that she liked it. After dinner, Deli mentioned that Charles Townes had given his wife something nice, namely a ruby and she asked Nico when he would make something nice for her. Nico replied: *"At the moment I'm working with cyanide"* causing much amusement with this remark [61]. Bloembergen was not actually working with cyanide, but with cyanate. A small amount of cyanide can kill humans and animals, while cyanate is much less toxic. This is an anecdote with a serious undertone, because according to Bloembergen, Townes had annexed ruby as a masing material, while the Kikuchi group who developed it had been totally forgotten.

In the late 1950s and early 1960s, there was a flood of publications on masers. The applications of the maser were still rather limited shortly after its discovery, but still attracted the attention of physicists and the military. Spectroscopy was the first application, as was immediately clear; Townes had already discussed it in his publications. The second was to use it as an atomic clock. Accurate clocks were important for physics research, but military personnel could also use them in intercontinental missile technology [62]. In 1965 Ramsey developed the hydrogen maser as a standard for frequency and time—this was the first 'atomic clock', an instrument taking advantage of the fact that the frequency of the vibrations is highly stable. It turned out that the radiation from the hydrogen maser is more stable than from any other source [63, 64]. This maser is now the international frequency standard for the *Bureau International des Poids et Mesures* in Sèvres, France. Centers broadcasting time signals (the 'beeps' that can still be heard on the radio) also use hydrogen masers. In Europe, the time-and-date station is located in Frankfurt am Main in Germany (DCF77) [65]. The station sends a signal with a carrier frequency of 77.5 kHz with information from local atomic clocks superimposed on it. Other broadcast stations in Europe are MSF located in Anthorn, Cumbria, and TDF in Allouis, France. In North America there is WWV in Fort Collins, Colorado, WWVH in Kauai, Hawaii, and CHU, Ottawa, Canada. Others can be found on Wikipedia. These local signals are connected to the mother clock in Braunschweig in Germany. The signal can be received over distances of up to 2000 km and is used, for example, to activate alarms [66]. In 1989 Ramsey received the shared Nobel Prize for Physics *for the invention of the separated oscillatory fields method and its use in the hydrogen maser and other atomic clocks.*

A fundamental application of the hydrogen laser is to the study of general relativity in astronomy [67].

Bloembergen patented a number of applications of the maser. For the oil industry, he first patented a device for the exploration of non-porous and porous earth layers using resonance techniques [68]. The patent was filed on 24 December 1954 and, remarkably, was not granted until 22 March 1966, almost twelve years later. A second patent for the oil industry was the quantum mechanical counter, described as follows: *This invention relates to radiation detecting systems and, more particularly, to systems for detecting electromagnetic quanta in the infrared or millimeter wave spectrum.* The intention was that the counter should be used in oil extraction; in the infrared, detection was improved significantly compared to the previously used photomultipliers and other detectors. Furthermore, microwave and infrared quanta would be converted into visible light. The patent was registered on 17 April 1959 and granted on 25 December 1962 [69]. Bloembergen was an advisor to the company Schlumberger Well Surveying Corporation in Houston, Texas, which was prospecting for oil in Texas, among other places.

With Philips, Bloembergen patented the three-level maser in a number of countries outside the United States, namely Switzerland, (West) Germany, France, Japan, Australia, Brazil, Austria, Argentina, Denmark, Norway, Canada, India, Italy, Sweden, and the Netherlands. These patents were granted in the period following 1956. In anticipation of the laser (then still called the optical maser), Bloembergen wrote to G. Oudemans at Philips on 31 March 1961 in connection with the Canadian patent (21 February 1961): *I would like to call your attention that the recent development in solid state optical masers is at least in part covered by the first 3 claims of the above mentioned Letters Patent* [70]. Here we anticipate Chap. 12 which is devoted to the laser.

Later, disagreement broke out with General Electric because it had allegedly violated the Philips patent for a number of European countries. Simons at Philips wrote to Bloembergen: *Our investigations in Brazil, Italy, and Japan so far lead to the result that they were using parametric amplifiers instead of masers. In France and Germany so far we found, that there were two masers delivered by Western Electric. We have approached the Post Offices in the countries about these Western Electric masers, but so far without result. In particular, the Post Offices stated, that they would contact Western Electric on this matter and, in Germany, the additional comment was made that the two masers were only for test purposes and not yet for commercial purposes. We are considering whether we shall reopen discussions in these countries; if you are aware of any further manufacture or use by others outside the U.S.A. and Canada, we would of course appreciate to be informed. We continuously watch publications in this respect* [71].

Bloembergen had an agreement with Bell Telephone that they could use the above patent freely in the United States and Canada. Bloembergen sold the rights outside North America to Philips. The only countries where Philips' patent was effective were France and the United Kingdom. In 1965 Bloembergen received a sum of money from Philips with which he could pay the study costs of his children. Bloembergen sold all rights except those for the government in the United States and Canada to Perkin Elmer. Bloembergen got stock options and when the shares went up, he could collect. Perkin Elmer never exercised the rights.

The French government bought masers from Bell Labs, but the company had no rights in France. Bell paid no fees to Bloembergen. Philips had the rights, but did nothing. Bloembergen had no time for the complications and left it at that [72].

Since ruby was initially a widely used material for both masers and lasers, a lot of research was done by Bloembergen. In Jerusalem in 1963 he presented a study on the influence of an electric field on the resonance of Cr(III) [73]. This was followed by ten more publications in this field, and Bloembergen noted: *The electric field effects provided excellent opportunities for both postdoctoral research fellows and graduate students* [74].

The maser found an important application in radio astronomy. This is the field in which the universe is studied by means of radio waves. In 1931, Karl Jansky (1905–1950), a Bell radio engineer, had discovered that the Milky Way emitted radio waves and that the strongest radiation came from the constellation of Sagittarius. Grote Reber (1911–2002), already a radio operator, heard about Jansky's discoveries and in 1937 built his own radio telescope with a diameter of almost 10 metres in his backyard in Wheaton, a suburb of Chicago. From 1938 to 1943, Reber published among other things contour maps of the sky which clearly showed that the brightest radiation came from the Milky Way [75]. Radio astronomy was only developed into a separate discipline of astronomy after the Second World War. In 1948, a German parabolic radar antenna from the Second World War was set up in the Netherlands and adapted to detect the 21 cm radiation by the then Radiodienst of the PTT in Kootwijk. In 1944 Henk van Hulst (1918-2000, Ph.D. with Minnaert) had predicted the 21 cm radiation for interstellar hydrogen, in Leiden. In May 1951, six weeks after Ewen and Purcell in Harvard, the 21 cm radiation of the Milky Way was found by Lex Muller in the Netherlands.

Jan Oort (1900–1992)
Oort studied at the University of Groningen in the Netherlands, and in 1924 moved to the Observatory at Yale University. In 1925, he went to Leiden University and in 1926 wrote a thesis entitled: *The stars of high velocity*. In 1952 Oort found the spiral structure for the Milky Way. This discovery was considered by some to be one of the most important discoveries in the history of astronomy, since the discovery of the moons of Jupiter by Galileo Galilei [76, 77].

In 1958, Thomas Gold, professor of astronomy at Harvard, arranged for John Jelley and Brian Cooper to develop a 21-cm maser in Bloembergen's laboratory [78]. With this maser a significant improvement could be achieved in the signal-to-noise ratio: the noise temperature[6] of the receiver was 85 K. This receiver was used for observations at Harvard College Observatory for years [79]. In 1959 H Scovill (1923–2010) succeeded in reducing the noise to 18 K at Bell [80].

Arno Penzias (1933) and Robert Wilson (1936)
Penzias and Wilson, both at Bell, were able to reduce the noise temperature of the three-level ruby maser to 2.3 K in 1964 [81]. With this maser as an amplifier they were able to use the Bell radio telescope in Holmdel, New Jersey, to determine the background radiation of the universe (3 K). They chose a wavelength of 7 cm; it was already known that the background radiation decreased with decreasing wavelength, and Penzias and Wilson expected a weak signal. To their surprise, the background radiation was high. First, they thought the signal came from the device itself, or possibly from the atmosphere. But in fact, the radiation came from space! Their conclusion was that the cosmos emits microwave radiation isotropically.[7] It transpired

---

[6]Thermal noise can be expressed in terms of the so-called noise temperature. The noise must be as low as possible. A good radio receiver would have a noise below 100 K.

[7]Isotropically means exhibiting the same properties or behavior in all directions. Wikipedia.

later from the distribution of the spectrum that the radiation was probably a remnant of the Big Bang [82]. In 1978 Penzias and Wilson received the (shared) Nobel Prize for their discovery.

A related application of the maser was for the Defense Advanced Research Projects Agency (DARPA): *DARPA's original mission, established in 1958, was to prevent technological surprises like the launch of Sputnik, which signaled that the Soviets had beaten the U.S. into space.* In the early 1960s, the Saturn rocket was developed, among others. However, many of the tasks at this time were taken over by the National Aeronautics and Space Agency (NASA), which was also established in 1958 [83].

DARPA played a major role in the development of the Internet and artificial intelligence. Another DARPA program, called DEFENDER, was involved in the prediction of ballistic atomic missiles and received the lion's share of the money [84]. Part of this was also the radar technology used by NORAD, the North American Aerospace Defense Command, a partnership between the United States and Canada, also founded in 1958. In the Distant Early Warning System, a radar system from Alaska to Baffin Island in Canada, a chain of 63 radar stations had already been built in 1957 and this system became the cornerstone of NORAD [85]. The installations, such as the long-range pulsed radars and omnidirectional radars, were equipped with maser amplifiers to improve the signal-to-noise ratio. This allowed earlier detection of invading enemy aircraft and missiles.

Solid-state masers are still in use for transatlantic microwave communications and in NASA satellites [86]. Another application was as for navigation systems, instead of celestial navigation in deep space [87].

The following quote explains why the maser did not occupy a very important position:

They [masers, RH] produced only a few nanowatts of power, severely limiting their usefulness. Because of this impediment, most in the field gave up on masers and moved on to lasers, which use the same principles of physics, but work with optical light instead of microwaves. [...] The poor maser lived on in obscurity. It found only a few niche uses, such as boosting radio signals from distant spacecraft—including NASA's Mars Curiosity rover [88].

However, it had been shown that masers could be constructed and used at room temperature, with a considerable improvement in amplification, by using pentacene as the crystal and a laser for excitation [89].

As can be seen from the above, the application of masers was somewhat limited and the maser was not therefore a commercial success. The most important scientific milestone was that stimulated emission could be achieved experimentally and that masers were relatively easy to build.

Perhaps the most important 'application' was the laser. This will be dealt with in the next chapter.

11.3 Solid-State Physics, a New Field [90, 91]

The NMR and maser research on solids belong to solid-state physics. In 1930, *Physical Abstracts* made the following classification of the different areas of physics: General Physics (including radioactivity), Heat, Sound, Electricity and Magnetism, and Chemical Physics. In all groups, however, there was research on solids. In 1940 a new classification included Solids as a separate group. This was divided into Theory of Solids, Crystals, Lattice Dynamics, and Quantum Theory. So from 1940 onwards we could speak of 'solid-state physics'. In 1937, F. Seitz and R. P. Johnson published three articles entitled *Modern Theory of Solids*, which emphasized the need to use quantum mechanical concepts to describe the properties of solids. In 1940, the first textbook saw the light, entitled *Modern Theory of Solids*, authored by Seitz. In 1930 the first research program on theoretical solid-state physics was set up at the University of Bristol (Department of Scientific and Industrial Research). The group in Bristol had close contacts with their colleagues in the United Kingdom and the United States. Before the Second World War, research in solid-state physics was also carried out in Germany and the Netherlands, in addition to the two countries mentioned above. After the Second World War, however, the research did not get off to a good start, even though there was generally close cooperation with industry.

A typical research area was (and still is) semiconductors. There are materials that do not behave as conductors like metals, but also do not behave as insulators like ceramics, glass, and many plastics. Such substances had been known since the mid-nineteenth century. In 1833, Faraday had determined that the resistance of silver sulfide (Ag_2S) decreases as a function of temperature, while the opposite happens with metals. Later in the nineteenth century, it was discovered that the element selenium (Se) provides power when light falls on it. Selenium was used for a long time as a light meter in photography. In 1874 Braun discovered that, when certain materials were brought into contact with each other, an alternating current was rectified. In 1911, the term *Halbleiter* was introduced, translated into English as 'semiconductor'. In the 1930s, the properties of semiconductors were successfully described by Bloch and Wilson using quantum mechanics [92]. Metals have overlapping valence and conducting bands and no 'forbidden zone' ('band gap'), allowing electrons to move easily to a higher level. In insulators, there is a wide forbidden zone and it is therefore practically impossible for electrons to move higher up, so no current flows. With semiconductors, there is only a narrow forbidden zone that is easily bridged using external energy (thermal energy or photons). During the Second World War, careful research was carried out on semiconductors as a possible detector for radar. In 1947, the transistor was invented by William Shockley of MIT and John Bardeen of Bell Labs.

Other areas of research are phonons (in crystal lattices), superconductivity, diamagnetism and paramagnetism, ferromagnetism and antiferromagnetism, magnetic resonances, and defects in crystal lattices.

Bloembergen's research on solids fell into some of these areas. At conferences he met people like Bardeen (who was the only person to receive the Nobel Prize for

Physics twice). For example, Bloembergen did research on simple crystal structures such as oxides, fluorides, and chlorides, but also on gallium arsenide, an important semiconductor [93].

By the end of the 1960s, solid-state physics had become the most widely pursued area of research in physics.

References

1. Bloembergen AR. Het gezin van Rie en Auke Bloembergen 1917–1956. Eigen Beheer (2003)
2. Einstein A. Zur Quantentheorie der Strahlung. Mitteilungen der Physikalischen Gesellschaft zu Zürich, 1916 en Physikalische Zeitschrift. 18: 121–128 (1917)
3. Tolman RC. Duration of Molecules in Upper Quantum States. Phys Rev 23: 693–709 (1924)
4. Kramers HA. The Law of Dispersion and Bohr's Theory of Spectra. Nature 113:673–674 (1925)
5. Ladenburg R. Dispersion in Electrically Excited Gases. Rev Mod Phys 5:243–256 (1933)
6. Lukishova SG. Valentin A. Fabrikant: Negative absorption, his 1951 patent application for amplification of electromagnetic radiation (ultraviolet, visible, infrared and radio spectral regions) and his experiments.. J Europ Opt Soc Rapid Publ 5: 10045S1–10 (2010)
7. nobelprize.org/nobel_prizes/physics/laureates/1966/kastler-bio.html
8. Brossel J and Kastler A. La détection de la résonance magnétique des niveaux excités: l'effet de dépolarization des radiations de résonance optique et de fluorescence. Compt Rend Acad Sci 229: 1213 (1949)
9. Lukishova SG. Valentin A. Fabrikant: Negative absorption, his 1951 patent application for amplification of electromagnetic radiation (ultraviolet, visible, infrared and radio spectral regions) and his experiments.. J Europ Opt Soc Rapid Publ 5: 10045S1–10 (2010)
10. Townes CH. How the Laser Happened. Oxford University Press, New York (1999)
11. Hecht J. Beam. The Race to Make the Laser. Oxford University Press, Oxford (2005)
12. Lamb WE Jr and Retherford RC. Fine Structure of the Hydrogen Atom by a Microwave Method. Phys Rev 72:241–243 (1947)
13. Lamb WE Jr and Retherford RC. Fine Structure of the Hydrogen Atom. Part 1. Phys Rev 79:549–572 (1950)
14. Pound RV. Nuclear Spin Relaxation Times in Single Crystals of LiF. Phys Rev 81:156–157 (1951)
15. Purcell EM and Pound RV. A Nuclear Spin System at Negative Temperature. Phys Rev 81:279 (1951)
16. Ramsey NF and Pound RV. Nuclear Audiofrequency Spectroscopy by Resonant Heating of the Nuclear Spin System. Phys Rev 81: 278–279 (1951)
17. Letter from NF Ramsey to RW Dixon, 14 February 2006
18. Ramsey NF. Nuclear Moments. Ann Rev Nuclear Science 1: 97–106 (1952)
19. Interview Rob Herber with RV Pound 21 September 2007
20. Interview Joan Bromberg and Paul L Kelley with Nicolaas Bloembergen, 27 June 1983. Niels Bohr Library & Archives, American Institute of Physics, College Park, MD USA, www.aip.org/history/ohilist/4511.html
21. Townes CH. How the Laser Happened. Oxford University Press, New York (1999)
22. Interview Rob Herber with N Ramsey, 20 September 2007
23. Myers RA and Dixon RW. Who invented the laser: An analysis of the early patents. Hist Stud Phys Biol Sci 34:115–149 (2003)
24. Hecht J. Beam. The Race to Make the Laser. Oxford University Press, Oxford (2005)
25. Townes CH. How the Laser Happened. Oxford University Press, New York (1999)

26. Cleeton CE and Williams NH, A Magnetostatic Oscillator for the Generation of 1 to 3 cm Waves. Phys Rev 44: 421 (1933)
27. Cleeton CE and Williams NH. Electromagnetic Waves of 1.1 cm Wave-length and the Absorption Spectrum of Ammonia. Phys Rev 45:234–237 (1934)
28. Townes CH. How the Laser Happened. Oxford University Press, New York (1999)
29. Townes CH. The Ammonia Spectrum and Line Shapes Near 1.25-cm Wave-Length. Phys Rev 70:665–671 (1946)
30. Gordon JP, Zeiger HJ and Townes CH. Molecular Device and New Hyperfine Structure in the Microwave Spectrum of NH_3. Phys Rev 95:282–284 (1954)
31. Gordon JP, Zeiger HJ and Townes CH. The Maser – New Type of Microwave Amplifier, Frequency Standard, and Spectrometer. Phys Rev 99:1264–1274 (1955)
32. Hecht J. Beam. The Race to Make the Laser. Oxford University Press, Oxford (2005)
33. Interview Rob Herber with N Ramsey 20 September 2007
34. Interview Jeff Hecht with N Bloembergen, 5 Nov 1984, Niels Bohr Library & Archives, American Institute of Physics, College Park, MD USA
35. Interview Joan Bromberg and Paul L Kelley with N Bloembergen, 27 June 1983. Niels Bohr Library & Archives, American Institute of Physics, College Park, MD USA, www.aip.org/history/ohilist/4511.html
36. Interview Rob Herber with N Bloembergen, 6 December 2006
37. Versitron Acts As Ultra Sensitive Amplifier And Thermal Detector. The Tech, February 8 (1957)
38. Interview Rob Herber with N Bloembergen, 6 December 2006
39. Hecht J. Beam. The Race to Make the Laser. Oxford University Press, Oxford (2005)
40. Interview Joan Bromberg and Paul L Kelley met NBloembergen, 27 June 1983. Niels Bohr Library & Archives, American Institute of Physics, College Park, MD USA, www.aip.org/history/ohilist/4511.html
41. Interview Rob Herber with N Bloembergen, 7 December 2006
42. Mattson J, Simon M. Pioneers of NMR and Magnetic Resonance in Medicine. The story of MRI. Dean Books, Jericho, 1996
43. Hecht J. Beam. The Race to Make the Laser. Oxford University Press, Oxford (2005)
44. Interview Arthur Guenther with D.N.G. Basov, 14 September 1984. Niels Bohr Library & Archives, American Institute of Physics, College Park, MD USA, www.aip.org/history/ohilist/4495.html
45. Interview Joan Bromberg J and Paul L Kelley with N Bloembergen, 27 June 1983. Niels Bohr Library & Archives, American Institute of Physics, College Park, MD USA, www.aip.org/history/ohilist/4511.html
46. Bloembergen N. Proposal for a New Type of Solid State Maser. Phys Rev 104:324–327 (1956)
47. Hecht J. Beam. The Race to Make the Laser. Oxford University Press, Oxford (2005)
48. Townes CH. How the Laser Happened. Oxford University Press, New York (1999)
49. Bloembergen AR. Het gezin van Rie en Auke Bloembergen 1917–1956. Eigen Beheer (2003)
50. Interview Joan Bromberg J and Paul L Kelley with N Bloembergen, 27 June 1983. Niels Bohr Library & Archives, American Institute of Physics, College Park, MD USA, www.aip.org/history/ohilist/4511.html
51. Townes CH. How the Laser Happened. Oxford University Press, New York (1999)
52. Artman JO, Bloembergen N and Shapiro S. Operation of a Three-Level Solid State Maser at 21 cm. Phys Rev 109:1392–1393 (1958)
53. Artman JO, Bloembergen N and Shapiro S. Operation of a Three-Level Solid State Maser at 21 cm. Phys Rev 109:1392–1393 (1958)
54. Makhov G, Kikuchi C, Lambe J and Terhune RW. Maser Action in Ruby Phys Rev 109:1399–1400 (1958)
55. Interview Jeff Hecht with N Bloembergen. 5 Nov 1984, Niels Bohr Library & Archives, American Institute of Physics, College Park, MD USA
56. Bloembergen N. Uninterrupted Amplification Key Stimulated Emission of Radiation from a Substance Having Three Energy States. United States Patent Office 2.909.654 (1959)

57. Townes CH. Production of Electromagnetic Energy. United States Patent Office 2.879.439 (1959)
58. Myers RA and Dixon RW. Who invented the laser: An analysis of the early patents. Hist Stud Phys Biol Sci 34:115–149 (2003)
59. Shapiro S and Bloembergen N. Relaxation Effects in a Maser Material, $K_3(CoCr)(CN)_6$. Phys Rev 116:1453–1458 (1959)
60. Bloembergen N. (ed) Encounters in Magnetic Resonances. World Scientific, Singapore (1996)
61. Interview Rob Herber with D and N Bloembergen, 7 December 2006
62. Hecht J. Beam. The Race to Make the Laser. Oxford University Press, Oxford (2005)
63. Ramsey N. The Atomic Hydrogen Maser. Metrologica 1:7–15 (1965)
64. http://tycho.usno.navy.mil/maser.html
65. www.timeanddate.com
66. en.wkipedia.org
67. Interview Rob Herber with N Ramsey, 20 September 2007
68. Bloembergen N. Paramagnetic Resonance Precession Method and Apparatus for Well Logging. United States Patent Office 3.242.422 (1966)
69. Bloembergen N. Quantum Mechanical Counters. United States Patent Office 3.070.698 (1962)
70. Letter from N Bloembergen to G Gloudemans, 31 March 1961
71. Letter from JP Simons to N Bloembergen, 12 May 1969
72. Interview Rob Herber with D and N Bloembergen, 7 December 2006
73. Bloembergen N, Royce EB. Electric Shift of the Cr^{3+} Magnetic Resonance in Ruby. In: Low W (ed) Low Symposium on Paramagnetic Resonance. Academic Press, New York, 1963
74. Bloembergen N. (ed) Encounters in Magnetic Resonances. World Scientific, Singapore, 1996
75. www.nrao.edu
76. Beekman G. Vijftig jaar Nederlandse radiosterrenkunde: een verjaardag zonder jarige. (Fifty years of Dutch radio astronomy: a birthday without a birthday) Zenit. April 1999
77. en.wikipedia.nl
78. Bloembergen N. (ed) Encounters in Magnetic Resonances. World Scientific, Singapore, 1996
79. Jelly JV and Cooper BFC. Operational Ruby Maser for Observations at 21 Centimeters with a 60-Foot Radio Telescope. Rev Sci Instrum 32:166–175 (1961)
80. Bloembergen N. (ed) Encounters in Magnetic Resonances. World Scientific, Singapore, 1996
81. Penzias AA and Wilson AA. A Measurement of Excess Antenna Temperature at 4080 Mc/s. Astroph J 142:419–421(1965)
82. nobelprize.org/nobel_prizes/physics/laureates/1978/press.html
83. Bridging the Gap. DARPA. Powered by Ideas. (2005)
84. Atta RH van, Deitchman SJ and Reed SG. DARPA Technical Accomplishments Volume III. Institute for Defense Analyses. Alexandria, Virginia (1991)
85. www.radomes.org
86. Mattson J, Simon M. Pioneers of NMR and Magnetic Resonance in Medicine. The Story of MRI. Dean Books, Jericho, 1996
87. Dong, Jiang. (2008). The Principle and Application of Maser Navigation. https://arxiv.org/abs/0901.0068
88. Brumfiel G. Microwave laser fulfills 60 years of promise. Nature News. August 12, 2012
89. Oxborrow M., Breeze JD, and Alford M. Nature 448:353–356 (2012)
90. Kragh H. Quantum generations: a history of physics in the twentieth century. Princeton University Press, Princeton, NJ, 1999
91. Kittel C. Introduction to Solid State Physics. John Wiley. New York (1996). 7th edn
92. Erwin SE. When is a metal not a metal? Nature 441:295-296 (2006)
93. Bloembergen N. Electrical Shift in Magnetic Resonance. In: Smith J (ed) Proc XIIth Colloque Ampère 39–57 (1963)

Chapter 12
The Wild West and Soviets—Laser Applications from Eyes to Cars

Bloembergen, however, subsequently filed a patent which neither he nor Harvard seems ever to have exploited, but which some of Bloembergen's early students have recently pointed out is very comprehensive and far-reaching and could well cover not only microwave masers but all optically pumped lasers as well [1].

12.1 From Maser to Laser: Harder Than It Seems

In 1951 Purcell, Pound, and Ramsey were able to demonstrate stimulated emission at a frequency of 50 kHz (a wavelength of about 6 km) [2, 3]. This is a frequency used for navigation, time signals, and long wave radio broadcasts. The maser developed by Townes worked at a frequency of 24 GHz (a wavelength of about 1.25 cm) [4], a frequency used for radar, kitchen microwaves, mobile phones, and WLAN.[1] The frequency of 24 GHz is two million times higher than 50 kHz, but still a frequency used for radio purposes. Townes first worked at 600 GHz (a wavelength of about 0.5 mm), a frequency above the radio frequency range, but found it too difficult to work with [5]. The microwave range goes up to 1 mm; shorter wavelengths are considered infrared. This range includes the wavelengths between 1 mm (300 GHz) and 780 nm (384 THz). So, there is another factor of a thousand from microwave to (near) infrared. Visible light lies between 780 nm and 400 nm.

When Townes and Gordon were able to show the first working maser, further developments of the maser followed each other quickly [6]. Bloembergen's three-level mechanism and 'pumps' formed the basis for practically all further research [7].

---

[1]WLAN is the abbreviation for Wireless Local Area Network, used for the wireless connection of computers, printers, and modems.

© Springer Nature Switzerland AG 2019
R. Herber, *Nico Bloembergen*, Springer Biographies,
https://doi.org/10.1007/978-3-030-25737-8_12

The electronics developed over the decades from the kilohertz to the megahertz range and on to gigahertz. The next challenge was to obtain stimulated emission at even higher frequencies, above 100 GHz, or in the mm range. But that was easier said than done because generators above those frequencies did not exist, and also the electronics from the lower frequency range could not be used. In fact, very little was known about the frequency range between 100 GHz and 100 THz. Working with even higher frequencies, however, was a well-known area, because we then encounter the frequencies of visible light [8].

As early as 1956, Dicke patented a method for obtaining resonance with ammonia in the microwave and infrared regions. The patent was granted in 1958 [9]. In the 'maser world', however, not much was done with it.

12.2   The Laser: A Genuine War Breaks Out

Townes and Gordon Gould (1920–2005): first together, then separately

Townes had already done a lot of research into the maser, and it was with him that the quest for the laser really began. Townes became interested in higher frequencies than those of the microwave range, but soon realized the problems that would have to be overcome above 100 GHz. After developing a ruby laser for radio astronomical purposes, he thought it would be better to shift the maser research to the optical field. In Paris, Kastler (1902–1984) had introduced a new method to obtain resonance. As a pump he did not use the usual microwave frequency, but an optical pump to achieve resonance in the maser [10]. Townes became aware of this method during a 'sabbatical' in Paris. He realized that this optical pump could also be used to generate radiation in the wavelength range of visible light, instead of microwave radiation, thereby producing an 'optical maser' [11]. Rabi, who was associated with Columbia University like Townes, introduced the doctoral student Gordon Gould to the optical pumping of masers. Under the supervision of Townes, Gould succeeded in 1957 in capturing thallium atoms in an atomic beam. The first part of the process, the optical pump, was now accessible. The hardest thing, however, was to generate optical radiation and that was still a long way off. The principle of generating light without electrodes was patented by General Electric in 1970 and the result was called an 'electrodeless discharge lamp' [12]. Such lamps with a high output of monochromatic light were and still are widely used in spectroscopy and spectrometry.[2] Gould then went his own way.

Townes and Ard Schawlow (1921–1999), keeping things in the family

Townes became the advisor at Bell Labs a few days a month and worked there

---

[2]Spectroscopy is a general term for all physical and physical-chemical techniques used to investigate the light interaction properties of composite molecular, atomic, or subatomic systems. Spectrometry is the technique of measuring the distribution of light across some part of the electromagnetic spectrum.

on the optical maser with his brother-in-law Ard Schawlow, also employed there. Columbia University and Bell Labs were not the only places the optical masers were being developed. This was also the case at Hughes Aircraft Company in California and the Lebedev Institute of Physics in Moscow, for example.

Townes and Schawlow first made theoretical calculations to identify the possibilities and found that at optical wavelengths with sufficient oscillator strength, a number of transitions between energy levels should be possible (Fig. 12.1). Then, on the basis of what they knew about the emission lines of elements, they chose not thallium, with which Gould had first experimented, but potassium. The latter has two important wavelengths in the visible spectrum. However, that turned out to be a mistake: potassium is extremely difficult to handle, because it is extremely reactive in air. Schawlow then came up with the idea of replacing the closed rectangular vibration cavity for microwaves with an open system with two plates (mirrors). This idea was inspired by the Fabry-Perot interferometer.[3] If the wavelengths of the incident radiation fit exactly between the plates, resonance will occur and the radiation will be amplified. However, many wavelengths fit between the mirrors, because the wavelengths of visible light are so short: less than 0.0007 mm. However, the wavelengths to be generated by the optical maser are limited in number. Townes realised that only the strongest resonant wavelengths would be amplified by the numerous reflections. In August 1958, Townes and Schawlow submitted their theoretical considerations for publication [13]. In it they described the use of potassium vapor as a material for Bloembergen's three-level mechanism and as an optical pump to produce the purple wavelength of 404.7 nm. Townes and Schawlow operated as authors at Bell Telephone Laboratories, which would later prove to be of great importance for patent claims. They had not actually built anything at that time, but Townes did not want to make the same mistake as with the maser, waiting for publication until he had theoretically and practically completed the matter, whereupon Prochorow and Basow could take the credit. The publication was far from being a blueprint for making an optical maser: too many practical issues had not yet been solved. Townes and Schawlow had already finished their work in February or March 1958, but the things at Bell were moving slowly and they lost valuable time. Townes also went to Bell to patent the invention, but at first the engineers didn't see anything in it. Townes, however, included applications for optical masers for communication purposes in the application and Bell filed the patent application at the end of July 1958 [14]. The patent was granted in March 1960 [15].

Townes and Schawlow's patent versus Bloembergen's patent

Until recently, the Lucent Bell Laboratories claimed on their website: *About 40 years ago we began the field of optoelectronics when we **invented the laser***. And: *The invention of the laser, which stands for light amplification by stimulated emission of radiation, can be dated to 1958 with the publication of the scientific paper **Infrared***

---

[3] A Fabry-Perot interferometer is an optical instrument used to measure and control the wavelength of electromagnetic radiation. It consists of two reflecting plates mounted parallel to each other.

Fig. 12.1   Arthur Schawlow. *Source* twitter.com

and optical masers, by *Arthur L. Schawlow, then a Bell Labs researcher, and Charles H. Townes, a consultant to Bell Labs.*

Myers and Dixon make the following remark: *When it suits, even a patent-oriented organization like Bell Laboratories claims inventorship priority using a scientific publication rather than a patent* [16].

Regarding the Nobel Prize of 1964, we can still read in 2010:

Charles H. Townes shared the Nobel Prize in Physics with A. Prokhorov and N. Basov of the Lebedev Institute in Moscow for "fundamental work in the field of quantum electronics which has led to the construction of oscillators and amplifiers based on the maser-laser principle." Townes invented the laser with Arthur L. Schawlow while a consultant at Bell Labs in 1958 [17].

According to Myers and Dixon, Schawlow and Townes' patent differed from Bloembergen's with regard to the three levels in that parallel mirrors were used as well as a more closely specified system for optical pumping. However, according to Myers and Dixon, the patent is based on Bloembergen's earlier work. While Bloembergen cited the work of Basow and Prochorow in his patent, there is no mention of Bloembergen in Schawlow and Townes'. Furthermore, although Bloembergen's patent had not yet been issued, it was well known to Townes and Bell. The patent writer at Bell (Torsiglieri) was also involved in Bloembergen's patent application! Torsiglieri figured out that Bloembergen's application would probably dominate the lasers in the optical field and in solid-state lasers. He avoided the problems with Bloembergens' request by including a section on optical communication.

In an official response from Torsiglieri to an action by the Patent Office to quote Dicke's patent application, Torsiglieri wrote: *Applicant's invention* [refers to the application from Townes, RH], *on the other hand, **includes a three level maser which can be used in the infrared, visible, or ultraviolet regions of the electromagnetic spectrum** [...]*

Neither patent was easily applicable in optics. Bloembergen used paramagnetic salts and these were inferior to doped ruby or yttrium aluminum garnet. Schawlow and Townes' optical maser with potassium vapor could only actually be made decades later. This optical maser with potassium vapor brought nothing fundamentally new compared to Bloembergen's three-stage system.

Bloembergen did not stand up for his rights [18]. Townes: *If the owner of the patent does not vigorously defend it, outsiders feel free to help themselves to the relevant technology* [19].

Torsiglieri later said about the period he was working on patents for Bell: *If Bloembergen had been aggressive, he probably would have made a lot of money prosecuting the patent.* Torsiglieri did not think that Bloembergen was such a person [20].

Gould, the inventor

According to Hecht, the race for the laser started with Townes and Schawlow's publication [21].

Gould was not a scientific researcher like Bloembergen, Townes, and Schawlow (Fig. 12.2). Gould's model was Thomas Alva Edison, the great inventor who had become rich through his patents. Gould saw the possibilities to build a 'laser' (as he called it) when he was working with Townes in Columbia. After Gould had found a way to achieve optical pumping, he wondered whether emissions in the optical field could also be achieved. In the summer of 1957 he came up with the idea that, when on one side of a tube a semi-permeable mirror is placed and on the other side parallel to it a normal mirror, light will be reflected between the mirrors. If something whose energy levels are inverted is placed between the mirrors, then stimulated emission will occur. Each photon will cause a cascade of emission from the other photons and a laser beam will exit through the semi-permeable mirror. However, reinforcement (the cascade effect) must take place and the mirrors must be aligned exactly parallel to each other in order for the effect to occur. The concentration at one point (the beam) causes a high intensity. In November 1957 Gould had his notes signed by a notary as proof. Gould based his theory on potassium, the same element Townes had chosen. Gould visited a patent lawyer, but understood from him that he first had to be able to show a working model of his arrangement. However, that was not actually the case, as we have seen with the patents filed by Bloembergen and Schawlow/Townes, which only described the theory of an arrangement. It would prove to be an expensive mistake. Gould had an advantage over Schawlow/Townes at that time. Hecht estimates in his book that if the patent had been granted to Gould in 1960, he would have earned several million dollars until 1977, when the patent

Fig. 12.2   Gordon Gould. *Source* https://www.aip.org

expired [22]. However, Gould did not have a laboratory at his disposal to develop a laser, unlike Schawlow/Townes at Bell Labs. On the other hand, Gould found money from a new company in Manhattan, the Technical Research Group (TRG). This company had defence contracts and employed Gould in March 1958. He worked on the laser in his spare time.

Townes and Schawlow: from potassium to solid matter

After their theoretical publication on a possible laser, Townes and Schawlow continued struggling with their experimental arrangement with potassium vapor. Schawlow switched to solid-state physics research at Bell and examined various materials for their suitability for the optical maser. Synthetic ruby seemed a good candidate and was abundantly present at Bell. Schawlow, however, had to find out how ruby absorbed and emitted light, which turned out to be a difficult task. In a solid, the atoms are close together and there are all kinds of interaction with the surrounding atoms, as we saw earlier in Bloembergen's NMR experiments of. Others at Bell came up with bad news: it had been experimentally determined that only 1 to 10% of the excited atoms gave the intended red fluorescence. That was too little. Schawlow then began to examine other solids for suitability.

Ali Javan (1926–2016): helium and neon gases

Ali Javan was another Columbia University employee, but he, too, moved to Bell. He heard from Schawlow at Bell that a pair of mirrors might be suitable as a laser resonator. Javan worked with gases, but it is not easy to transfer energy to gases, because the molecules or atoms are far apart. Javan tried a mixture of the noble gases helium and neon. Helium was excited by an electrical discharge and the energy was then transferred to the neon by collisions. There is an energy level of helium exactly at the same level as an energy level of neon. Neon falls back to a lower energy level and emits radiation with a wavelength of $1.153\ \mu m$. This emission lies in the infrared.

Gould: Killing rays?

Gould had fewer resources at his disposal at TRG than the researchers Schawlow and Javan at Bell, so his research lagged behind theirs. Gould's idea was that light would normally be scattered over a relatively short distance, but that would not be the case with lasers (it turned out to be a correct assumption later on), whence laser beams would be suitable for communication purposes. Since laser beams would be very narrow in diameter, it would be difficult for an enemy to intercept them. The US government had set up a special research agency for difficult projects, the Advanced Research Projects Agency (ARPA). While the others at Columbia saw the laser as a research tool, Gould already dreamt of applications such as intercepting nuclear missiles with powerful lasers, and clearly saw their enormous potential. The military scientists were impressed by this vision. There was a long tradition in science fiction about the serious potential of 'death rays'. As early as 1934, Tesla (1856–1943) claimed without proof that he had invented a killing beam that could kill millions of soldiers. In 1939, the U.S. Army pledged $10,000 to anyone who could kill a goat with deadly rays. The prize had not yet been collected in 1959, but this background explained the army's interest in the things that Gould predicted, although he himself never talked about killing rays. TRG received a contract from the army for almost one million dollars, an immense amount of money at the time. According to the army, a number of groups within TRG should work on the various projects to develop the laser. However, in view of the possible military applications, the army wanted to keep the investigation secret, and researchers at TRG had to undergo a confidentiality procedure, so-called 'clearing'. Culver of the Rand Corporation had the task of finding interesting developments; he first went to ARPA with his copy of the TRG proposals, and then to TRG. He subsequently went to see Townes, who to his surprise saw that there was also a proposal for a gaseous laser, the research that Javan was doing. Culver was informed that the proposals were secret and that they could not be copied or seen by unauthorized persons. A complication for Gould was the fact that no secret material could be used in his patent. The patent was filed on 6 April 1959, but it would take a long time before it was granted. The procedures would take until the mid-1980s!

In May 1959 TRG got the contract. To his great dismay, Gould, who had submitted the proposals, was not allowed to carry out the investigation, and this caused great consternation at TRG. Gould did not get through the security procedures. This was in the middle of McCarthy era and there were several things that spoke against Gould. Gould had been quite left wing, had a somewhat dissipated lifestyle, and had also been divorced [23].

Bloembergen's Role

What was Bloembergen's role in these developments? He developed the solid state laser in 1956 (Chap. 11). The basis for this was a three-level mechanism with a continuous wave. This principle seemed to be applicable not only to solid-state masers, but to all masers, as described in the previous chapter. Since then, research on the maser has been carried out almost exclusively with the three-level system.

Tony Siegman, who wrote a standard history of laser development, makes the following remark: *Bloembergen, however, subsequently filed a patent which neither he nor Harvard seems ever to have exploited, but which some of Bloembergen's early students have recently pointed out is very comprehensive and far reaching and could well cover not only microwave masers but all optically pumped lasers as well* [24].

Moreover, the initial research on lasers was done exclusively with Bloembergen's three-level system.

Bloembergen did not do any research on the laser at that time. In an interview he stated: "*I didn't have the courage to set up a new optical lab myself. I had enough to do with my continuation of my magnetic resonance, the microwaves and the radio frequencies. So, I didn't have the guts to jump into optics just like that*" [25].

Townes, Schalow, and Javan had Bell behind them and Gould was working at TRG. The development of devices was all part of their normal business in these companies. Bloembergen did not have that at Harvard. In an interview, he commented: "*I thought that it would be very hard to build a laser, an optical maser (because the word laser hadn't been invented yet), and that even if somebody succeeded, it probably wouldn't be done at a university. And certainly not by me. So maybe a defeatist attitude. But in retrospect, it's very remarkably true, because most types of lasers were first made to work in industrial research labs, in the United States. And the point is that the U.S. universities are not set up to do that work, and bring together to a common focus very different techniques—optics, and microwaves, and all this together. And I had no experience really in optics, except for my early work in photoelectric detection, and I'm sure I would never have been the first to make it work*" [26].

In another interview, in response to a question from Jeff Beck about the support he was given in Harvard, Bloembergen also commented: "*Well, all the support is government contracts. Harvard only provides buildings. Nothing else. No support, whatsoever, from Harvard [...] You need technicians, you know, glass blowers, opticians and whatnot. At any rate, history clearly shows that whoever developed the laser first, then you could try to duplicate them, but even that's hard*" [27].

If Bloembergen had no chance of winning, he wouldn't take part in the race [28].

Valentin Fabrikant and Fatima Butayeva in the Soviet Union: mercury?

The Soviets had good ideas, as the development of the maser had already shown, but lacked good equipment. Fabrikant and Butayeva claimed in 1959 that a mixture of mercury vapor and hydrogen would strengthen the blue and green mercury lines by 1% and that population inversion would occur. However, this was based on a measurement error, which was only admitted after the first laser had been demonstrated elsewhere [29].

Swawanga Lodge Meeting, 14–16 September 1959

Early in 1959 Irving Rowe, a representative of the Office of Naval Research (ONR), which also provided grants to Bloembergen, asked Townes if it would be a good idea to organize a conference on masers. The ONR would be ready to subsidize such a conference and Townes could be the chairman. Townes agreed and set up an organizing committee of eleven people. Apart from Rowe, the members of the committee were Bloembergen (Harvard), Dicke (Princeton), Siegman (Stanford), Birnbaum (Hughes Research), Kittel (Berkeley), Strandberg (MIT), and Kompfner (Bell Labs) [30]. Townes left Columbia in the summer of 1959 to work at the Institute for Defense Analysis in Washington, DC—to the great sorrow of his students who suddenly found themselves without a supervisor. Townes suspected that the maser/laser would probably yield a Nobel Prize. He also observed that the development of an idea would usually win prizes, even if the idea wasn't applied, and he felt that it no longer mattered who built the first laser: the idea had been developed and he could leave university in 1959 with peace of mind (Fig. 12.3).

This conference is viewed as the beginning of the laser age. The title of the conference became: *The Conference of Quantum Electronics—Resonance Phenomena.* The term *Quantum Electronics* was used here for the first time. The location was Shawanga Lodge in the Catskill Mountains in the northwest of New York State. There were 66 lectures, of which only two (those of Schawlow and Javan) were about the laser, yet everyone talked about it during the meals and in the corridors. Although there was a lot of discussion, nobody had yet managed to develop a laser,

Fig. 12.3 Members of the Steering Committee of the First Quantum Electronics Conference, Shawanga Lodge, NY, 1959. In front, in the light suit, Nico Bloembergen, and moving clockwise, Benjamin Lax (MIT), Robert Dicke (Princeton), Tony Siegman (Stanford), unknown, Irving Rowe (Office of Naval Research), George Birnbaum (Hughes), Rudolf Kompfner (Bell Labs), Charles Townes (Columbia), Charles Kittel (U.C. Berkeley), and Malcolm (Woody) Strandberg (MIT). *Source* Library of Congress

to use Gould's term for such a system. A lot of attention was paid to the lecture by Schawlow, who was struggling at the time to develop a ruby laser. His conclusion was that it would not be possible to build such a laser.

There was one listener who disagreed: Maiman. This was a new athlete in the race to the laser. Maiman comments: *I attended the first IQEC meeting; there was all kind of talk about lasers, gas discharge proposals, in addition to the alkali vapor. Schawlow gave a paper in which he made his famous comment on how ruby couldn't work because the ground state would have to be depopulated. Another one of his comments was, "Nobody knows what form the laser will take." What I mean is, some frustration already set in* [31].

Both at Bell and at TRG and Hughes Research Laboratories, people were working feverishly to be the first to demonstrate a working laser. It was clear to the participants of the conference that constructing a laser would not be an easy task.

Theodore (Ted) Maiman (1927–2007): A maverick constructs the first laser but doesn't get the honor

Ted Maiman worked at Hughes Research Laboratories[4] in Malibu, California, on a compact ruby laser, cooled with dry ice.[5] That was no news in 1959, because maser research was commonplace at that time.

Maiman didn't want to be the umpteenth to build a maser, but was investigating fundamental issues. He had worked on optics for his dissertation. At the Swawanga Lodge Meeting, the general idea was that the laser should involve four levels (instead of Bloembergen's three levels), one of which was metastable. Metastable means that it is difficult for the atoms to fall back to a lower level. In the case of ruby, the temperature should be close to absolute zero. Maiman did not see any future for such a design. He wanted a laser that was easy to build and that would work at room temperature. Charles Asawa, Maiman's assistant, calculated that ruby was a good candidate as a material for the laser. However, there remained the problem of how to achieve emissions. Somehow energy was leaking away and the two of them couldn't find out what was going on. It was clear that an extremely bright light source was needed to excite ruby. One candidate was the arc, in which a strong current between two graphite electrodes generates a plasma that provides intense light.[6] These arcs had been known for a hundred years or more. An example is the electric arc used for welding. This immediately indicates the problem: the arc becomes too hot for a laser and a lot of smoke is produced. The lamp of a film projector was the next candidate (Fig. 12.4). There was a special compact mercury lamp that produced the necessary ultraviolet wavelengths and Asawa recommended it to Maiman. As an alternative, Asawa also discovered the xenon flash lamp used by professional photographers at

---

[4]Hughes Research Laboratories was founded in 1960 by Howard Hughes (1905–1976). He was known as a film producer, playboy, pilot, and owner of Hughes Aircraft Company, of which Hughes Research Laboratories was part. Hughes was the richest person in the world for some time.

[5]Dry ice or carbon dioxide snow is the solid form of carbon dioxide, CO_2. It has a temperature of 194 K (-78 °C).

[6]Plasma is a gas phase, in which the atoms have lost their electrons due to the high temperature. That temperature can be as high as 8000 K.

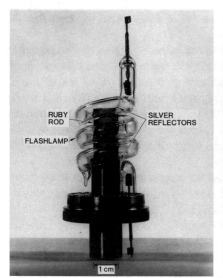 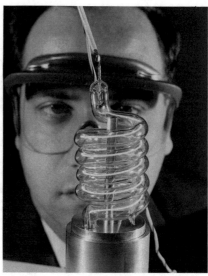

Fig. 12.4 Theodore Maiman with the first laser (left) in 1960. The right-hand picture with a much larger laser was shown to the press. *Source* Hecht J Opt Eng **49**(9) 091002 (1960)

the time, which produces an intense flash for a very short time. Maiman calculated that the mercury lamp produced too little energy to make a laser work efficiently. This left the flash lamp as the only candidate. A laser that would work with a flash lamp had to be a pulsed laser as well. A pulse is a short-lived (electromagnetic) phenomenon. The flash of the photo lamp was also short-lived. No one had worked with anything like that before. Maiman was the first to experiment with a flash lamp. The construction that Maiman and his assistants Asawa and Irnee D'Haenens came up with was very simple compared to the one designed by their competitors at Bell, who were working on the helium-neon laser. Their system consisted of a ruby bar about 15 mm long and 10 mm across, whose ends were polished and parallel to each other and then coated with a layer of silver. In the middle of one of the ends a small opening was made in the silver layer to let the radiation escape. The rod was enclosed in a flashlight and the whole thing was mounted in an aluminum cylinder [32].

The big day was 16 May 1960: Ray Hoskins and Asawa saw the deep red laser radiation for the first time. D'Haenens, who was there too, only saw white light: it turned out that he was color blind and could not see red [33].

Maiman wanted his work to be published as soon as possible and *Physical Review Letters* was perfectly suited for that. He posted his manuscript on 22 June 1960.

Physical Review Letters was launched on 1 August 1959. The editor-in-chief was Samuel Goudsmit, whom we already met in Chaps. 5, 7, and 8. Goudsmit wanted it to be a journal that would publish the latest developments in physics, but instead he was often sent applications, including many applications of the maser. As a theoretical physicist, Goudsmit was not very interested in equipment and applications. In the spring of 1960, Goudsmit had received few manuscripts about the laser that the editors considered worth publishing.

In an attempt to connect it to Townes and Schawlow's terminology, Maiman had used the term 'optical maser' in his manuscript: *Optical Maser Action in Ruby*. That turned out to be a fatal mistake. Goudsmit only saw the word 'maser', so he decided to return the manuscript without a proper evaluation. Goudsmit had clearly not understood what Maiman had done. Goudsmit did not even send Maiman's manuscript to the referees, not the standard practice at all.

Forty years after the event, Maiman was still furious whenever he recalled this refusal.

After Goudsmit's rejection, Maiman sent the manuscript to the British magazine *Nature*, which immediately accepted it and published it under the title *Stimulated Optical Radiation in Ruby* on 6 August 1960 [34].

Hughes Research Labs organized a press conference in Manhattan, accompanied by a press release about *atomic radio light*. When Maiman, at a press conference, could not deny that the laser could ever be used as a weapon, headlines appeared in the newspapers such as *LA man discovers science fiction death ray* and *Death ray possibilities by scientists*.

Maiman also sent a manuscript to the *Journal of Applied Physics*, a magazine for applied physics. Although the article could only be published six months later, it was a reliable magazine. Maiman was hit again when the publication appeared in September 1960 in the completely unknown journal *British Communications and Electronics*. The manuscript had been sent to the British journal without Maiman or the Hughes Research Labs knowing about it. The law firm Byoir Associates had organized a press conference for Hughes Research Labs and had sent a press statement and manuscript to the aforementioned British magazine. An editor of the article had published the article without asking Maiman, Hughes, or Byoir anything. This meant publication in the *Journal of Applied Physics* was no longer possible. The publication in *Nature* (twelve lines!) was the only thing Maiman had [35].

In 1961, Maiman's group published further research on methods for improving the efficiency of the laser [36, 37].

The battle for the lasers: Bell (Schawlow, Collins, and Nelson) versus Hughes Research (Maiman)

Soon TRG and Bell Labs succeeded in reproducing Maiman's laser [38]. They used Bloembergen's three-level principle and Maiman's ruby laser and came to the same result as Maiman. Schawlow's group at Bell soon published their contribution in *Physical Review Letters*. They probably knew about the difficulties Maiman had had with this journal, so they avoided the word 'maser' and thus the difficulties associated with it [39].

Hecht discusses the paper published by Schawlow's group in detail in his book. He blames the Bell group for their short-sightedness. The paper began with a reference to the article *Infrared and Optical Masers* by Schawlow and Townes [40], in which the laser concept was presented. Next, it considered the paper by Schawlow's group: *The use of a ruby rod for the observation of these effects has been proposed by*

Schawlow, and referred to the Swawanga Lodge Meeting [41]. However, at this conference, Schawlow had stated that Maiman's concept could not work! Maiman himself is hardly mentioned, and neither is Bloembergen, while Kikuchi (who had the ruby idea) is cited in the paper. Maiman is thus mentioned as having *observed a decrease in the lifetime and a narrowing of the line shape for ruby fluorescence.* Weeks before their article was sent in, Bell had received a preprint of Maiman's article for the *Journal of Applied Physics* and should have cited it as 'in print', or else cited his article in *British Communications and Electronics.* According to Hecht, the article by the Bell scientists comes close to falsification: by secretly sending their work past Goudsmit and ignoring Maiman's experiments, Bell made sure that they appeared to be the ones who invented the laser, whereas that was clearly Maiman's work.

As soon as Bell's group had the ruby laser working and it was clear that it could be published, Bell organized an experiment in which the laser covered a distance of about 40 km. In daylight an expanding beam of 6 m diameter could be seen at the place of 'arrival'. Bell organized a press conference and there was talk of '*death rays*' once again [42].

TRG (Gould) versus Bell (Townes and Schawlow)
Gould's publications were broader in scope than those of Townes and Schawlow. Gould had been left behind for years due to his problems with the Ministry of Defense and the conflicting claims with Bell's patents. Four patents were finally granted: this was only a fraction of his work. Gould won the patent wars after almost thirty years: none of the materials proposed by Townes and Schawlow turned out to work [43]. Some consider Gould to be the inventor of the laser [44, 45]. The problem here is that it is not made clear what is meant by 'inventor'. This can have a scientific meaning (the theory, the first working laser), a practical meaning (the first usable laser), or a commercial meaning (the first patent).

Hughes Research (Maiman) versus TRG (Gould)
Maiman was disparaging about Gould: alkali vapor had never worked as a medium for the laser, in the way that Gould had described it [46]. However, TRG succeeded in 1962 in building a working cesium laser.

Further developments
After Maiman had succeeded in demonstrating his working ruby laser, developments followed each other in rapid succession. The Bell group has already been mentioned, with work by the researchers Collins and Nelson.

Townes made the following remark about these developments: *When the goal is clear, industry can indeed be effective. Yet the first lasers, though built in industrial laboratories, were invented and built by young scientists recently hired after their university research in the field of microwave and radio spectroscopy—students of Willis Lamb, Polycarp Kusch, and myself, who were together in the Columbia Radiation Lab, and of Nico Bloembergen at Harvard* [47].

A student of Bloembergen's at Harvard, Robert Myers, had copied the ruby laser and found that *it was almost ridiculous easy to repeat* Maiman's experiment. This was the same Myers who later, together with Dixon, produced a critical review about the development of patents [48].

Schawlow and George Devlin built a cooled laser at Bell, a four-level laser which Schawlow had claimed would not work at the Swawanga Lodge Meeting. Irwin Wieder at Varian Associations (Palo Alto, California) even built a cooled laser a little earlier than Schawlow [49].

The two groups published the result in the same issue of Physical Review [50, 51].

Peter Sorokin (a former Bloembergen coworker) and Mirek Stevenson of IBM built their own ruby laser, but then successfully built a four-level solid-state uranium-doped calcium fluoride laser, equipped with a flash lamp [52]. This was the first working laser that did not use ruby.

After difficult experiments, Herriott, Bennett, and Javan at Bell succeeded in making the first continuous wave gas laser (ruby and calcium fluoride lasers were pulsed). They developed the helium-neon laser [53].

The TRG group associated with Gould (who was still not allowed to enter the laboratory due to Pentagon security restrictions) developed a gas laser with cesium [54].

Bloembergen's research

When Maiman built the first laser in 1960, Bloembergen was working on the modulation of light beams using microwaves. In 1961 he bought his first commercial laser from Trion. Then Bloembergen decided to modulate light with light, and it was with this that he actually began his research on non-linear optics (Chap. 13) [55].

In 1962, Bloembergen visited Israel for the first time. At this point he was investigating ruby. For his lecture at the congress, Deli found two appropriate Old Testament proverbs. The first was from Job 28:18 which reads in the *The King James Bible*, according to Bloembergen: *The price of wisdom is above rubies.* **In reality, the State translation is as follows: *Ramoth and Gabian shall not be remembered: for the migration of wisdom is greater than that of the Rubies. The Ramoth: Some understand by this word coral, others a precious stone called sandastros or garamantites. Gabisch: The name of a pearl, growing in the shell of a fish (...) Others therefore understand a certain gemstone. Not thought, Know, when one mentions the value of wisdom. Rubies: The word means stone, which has been red in colour.*

The second proverb was, according to Bloembergen, from Proverbs 8:11: *Wisdom is better than rubies.*

When Bloembergen quoted these, the Orthodox Jewish scientists recognized them and found it amusing. In the evening, there was a dinner and Deli asked a rabbi what the what the two proverbs were referring to. The rabbi replied that they were about "*crystalized stones*", and not about rubies. Nico checked all the sources, including the Dutch official translation, and a British and a French Bible, but everywhere he found 'rubies'. According to the rabbi, everything was based on an erroneous Greek translation from Aramaic [56]. Bloembergen's knowledge of physics at the congress is in any case much better appreciated than his knowledge of the Old Testament [57].

Who invented the laser?

The 1964 Nobel Prize in Physics honored the development of the maser-laser. Townes received half of the shared prize: *for fundamental work in the field of quantum electronics, which has led to the construction of oscillators and amplifiers based on the maser-laser principle* (Fig. 12.5). Basow and Prochorow got the other half: *for fundamental work in the field of quantum electronics, which has led to the construction of oscillators and amplifiers based on the maser-laser principle.*

It may well be argued that Basow and Prochorow should not have been awarded the prize. They owed their description of the theory solely to the fact that Townes and Schawlow had published an incomplete description, after which Basow and Prochorow merely completed the work (Fig. 12.6). According to Myers and Dixon, Basow and Prochorow never demonstrated that their theoretical description would work in practice [58].

Frances Townes, Charles' wife, said: *"We are robbed."* By this she probably meant that Schawlow should have been awarded the prize instead of the two Soviets

Fig. 12.5 Alexander Prochorov. *Source* https:// www.kp.ru

Fig. 12.6 Nicolas Basov. *Source* https://www.kp.ru

[59]. Townes himself thought that the Soviets did not deserve to get the prize [60]. Someone who felt sure of being robbed of the prize was Maiman.

According to Hecht, the choice of the Nobel Committee was not the worst. He did think that Bloembergen should have received the prize on the basis of his three-level principle, but the fact is that the prize is awarded to a maximum of three people.

Hecht felt that the prize should not have been awarded for the maser, but for the much more important laser. According to him, Townes and Maiman should have been awarded the prize in that case [61].

Siegman said in an interview: *"I believe that Townes invented the maser. He knew how microwave hardware looked like. He had both sides* [both theory and practice, RH] *of those."* [62]

Myers and Dixon write that nobody could beat their chest as 'the' inventor of the laser [63].

In an interview in 2007, Townes, on the other hand, says: *"Even when I discovered the laser…"* [64] [emphasis of RH].

In a 2005 publication, Townes mentions Weber, Basow, and Prochorow and himself: *three independent ideas for amplification by stimulated emission were generated about 1950.* However, in this article Townes also clearly gives a positive opinion of Bloembergen's contribution: *Nicolaas Bloembergen, who had been working with paramagnetic crystals, recognized that the unequally spaced electron spin*

levels in such crystals would allow a very convenient method for population inversion, his famous 3-level pumping system. This rapidly became the most popular type of maser and helped the field to blossom fast [65].

Bloembergen states that the main function of the solid-state maser is that it is the precursor of the laser [66].

Townes' patent of 1959, viz., *Production of Electromagnetic Energy* [67], describes the population inversion amplification between two energy levels and applies to both masers and lasers and both solids and gases. However, this broad interpretation does not follow so clearly from the specifications and claims described in the patent.

Bloembergen's patent, also from 1959, viz., *Uninterrupted Amplification Key Stimulated Emission of Radiation from a Substance Having Three Energy States* [68], describes the possibilities for stimulated emission with solid-state lasers. The three-level, or more generally multi-level principle lies at the heart of such lasers. The two patents, that is, Townes' and Bloembergen's, could not work without each other. While Bloembergen's was the handle, Townes' was the basket [69].

The Schawlow and Townes' patent of 1960, viz., *Masers and Maser Communications System* [70], was largely based on the earlier patents filed by Townes and especially Bloembergen in 1959.

In the patent war between Gould and Townes, the American court ruled that Townes' patent prevailed over Gould's [71].

According to Myers and Dixon, the question of why Bloembergen's patent was not recognised as the basic laser patent can be answered as follows [72].

The patent office classified the patent under a title that caused confusion. Bloembergen's patent, although filed long before those of Townes and Schawlow, was only granted after the latter two.

In an interview in 2006, Bloembergen remarked: *"Those lawyers, patent attorneys, who should have seen that. They should have been after it and then I would have become rich"* [73]. His brother Evert Bloembergen, a lawyer himself, also thought that the lawyers should have done better [74].

Bloembergen's three-level system was immediately recognized as being of great practical significance for military radar, satellite communications, and scientific purposes such as radio astronomy. It was even used so widely that nobody referred to it anymore, for the same reason that there is no longer any reference to Newton or Maxwell [75].

Siegman said in an interview about this: *"His* [Bloembergen's, RH] *patent was never mentioned, as if it didn't exist. I cannot understand that. That is beyond belief"* [76].

As always, more interested in scientific development than the question of who had the patent rights, Bloembergen observed: *First masers were (Townes, Ramsey) based on separation in space of atoms in the higher state from those in the lower state. A more powerful and widely applied method is the pumping method. The precursors of that principle go back to magnetic resonance* [77].

12.3 From Eyes to Cars: Applications of Lasers

What should the world do with a laser? Irnee D'Haenens, Maiman's associate, stated that the laser was *"a solution looking for a problem"* [78]. Since then we have learnt otherwise! The number of applications of lasers is countless and it would be impossible to list them all.

Of the pioneers, only the ruby and helium-neon lasers are still used, because they are relatively simple to manufacture [79]. In 1969, the physics laboratory of the University of Amsterdam was already using a helium-neon laser for experiments with an interferometer. This type of laser is often used as a barcode scanner, in land surveying, and for holography.

The compact disc (CD) works with a diode laser that is only a few mm in size and has a wavelength of 780 nm (far red). The material used is aluminum gallium arsenide. The DVD works with a slightly shorter wavelength of 650 nm (red), while the Blue Ray disc works with a wavelength of 405 nm (violet). The material of the DVD laser is aluminum gallium indium phosphide and the Blue Ray works with indium gallium nitride. These are examples of semiconductors and thus solid-state lasers. Robert Hall of General Electric succeeded in 1962 in making the first gallium arsenide laser [80].

Previous applications were limited to low power ratings: mostly reading data and sometimes (with CD, DVD, or Blue Ray) also burning data in a polymer layer.

Industry also makes extensive use of lasers. Here we find applications such as burning the sell-by date in labels, as well as engraving ballpoint pens and other materials. High power applications include cutting all kinds of materials from textiles and plastics to centimeter thick stainless steel. It can also be used for drilling and welding. A well-known high-power laser is the neodymium-doped yttrium aluminum garnet laser (Nd:YAG) with a wavelength of 1064 nm (infrared). This laser is also used in dentistry and medicine and is a solid-state laser. An application in ophthalmology has become so popular that it has even been given its own name: *laser eye surgery*. This changes the focusing power of the eye, so that the image falls back on the retina and not in front of it. Another well-known laser is the carbon dioxide (CO_2) gas laser, which uses a mixture of nitrogen, carbon dioxide, hydrogen and/or xenon, and helium. The CO_2 laser is also used in industry and medicine and works in the infrared. It can generate a power of up to several hundred kilowatts. This laser is used for cutting and welding large objects, as in the automotive industry, for example. In the graphics industry, the laser has completely replaced printing with lead, but later also working with films.

In 1958, Gould proposed to pulse the laser beam, which is called Q-switching [81]. This allows extremely high (gigawatt) peak powers to be reached, much higher than without pulses. Nd:YAG lasers are often equipped with this.

With the dye laser, an organic dye such as rhodamine 6, usually in a solution, provides the laser effect. A dye laser is 'fired' with a flash lamp or an external laser such as a ruby laser. The advantage of dye lasers is that a much broader wavelength range is possible compared to gases and solid matter. This type of laser was simultaneously developed by Fritz Schäfer of the University of Marburg [82] and Peter Sorokin of IBM [83].

According to Bloembergen, optical communication with erbium-doped fiber amplifiers is common for the fiber optic network [84].

Defense ministries have always been interested in lasers; the killing beam in particular appeals to their imagination. At the presentations by Maiman and the Bell Labs, there was talk of 'death rays'. The James Bond Film *Goldfinger* of 1964, like countless earlier films and comic books, was full of it. In 1983, President Ronald Reagan inaugurated the *Strategic Defensive Initiative* (Star Wars), whereby a protective (laser) shield was supposed prevent space-based attack on the United States. This is discussed in Chap. 15.

An important application of the laser was also for further research. These coherent, intense light sources are the basis for numerous new research projects. Bloembergen already saw this in 1961 and was investigating a number of possibilities in a new field: non-linear optics.

References

1. Siegman AE. Masers and Lasers: Looking back over 50 years. CLEO 2004 Plenary Presentation, San Francisco, 2004
2. Purcell EM and Pound RV. A Nuclear Spin System at Negative Temperature. Phys Rev 81:279 (1951)
3. Ramsey NF and Pound RV. Nuclear Audiofrequency Spectroscopy by Resonant Heating of the Nuclear Spin System. Phys Rev 81: 278–279 (1951)
4. Gordon JP, Zeiger HJ and Townes CH. The Maser—New Type of Microwave Amplifier, Frequency Standard, and Spectrometer. Phys Rev 99:1264–1274 (1955)
5. Hecht J. Beam. The Race to Make the Laser. Oxford University Press, Oxford (2005)
6. Gordon JP, Zeiger HJ and Townes CH. The Maser—New Type of Microwave Amplifier, Frequency Standard, and Spectrometer. Phys Rev 99:1264–1274 (1955)
7. Bloembergen N. Proposal for a New Type of Solid State Maser. Phys Rev 104:324–327 (1956)
8. Hecht J. Beam. The Race to Make the Laser. Oxford University Press, Oxford (2005)
9. Dicke RH. Molecular Amplification and Generation Systems and Methods. United States Patent Office 2.851.652 (1958)
10. Kastler A. Optical methods for studying Hertzian resonances. Nobel Lecture, 12 December (1966)
11. Townes CH. How the Laser Happened. Oxford University Press, New York (1999)
12. Anderson JM. The Electrodeless Gaseous Electric Discharge Devices Utilizing Ferrite Cores. United States Patent Office 3.500.118 (1970)
13. Schawlow AL and Townes CH. Infrared and Optical Masers. Phys Rev 112:1940–1949 (1958)
14. Hecht J. Beam. The Race to Make the Laser. Oxford University Press, Oxford (2005)
15. Schawlow AL and Townes CH. Masers and Maser Communications System. United States Patent Office 2.929.922 (1960)
16. Myers RA and Dixon RW. Who invented the laser: An analysis of the early patents. Hist Stud Phys Biol Sci 34:115–149 (2003)
17. www.alcatel-lucent.com
18. Myers RA and Dixon RW. Who invented the laser: An analysis of the early patents. Hist Stud Phys Biol Sci 34:115–149 (2003)
19. Townes CH. How the Laser Happened. Oxford University Press, New York (1999)
20. Myers RA and Dixon RW. Who invented the laser: An analysis of the early patents. Hist Stud Phys Biol Sci 34:115–149 (2003)
21. Hecht J. Beam. The Race to Make the Laser. Oxford University Press, Oxford (2005)
22. Hecht J. Beam. The Race to Make the Laser. Oxford University Press, Oxford (2005)
23. Hecht J. Beam. The Race to Make the Laser. Oxford University Press, Oxford (2005)
24. Siegman AE. Masers and Lasers: Looking back over 50 years. CLEO 2004 Plenary Presentation, San Francisco, 2004
25. Interview Rob Herber with N Bloembergen, 6 December 2006

26. Interview Joan Bromberg J and Paul L Kelley with Nicolaas Bloembergen, 27 June 1983. Niels Bohr Library & Archives, American Institute of Physics, College Park, MD USA, www.aip.org/history/ohilist/4511.html
27. Interview Jeff Hecht with N Bloembergen, 4 Nov 1984. Niels Bohr Library & Archives, American Institute of Physics, College Park, MD USA
28. Hecht J. Beam. The Race to Make the Laser. Oxford University Press, Oxford (2005)
29. Hecht J. Beam. The Race to Make the Laser. Oxford University Press, Oxford (2005)
30. Townes CH. How the Laser Happened. Oxford University Press, New York (1999)
31. Laser Pioneer Interviews: Theodore H. Maiman. A Candid Discussion with the Man Who Built the First Laser. Laser and Applications. May 1985 85–90
32. Hecht J. Beam. The Race to Make the Laser. Oxford University Press, Oxford (2005)
33. Interview Joan Bromberg with I D'Haenens, 5 February 1985. Niels Bohr Library & Archives, American Institute of Physics, College Park, MD USA, www.aip.org/history/ohilist/5015.html
34. Maiman TH. Stimulated Optical Radiation in Ruby. Nature 187:493–494 (1960)
35. Hecht J. Beam. The Race to Make the Laser. Oxford University Press, Oxford (2005)
36. Maiman TH. Optical and Microwave-optical Experiments in Ruby. Phys Rev 123:1145–1151 (1961)
37. Maiman TH, Hoskins RH, D'Haenens IJ, Asawa CK and Evtuhov V. Stimulated Optical Emission in Fluorescent Solids II. Spectroscopy and Stimulated Emission in Ruby. Phys Rev 123:1151–1157 (1961)
38. Hecht J. Beam. The Race to Make the Laser. Oxford University Press, Oxford (2005)
39. Collins RJ, Nelson FJ, Schawlow AL, Bond W, Garrett CGB, and Kaiser W. Coherence, Narrowing, Directionality, and Relaxation Oscillations in the Light Emission from Ruby. Phys Rev Lett 5:303–304 (1960)
40. Schawlow AL and Townes CH. Infrared and Optical Masers. Phys Rev 112:1940–1949 (1958)
41. Schawlow AL. In: Townes, CH (ed) Quantum Electronics. Columbia University Press, New York (1960)
42. Hecht J. Beam. The Race to Make the Laser. Oxford University Press, Oxford (2005)
43. Hecht J. Beam. The Race to Make the Laser. Oxford University Press, Oxford (2005)
44. McPartland S. Gordon Gould. Rourke, Vero Beach (1993)
45. Taylor N. Laser: The Inventor, the Nobel Laureate, and the Thirty-Year Patent War. Simon and Schuster, New York (2000)
46. Maiman T. The Laser Odyssee. Laser Press, Blaine (2000)
47. Townes CH. How the Laser Happened. Oxford University Press, New York (1999)
48. Myers RA and Dixon RW. Who invented the laser: An analysis of the early patents. Hist Stud Phys Biol Sci 34:115–149 (2003)
49. Hecht J. Beam. The Race to Make the Laser. Oxford University Press, Oxford (2005)
50. Wieder I and Sarles LR. Stimulated Optical Emission from Exchange-Coupled Ions of Cr+++ in Al_2O_3. Phys Rev Let 6:95–96 (1961)
51. Schawlow AL and Devlin GE. Simultaneous Optical Maser Action in Two Ruby Satellite Lines. Phys Rev 6:96 (1961)
52. Sorokin P and Stevenson MJ. Stimulated Emission from Trivalent Uranium. Phys Lett 5:557–559 (1960)
53. Javan A, Bennet WR Jr and Herriott DR. Population Inversion and Continuos Optical Maser Oscillation in a Gas Discharge Containing a He-Ne Mixture. Phys Rev Let 6:106–110 (1961)
54. Jacobs S, Gould G and Rabinowitz P. Coherent Light Amplification in Optically Pumped Cs Vapor. Phys Rev Let 7:415–417 (1961)
55. Myers RA and Dixon RW. Who invented the laser: An analysis of the early patents. Hist Stud Phys Biol Sci 34:115–149 (2003)
56. Interview Rob Herber with N Bloembergen and Deli Bloembergen-Brink, 7 December 2006
57. Bloembergen N. (ed) Encounters in Magnetic Resonances. World Scientific, Singapore (1996)
58. Myers RA and Dixon RW. Who invented the laser: An analysis of the early patents. Hist Stud Phys Biol Sci 34:115–149 (2003)
59. Hecht J. Beam. The Race to Make the Laser. Oxford University Press, Oxford (2005)

60. Interview Rob Herber with C Townes, 26 November 2007
61. Hecht J. Beam. The Race to Make the Laser. Oxford University Press, Oxford (2005)
62. Interview Rob Herber with AE Siegman 13 June 2007
63. Myers RA and Dixon RW. Who invented the laser: An analysis of the early patents. Hist Stud Phys Biol Sci 34:115–149 (2003)
64. Interview Rob Herber with CH Townes, 26 November 2007
65. Townes CH. Early history of quantum electronics. J Mod Optics 52:1637–1645 (2005)
66. Bloembergen N. Historical comments on the pumping of masers and lasers. J Mod Optics 52:1653–1655 (2005)
67. Townes CH. Production of Electromagnetic Energy. United States Patent Office 2.879.439 (1959)
68. Bloembergen N. Uninterrupted Amplification Key Stimulated Emission of Radiation from a Substance Having Three Energy States. United States Patent Office 2.909.654 (1959)
69. Myers RA and Dixon RW. Who invented the laser: An analysis of the early patents. Hist Stud Phys Biol Sci 34:115–149 (2003)
70. Schawlow AL and Townes CH. Masers and Maser Communications System. United States Patent Office 2.929.922 (1960)
71. Myers RA and Dixon RW. Who invented the laser: An analysis of the early patents. Hist Stud Phys Biol Sci 34:115–149 (2003)
72. Myers RA and Dixon RW. Who invented the laser: An analysis of the early patents. Hist Stud Phys Biol Sci 34:115–149 (2003)
73. Interview Rob Herber with N Bloembergen and Deli Bloembergen-Brink, 7 December 2006
74. Interview Rob Herber with E Bloembergen, 20 November 2006
75. Myers RA and Dixon RW. Who invented the laser: An analysis of the early patents. Hist Stud Phys Biol Sci 34:115–149 (2003)
76. Interview Rob Herber with AE Siegman, 13 June 2007
77. Hecht J. Voorzitter, chairman of the Panel Session of Development of the Laser and Non Linear Optics. Optical Society of America. 20 October 1982. Magnetic tape recording. Niels Bohr Library & Archives, American Institute of Physics, College Park, MD USA
78. Hecht J. Beam. The Race to Make the Laser. Oxford University Press, Oxford (2005)
79. Hecht J. Beam. The Race to Make the Laser. Oxford University Press, Oxford (2005)
80. Hall RN, Fenner GE, Kingsley JD, Soltys TJ and Carlson RO. Coherent Light from GaAs Junctions. Phys Rew Lett 9:366–368 (1962)
81. Taylor N. Laser: The Inventor, the Nobel Laureate, and the Thirty-Year Patent War. Simon and Schuster, New York (2000)
82. Schäfer FP, Schmidt W and Volze J. Organic Dye Solution Laser. Appl Phys Lett 9:306–309 (1966)
83. Sorokin PP and Lankard JR. Stimulated emission from an organic dye, chloro-aluminium phthalocyanine. IBM Res J Develop 10:162–163 (1963)
84. Bloembergen N. Historical comments on the pumping of masers and lasers. J Mod Optics 52:1653–1655 (2005)

Part V
Non-linear Optics, the Icing on the Cake

Chapter 13
From Lasers to Nonlinear Optics

He grasped its significance immediately, and I've always thought that right there before my eyes he changed the direction of his own research career. Spin resonance was pushed into the background, and he moved strongly into what was to become nonlinear optics and a whole new chapter in his contributions to physics. John Armstrong, PhD student supervised by Bloembergen from 1958 to 1963 [1].

13.1 Of Radio Waves, Microwaves, Lasers, and Non-linearity

Maiman's creation of the laser in 1960 made it possible to produce high-intensity monochromatic coherent light with a wavelength of 694.3 nm. The number and type of lasers increased explosively after 1960.

As described in Chap. 12, Bloembergen had not yet designed a laser in 1960, but he found the laser interesting enough to want to build one right away:

And then, Maiman got the laser going. Then I said, aha, we have to get in, but we cannot work on new laser types or improve them. We just have to do something with them, to use them. [...] In '60, I asked Peter Pershan to build us a duplicate ruby laser, just copying what Maiman did, and we had a hell of a time, you know. If you have no optical experience, it just isn't that easy. It isn't that easy, in the first year. So, we really got going in '62, when we could use the so-called Trion ruby laser. And because we had worried about interaction of microwaves and light, here was an interaction of light on light and general nonlinear phenomena, which I knew from ferromagnetic resonance and what not [2].

Bloembergen was referring here to the publications of his group in 1952, 1953, and 1954. Non-linear behavior of components in radio tubes, for example, had long

© Springer Nature Switzerland AG 2019
R. Herber, *Nico Bloembergen*, Springer Biographies,
https://doi.org/10.1007/978-3-030-25737-8_13

been known in radio and microwave technology [3]. The paper *Relaxation Effects of Ferromagnetic Resonance* by Bloembergen and Damon describes the saturation of the electron spins in a nickel ferrite crystal, $NiO.Fe_2O_3$ [4]. Bloembergen made the following remark about this in an interview:

We found the first evidence for nonlinear effects in ferromagnetic resonance, which we didn't clarify completely, but at least we found significant nonlinear effects experimentally [5].

In a follow-up article by Damon called *Relaxation Effects in the Ferromagnetic Resonance*, it is observed that the usual theory cannot explain the spin-lattice relaxation time in a strong magnetic field. Damon noted that the deviation from the theory depended on the power of the applied microwave frequency [6].

In *Relaxation Effects in Para- and Ferromagnetic Resonance* by Bloembergen and Wang, it is stated that the theory above the Curie temperature[1] gives a good description of experimental results. Below the Curie temperature, however, the theory is not correct: *The present experiments, which were undertaken to clarify some of the issues of ferromagnetic relaxation, have only created new problems in this respect* [7].

Hans Dehmelt of the University of Washington showed that the orientation of atoms can be measured by the absorption of a polarized light beam, with the precession of atoms causing the absorption. This precession is generated by radio frequencies [8].

According to Bloembergen, the first International Quantum Electronics Conference (IQEC) in High View, New York, in September 1959 was largely dedicated to paramagnetic resonance and masers. From conversations Bloembergen had at the conference, he expected a wave of optics research to arrive in the near future [9].

A next step made by Bloembergen's group (Bloembergen, Peter Pershan, and Lee Wilcox) made after studying ferromagnetic resonance was the modulation of light in a solid by means of microwave radiation [10]. Subsequently, one of Bloembergen's Ph.D. students, Yuen-Ron Shen, investigated the Faraday rotation in rare earth metals. This effect occurs when the polarization of light is changed, for example, by a magnetic field. The work described in the dissertation was later partly published in three publications [11–13].

In November 1961, Bloembergen filed a patent application entitled: *Microwave Modulation of Optical Radiation in a* Waveguide, which was granted in March 1966 [14]. This method, which would later become known as quasi-phase matching, is widely used in optical waveguides (Fig. 13.1).

In 1961 a publication by Peter Franken et al. of Michigan University was published about the effect of the laser beam in solid matter. In the article *Generation of Optical Harmonics*, the beam of a ruby laser was filtered with a red filter to eliminate the background radiation of the xenon flash. The beam was then focused in a quartz crystal and the emitted radiation analyzed with a spectrograph using a photographic plate. The main wavelength of the beam at 694.3 nm appeared on the plate along

---

[1]The Curie temperature is the temperature above which ferromagnetic materials such as iron lose their permanent magnetic field.

Fig. 13.1 Peter Franken. *Source* michiganphysics.wordpress.com Jens Zorn

with a weak image of the second harmonic[2] radiation at 347.2 nm. Franken et al. set out in the article the conditions that optical material must meet in order to produce harmonics. These are a nonlinear dielectric[3] coefficient and transparency for both the main wavelength and the overtones to be measured. If the electric field is strong enough, all dielectrics meet the condition of producing harmonics. Glass and quartz would be suitable materials, but glass is isotropic, i.e., its properties do not depend on the direction in space. Gases and liquids are isotropic. Their properties, for example, conductivity or permittivity, can be described by a single number which is then called a *scalar*. Quartz is anisotropic: its properties depend on the direction in space, in this case the directions of the crystal axes. A quantity with a direction, such as velocity, is described by a *vector*. With a crystal, there are a number of directions and certain physical quantities have to be described by a *tensor*. In a crystal, the properties perpendicular to a crystal axis are different from those parallel to that axis. However, a crystal has three axes and the situation is therefore complex [15].

Lee Wilcox, present at the Second Quantum Conference in Berkeley, received a preprint (a copy of a scientific article that appears before it is officially published) from Franken and showed it to Bloembergen. This was the first paper in which

---

[2]A harmonic is a frequency that is an integer multiple of the fundamental frequency. In music, the harmonics are called overtones.

[3]A dielectric material is a material which is a poor conductor (insulator), such as glass, quartz, or porcelain, and which can be polarized by an electric field. The dielectric coefficient determines the strength of the interactions between charged particles. It is now called the permittivity and is a parameter that describes how an electric field influences a medium and is influenced by a medium.

the phenomenon of non-linear radiation was described in optics. Bloembergen's interest was immediately aroused and he decided to turn his attention to this new phenomenon. He noted that Franken's article described an elementary process in which two photons come together and mutually destroy each other, producing a new photon that is released with twice as much energy, and thus twice the frequency.

Another publication from 1961 that Bloembergen found particularly interesting was by Kaiser and Garrett. They describe a two-photon process in calcium fluoride. This crystal was irradiated with the light of a ruby laser (wavelength 694.3 nm, red light) and the authors observed a wavelength of 425 nm, which lies in the blue. They pointed out that it was essential that the crystal have no center of inversion, that is, no center of symmetry. Kaiser and Garrett also observed the absorption of two photons [16].

13.2   Maxwell—From Magnetic Fields to Electric Fields

When he turned to non-linear optics Bloembergen entered a new area largely based on Maxwell's equations.

Research in the theory of electricity has resulted in a large number of physical laws that we can now recognize mainly by the physical units that are used. The volt, the ohm, the ampere, the coulomb, the gauss, the faraday and many other units refer to different phenomena in electricity and magnetism. In 1865 James Clerk Maxwell formulated a number of equations (in fact, twenty), which could give a general description of various electromagnetic phenomena. These twenty equations were reduced to a very compact form in 1884 by Oliver Heaviside and Heinrich Hertz, requiring only four basic equations. Their great significance is that they describe the electrical and magnetic forces in matter for the entire wavelength range from radio waves to gamma radiation.

According to Ad Lagendijk, quantum mechanics is a rare phenomenon on the periphery of an especially electromagnetic reality, as described by the Maxwell equations [17].

Maxwell's equations simplify in isotropic linear media, for example, when describing the refraction visible light in a lens or prism. The vacuum is a special case of such an isotropic medium. In vacuum there are no charges or currents; moreover there can be no polarization or magnetization. In vacuum, the electromagnetic field can be described by sinusoidal waves, in which the directions of the electric and magnetic fields are perpendicular to each other and to the direction of propagation [18].

In anisotropic media, as described earlier by Franken et al., the constants describing the permittivity and permeability must be replaced by tensors (which are effectively like sets of vectors). The constants are therefore replaced by arrays of numbers, or matrices, that depend on the directions characterizing the anisotropy, for example, the directions of the crystal axes.

Bloembergen realized that Franken et al.'s research was an extension of the so-called Pockel effect [19]. Friedrich Pockel found in 1893 that when an optical medium produces a constant or varying electric field, birefringence can occur. This birefringence is proportional to the electric field and occurs only with certain crystals that do not have inversion symmetry, such as lithium niobate or gallium arsenide, i.e., the properties in one direction are not always the same as those in the opposite direction. Bloembergen recognized that, in Franken et al.'s research, the refractive index of the material (quartz) is changed by laser radiation. As mentioned above, this meant that it was an extension of the Pockel effect, which occurs with microwaves. Even before Franken's article appeared in July 1961, Bloembergen had convened a discussion group in his laboratory. He was aware of the fact that non-linearity stabilizes vibrations. In his Nobel lecture in 1981, he put this as follows: *The amplitude of the laser output is limited by a nonlinear characteristic, as for any feed-back oscillator system. It is the onset of saturation by the laser radiation itself which tends to equalize the populations in the upper and lower lasing levels* [20]. Much later, he explained this as follows: *Actually the ruby laser itself is an example of nonlinear optics. Maiman achieved stimulated emission, in a negative imaginary part[4] of the susceptibility at the laser frequency $_L$ by pumping at $_P$, where $_P$ stands for the blue and green absorption bands by the ruby* [21]. Maiman was the one who demonstrated non-linearity for the first time, without realizing it himself (and before Bloembergen realized it either).

John Armstrong (Born in 1934)

Armstrong joined the Bloembergen group as a postdoc. He gives a slightly different reading of the way Bloembergen viewed Franken's article:

I remember looking at the Phys. Rev. Letter preprint from Michigan on second harmonic generation in quartz and saying, "Well now there's a curiosity," but I didn't see at the time why anyone should think it should be a Physical Review Letter. I had seen the preprint because it had been addressed to Lee Wilcox, whose mail I was handling while he was in Beirut. Nico came into the lab that afternoon and I showed him this preprint and he read it on the spot. He grasped its significance immediately, and I've always thought that right there before my eyes he changed the direction of his own research career. Spin resonance was pushed into the background, and he moved strongly into what was to become nonlinear optics and a whole new chapter in his contributions to physics [22] (Fig. 13.2).

In an interview with Bromberg, Armstrong remarked: "*Nico saw through the very modest beginnings of that* [referring to Franken's paper, RH] *immediately, in my perception, and within a day or two, he had derived the very simplest form of nonlinear equations [...] And I remember, Ducuing and I had these equations, and Nico didn't know how to solve them right off the bat* [23].

---

[4]In mathematics there is an imaginary number i whose square is a negative real number. Indeed, by definition, $i^2 = -1$, or $\sqrt{-1} = i$. Imaginary numbers are useful in the practice of mathematics and physics, especially in electricity and wave theory. In general, a complex number consists of a real and an imaginary part, for example, $5 + 3i$.

Fig. 13.2 John Armstrong. *Source* www.aip.org

Jacques Ducuing (Born 1932)

Ducuing came to Bloembergen in 1960 as a postdoc with a NATO scholarship. He actually wanted to work on masers, but agreed to come and explore Bloembergen's newly formed optics group [24]. In 1964 Ducuing obtained his doctorate at Harvard and later at the University of Paris, with the thesis *Contribution à l'étude des non-linéarités optiques* (*Contribution to the study of optical non-linearities*), work he had carried out with Bloembergen.

Peter Pershan (Born 1934)

Pershan started working on his thesis in 1956 under the supervision of Bloembergen. In 1960 he obtained his Ph.D. with Bloembergen for a thesis on NMR in lithium fluoride. He stayed in Harvard for a postdoc period and became Assistant Professor. After three years at Harvard, he worked for one year at Bell. He was then asked to come back to Harvard and was given a permanent tenure as Associate Professor. Pershan never left Cambridge (Fig. 13.3).

Pershan: *"When I was studying for my PhD, Bloembergen changed fields [...] In those years, there was little money and it was hard to compete. Bloembergen wanted to do something with lasers and he looked first to high-speed modulation of light. Microwave modulation of light. We made some attempts to study that. So when non-linear optics came along, it was an extension, not with microwaves, but with light. The physics studied by Peter Franken was very similar to the physics I had learned as an undergraduate. So I knew immediately how to approach the problem. Bloembergen started to work on that with Jacques Ducuing, and John Armstrong. But then Bloembergen came down and asked me to help with the paper, and I was very happy to join."* [25]

Fig. 13.3 Peter Pershan. *Source* liquids.seas.harvard.edu

Bloembergen asked Pershan to build his own laser. However, the self-made laser did not meet their requirements and Bloembergen's group decided to postpone experimenting with lasers until they could obtain a commercial device in the lab. As long as they did not yet have their own laser, the group (Armstrong, Bloembergen, Ducuing, and Pershan) worked on theoretical aspects of the interaction of light waves with a non-linear dielectric [26].

Pershan notes: *"Then we worked on two other papers. But I wanted to make a career of myself and I did my own things. We remained friends and I followed his work."* [27]

13.3 Theory of Nonlinear Optics

Bloembergen based his theoretical study on Léon Rosenfeld's lectures, attended during his studies in Utrecht (see Chap. 4) and on the book *Theory of Electrons* by Rosenfeld [28]. The starting point for understanding electromagnetism was the microscopic equations of Maxwell in vacuum together with the expression for the Lorentz force.[5] These microscopic equations are based on the total electric charge and current distributions in matter, including the distributions in the atoms, which are difficult to calculate. Bloembergen further examined the relationship between the microscopic equations in vacuum, and the macroscopic equations in media according

---

[5]The Lorentz force is the force exerted on a charge by an electromagnetic field.

Fig. 13.4 Nico Bloembergen with graduate student Pierre Lallemand. *Source* Nico Bloembergen, Cruft Laboratory, Harvard University

to Maxwell's equations [29]. The macroscopic equations make use of two new 'auxiliary fields' and free charge and current that circumvent the problems of calculating the atomic scale charge distributions [30]. Bloembergen emphasized these relationships at lectures he gave in Harvard (Fig. 13.4). He warned the students that the introduction of electrical and magnetic permeability in vacuum obscures the fundamental physics of the relationship between fields and matter. Although permeability is easy to calculate (such as the derived refractive index of glass), it says nothing about the underlying physics [31].

In the aforementioned 1962 article by Armstrong, Bloembergen, Ducuing, and Pershan, the theory is unfolded comprehensively [32].

A light wave is a variable electromagnetic field that propagates through space as a travelling wave. When it passes through a transparent medium (a dielectric), the medium is polarized by the electromagnetic field. The changing polarization of the medium itself acts as a source of secondary electromagnetic waves that interfere with the original wave. The result is a change in the speed of the resultant wave and it is this that leads to the phenomenon of light refraction. As a rule, the polarization is proportional to the electric field. In that case, nothing changes in the frequency of the incident light. However, this is not true for birefringent media such as quartz, as a result of which the polarization due to two different electric fields is no longer equal to the sum of the individual polarizations due to each of those fields.

Bloembergen himself compared the phenomena of nonlinear optics with effects that occur with sound. If one blows too loudly into a musical instrument such as a flute or an organ pipe, distortion occurs: tones with higher frequencies become audible. This occurs when the sound intensity becomes too high [33].

The above-mentioned classic article by Armstrong, Bloembergen, Ducuing, and Pershan has 22 pages and forms the basis of non-linear optics. It has been quoted 1172 times up to the time of writing. It is therefore included in the series *This Week's Citation Classic* [34].

Bloembergen assumed that the nonlinear polarizations caused by nonlinear behavior could serve as additional sources in the Maxwell equations. The nonlinear behavior can then be expressed as dependent on the nonlinear susceptibility [35].[6] The electrical susceptibility of a dielectric material is a measure of how fast the material polarizes as a result of an electric field. The dielectric constants only apply to linear behavior of materials and, as Franken stated, if the electric field is large enough, all dielectrics exhibit non-linear behavior. The dependence on susceptibility can be expressed in a certain mathematical way (second and third order non-linearity) which can be understood physically in terms of the mathematical account given by perturbation theory in quantum mechanics. This theory aims to find approximate solutions to problems that cannot be solved exactly [36].

In the case of nonlinear susceptibility, this approach leads to a set of coupled electromagnetic wave equations, for which solutions must be sought. Bloembergen explained later that the mathematical derivation was mainly done by his co-authors, because they were much more skilled than himself in the laborious, detailed elaboration of mathematical equations [37].

The non-linear polarization terms, as measured by Franken, were calculated for non-dissipative media. These are media from which energy cannot escape. These non-linear polarization terms then lead to electromagnetic waves with different frequencies.

According to the theory they developed, it appeared to be possible to convert a light frequency completely without loss into another light frequency, assuming that the phases were the same. If the latter was not the case, corrections would arise that would compensate for the difference in phase. It would take more than ten years before experiments could confirm this theory.

The first publication of the theory of nonlinear optics was followed by more than two hundred Harvard publications in this field [38].

To find out about the optics that are new to him, Bloembergen consulted the standard textbook for his first article on non-linearity, namely, *Principle of Optics* by Born and Wolf [39]. It began to dawn on him that many chapters in the book could be expanded and generalized to the case when high-intensity light causes non-linear effects. An obvious question, according to Bloembergen, was the effect that occurs when a high-intensity laser beam in vacuum strikes the interface with a non-linear crystal. He became interested in the history of the laws of reflection and refraction in optics. The first to describe reflection was Hero of Alexandria (10–70 years after Christ). Hero stated that, for reflection in a mirror, the angle of incidence is equal to

---

[6]The susceptibility of a material or substance describes its response to an applied field. This may refer to a magnetic or electric field. en.wikipedia.org.

the angle of reflection. Refraction was not described mathematically until Snellius in 1621, almost 1600 years later. The law of refraction says that the ratio of the sine of the angle between the incident beam and the normal (vector perpendicular to the plane) and the sine of the angle between the refracted beam and the normal is equal to the index of refraction of the medium (for example, glass) to the index of refraction of the other medium (for example, air). This law is in agreement with Fermat's principle of 1657, which says that: "The path taken by a beam of light between two points is the one that can be travelled in the shortest time". August-Jean Fresnel (1788–1827) generalized the laws in the linear case. At a certain angle, a polarized incident beam is completely refracted without reflection. This is called the Brewster angle, after David Brewster (1781–1868). For the interface between glass and air, it is 56°.

Bloembergen and Pershan determined the intensities of the second harmonic at high intensities. They also determined the intensities for refraction of the second harmonic of two wave fronts at high intensities. Other properties studied were polarization, the Brewster angle, and total reflection of harmonics. The conditions for non-linear reflection in parallel plates and prisms were also determined [40]. Although Bloembergen's principles were simple, the elaboration of the theory required laborious and tough algebra, taken care of by Pershan. The seven page publication attracted less interest than the first in the field of non-linear optics, but later turned out to be of great importance for the physics and chemistry of interfaces [41]. The publication was cited 328 times up to 2011, a considerable number.

In the autumn of 1962, Bloembergen realized that the calculations that were used for the complex susceptibility in magnetic resonance could also be used in optics. The model previously used in the first publication on non-linear optics for closed thermodynamic processes could be extended to more general situations.

Ron Shen, who obtained his Ph.D. at Harvard in 1962, stayed on as a postdoc for another year to develop this subject. The resulting paper described the quantum mechanical one- and two-photon transitions from the same ground levels to the same excited levels.

The equations take into account the absorption process of both the two-photon transition and the Raman transition.[7]

Although Bloembergen claimed that the article did not receive much attention and that it had not been often cited, it didn't do that badly. Up until 2011 it was cited 118 times and that was more than many of Bloembergen's other publications [42].

In February 1963, the Third International Symposium on Quantum Electronics was held in Paris. Bloembergen chaired the Scientific Program Committee of the conference. As chairman, he received a contribution from Malcolm Stich of Hughes Research, where Maiman had developed the laser. The authors Woodbury and Ng in Stich's working group measure infrared radiation from a ruby laser. However, they were not sure if this was infrared fluorescence or a Raman effect. The laser was

---

[7]Raman spectroscopy, like infrared spectroscopy, is an important tool for studying vibrations and rotations in a system. Both techniques are applied in solid-state physics and chemistry. In chemistry, infrared spectroscopy is mainly used for the structural analysis of organic compounds. Raman spectroscopy (named after Chandrasekhara Raman, discoverer of the effect in 1928) is used in inorganic chemistry and in organic chemistry. Raman spectroscopy uses lasers as light source.

pulsed with a so-called Q-switch. Continuous laser light (usually with an acousto-optic modulator) was converted into short pulses of high power. Each pulse typically had a duration of 100 ns. According to Bloembergen, the phenomenon could indeed be explained with his theory as stimulated Raman scattering. In October 1962, the work at Hughes Research was published in the journal of the Institute of Radio Engineers without Bloembergen's knowledge [43].

Michael Bass from the University of Southern California comments:

By this time the Bloembergen analysis relating the various coefficients had appeared [...] We then decided by looking at the Pockel's coefficients in different materials that it would be possible to measure the coefficients of optical rectification, which is the parallel process to frequency doubling.

Bloembergen and his group were ready to move in that direction and they moved beautifully. They did the analysis as well. Bloembergen, (Peter) Pershan, I guess it was (Jacques) Ducuing was visiting with them at the time, and (John) Armstrong - that became a classic paper, because they got the analysis right [44].

At that Paris conference, Bloembergen gave a talk in French: *Optique non-linéaire* (Non-linear optics), which was basically a more detailed version of the two publications mentioned above. Ducuing translated the lecture into correct French and gave Bloembergen some tips on the pronunciation. However, there was simultaneous translation into English and Russian and the audience, expecting Bloembergen to give a lecture in English, got confused. There was some hassle with microphones, headphones, and switches at the beginning of Bloembergen's lecture [45]. And to thank him for his effort to give the lecture in French, he was told by some of the French members of the audience that his French was not good [46].

At the conference there were almost no more lectures about microwaves. Almost everything was about optics. This was indicative of the changed interests of the physicists involved [47].

Bloembergen describes how, during a particularly boring lecture at the above conference, he penned a limerick about non-linearity:

There was a young lady who radiated
When, in a mirror, her figure she contemplated
The nonlinear way
Induced her to say
By Jove, I am ultraviolated.

A year later Bloembergen went on a sabbatical to Berkeley. He gave a lecture for students at a colloquium and asked his good friend and colleague Erwin Hahn if he (Bloembergen) should read the limerick at the lecture on non-linear optics. Hahn believed that most of his colleagues and the students would not appreciate the limerick. Bloembergen still read out the limerick and Hahn was right [48].

With these three publications—Armstrong, Bloembergen, Ducuing, and Persan (1962), Bloembergen and Pershan (1962), Bloembergen and Shen (1964)—the foundations of the field of non-linear optics were laid by Bloembergen. His many later publications on non-linear optics were basically partial elaborations and experiments building on this foundation.

Although Bloembergen received a grant from the Joint Services Electronic Program in 1961 and subsequent years for doing research into magnetic resonance and masers, he didn't say in his research proposal that the money would be used to do research in optics. He reported regularly in quarterly research reports and the publication appeared in 1962 as JSEP Cruft Laboratory Technical Report 358 which was widely distributed.

In 1964, after the three previous publications, Bloembergen worked on the standard textbook *Nonlinear Optics*, which only appeared in 1965, published by W.A. Benjamin. In 1992 the third edition was published by Addison Wesley and in 1996 the fourth by World Scientific.

Further Research in Nonlinear Optics: Reflection, Refraction, Dispersion, and Polarization

The research into these phenomena covers the period from 1963 to 1971. Ducuing participated in the first four publications, Pershan had already gone his own way, as mentioned earlier.

The non-linear optics that Bloembergen and others developed soon had applications. In an interview of Hecht, we find the following exchange:

Hecht: In other words, some of the other [he was not referring to conical refraction, which remained academic in 1984 at the time of the interview, RH] non-linear optics established applications very rapidly?

Bloembergen: Oh, many. And that's what people worked on - to get better harmonic generators, light mixing devices, parametric downconverters, and so on. A lot of technical development. [...] We were, in many cases, the first to study that (properties of non-linearity) in detail. Various classes, metal semiconductors, dielectrics, and then we got into the very high intensities where even dielectrics break down and get optically damaged, and that has turned out to be a very important field, of course, for obvious reasons. Because if you want to build a high power laser you don't want your components to be damaged [49] (Figs. 13.5 and 13.6).

In two publications, the nonlinear reflection of piezoelectric crystals such as gallium arsenide was determined. These experiments were the first in which the Bloembergen group used a (commercial) laser.

An interesting development was generation of the third harmonic. While the second harmonic is already weak (in 1961, Franken saw a weak image of the second harmonic on the photographic plate), the third is not visible at all with a continuous laser. This requires a pulsed laser with a very short pulse time, on the order of picoseconds (1 picosecond is 10^{-12} s or one millionth of a millisecond), for highly absorbent media such as silicon, germanium, silver, and gold. With such a laser, a power of 1 Terawatt (1 million Megawatt) or more can be generated. A laser with full continuous power would damage the surface and could not be used. Bloembergen used a neodymium-glass laser with a wavelength of 1.06 μm, which lies in the infrared. The third harmonic has a wavelength of 353 nm in the UV [50].

Conical refraction, as mentioned in the above interview with Jeff Hecht, was predicted by Hamilton in 1832 and was confirmed experimentally by Lloyd shortly

Fig. 13.5 Single calcite crystal showing double refraction. *Source* Rob Herber

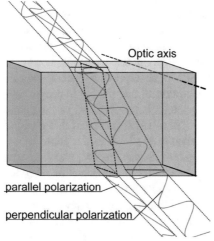

Optic axis

parallel polarization

perpendicular polarization

Fig. 13.6 Double refraction and polarization. *Source* en.wikipedia.org Mikael Häggström

afterwards.[8] Bloembergen and Hansen Shih formulated a theory of nonlinear conical refraction and published the result in 1969 [51]. In 1977, experimental work by Jane Schell and Bloembergen confirmed the theory [52].

Further Research in Nonlinear Optics: Coupled Light Beams

Three beams of light in a non-linear medium can produce a fourth beam by interaction. The intensity of the fourth beam can be increased by resonance. All kinds of scatterings can cause this resonance, including stimulated Raman scattering, mentioned previously. In 1964, Shen and Bloembergen published their first article about the coupling involved in such a phenomenon [53].

In 1966, the dye laser was developed independently by Peter Sorokin, a former student of Bloembergen, now at IBM, and Fritz Schäfer of the University of Marburg. A dye dissolved in a liquid, such as Rhodamine B (red to purple) is optically pumped by, for example, another laser. This can involve a large number of stimulated emissions, which in turn can be selected using a diffraction grating. In 1971, the Bloembergen group built its own dye lasers with frequency doubling, pumped with a neodymium-glass laser or a nitrogen laser. These lasers were used by the Bloembergen group for experiments with coupled light beams. First, resonance and non-resonance were demonstrated in diamond [54]. In a further experiment, the coupling between three light beams was studied in a copper (I) chloride crystal [55], and then the interference between Raman resonances for four coupled beams [56].

Further Research in Nonlinear Optics: Two-Photon Absorption Spectroscopy

Two-photon absorption is the simultaneous absorption of two photons with the same or different frequency. According to Einstein, a threshold energy is needed to cause the photoelectric effect. It turns out that two photons can also reach a higher energy level together. This non-linear absorption is much weaker than the linear absorption of one photon. The absorption of one photon is proportional to the light intensity; the absorption of two photons is proportional to the square of the light intensity. Maria Goeppert-Mayer (1906–1972) predicted this phenomenon in 1931 [57], but it was only with the arrival of the laser that it could be confirmed experimentally [58]. Normally in the case of absorption, Doppler line broadening occurs because a number of atoms will be moving in the detection direction and others away from it. Now it turned out that Doppler broadening was sometimes absent in two-photon absorption. When the Soviet scientist Chebotaiev predicted this absence in 1970, the Bloembergen and his group began researching it and made an incursion into atomic spectroscopy [59, 60]. These studies are carried out by Marc Levenson as postdoc and Michael Salour as Ph.D. student.

---

[8]In a biaxial crystal, a beam of light, which runs along one of the two optical axes, produces a normal beam for polarization components in the direction of the central main refraction axis. For all other polarization components, an extraordinary beam is created, which has a different propagation direction but the same refractive index for each polarization component. Therefore, all polarization directions are equal, and there is no discrete splitting into two beams. Instead, conical refraction of the abnormal beam occurs. This means that, for unpolarized light, one sees a beam cone containing the optical axis in its lateral surface.

Further research by Bloembergen on two-photon absorption concerned ions of rare earth metals. This was done by the postdocs Mario Dagenais, Reinhard Neumann, and Albert Bivas, together with one of Bloembergen's last Ph.D. students, Michael Downer. In 1984 Bloembergen wrote an overview article of the research in this field in Harvard [61].

Further Research in Nonlinear Optics: Multi-Photon Processes
It appears that a higher energy level can be achieved, not only by two photons, but also by even more photons that can be simultaneously (on each other's shoulders, as it were) excited to a higher energy level. However, this process requires even higher energy than with two-photon absorption [62].

Bloembergen's group carried out these studies from 1975 to 1980, but in 1959 Bloembergen had already published a study that could be seen as a precursor. The theoretical article describes the possibility of using a maser to develop a quantum counter for infrared radiation. The counter was never realized in practice [63].

In 1975 Bloembergen gave a lecture in Les Houches, France, about theoretical multi-photon dissociation[9] in the infrared under the influence of intense electromagnetic radiation. After the lecture Vladilen Letochov (1939–2009), leader of a research group in the Soviet Union, came over to him, furious, with the announcement that he [Letochov, RH] had described that phenomenon years before. Bloembergen asked Letochov, where he had published that, to which the latter answered that the phenomena were *"too simple, obvious, and unimportant"*, but that he would show Bloembergen his laboratory diary. Bloembergen told him that he was unable to read Russian and that Letochov shouldn't be so excited if it was such an unimportant problem. Khohklov, the Soviet delegation leader, who had been following the heated discussion, tried to smooth things out and told Letochov to make up. Letochov apologized at lunch. The Letochov and Bloembergen groups remained fierce competitors for many years to come, until Letochov and Bloembergen were finally reconciled in 1980. Bloembergen stated that the Letochov group was the one that had done most work on the subject and cited two books on this topic that Letochov had published [64, 65] (Fig. 13.7).

The fact that this research was carried out on molecules, in fact investigating the interaction of electromagnetic radiation and molecules, shows that this type of research was actually at the interface between physics and chemistry. This is also apparent from a 1977 publication involving the chemist Shaul Mukamel at MIT in Cambridge, Massachusetts [66].

In the 1980s, the research was deepened and expanded. Eric Mazur arrived as a postdoc in the Bloembergen group [67].

---

[9]Dissociation is the process by which a particle disintegrates into two or more smaller parts (nl.wikipedia.org).

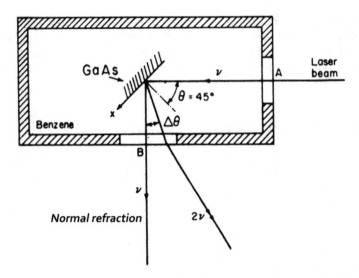

Fig. 13.7 A nonlinear mirror of gallium arsenide. The nonlinear beam makes an angle of $\Delta\theta$ with the normal and has a frequency of 2ν. The benzene enlarges the angle $\Delta\theta$. *Source* Bloembergen N and Ducuing J. Experimental verification of optical laws of nonlinear reflection. Phys Lett **6**:5–6 (1963)

Eric Mazur (Born 1954)

Eric Mazur is the son of Peter Mazur (1922–2001). The latter was born in Austria and joined Leiden University in 1954 as a lecturer. Together with S. R. de Groot, he founded the Lorentz Institute of theoretical physics [68]. His son Eric had been spoon fed with physics from an early age, and then went on to study physics and astronomy in Leiden, graduating in 1977. He obtained his doctorate in 1981 with Jan Beenakker (1926–1998), professor of physics since 1963. After his doctorate, Mazur wanted to work at Philips in Eindhoven, because he was offered a job there and wanted to leave the university world. There he would work with the group that was designing the compact disc. However, his father convinced him to work as a postdoc with Bloembergen for one year in 1982, before spending the rest of his life at Philips. Eric had never been to the United States before and thought it was not such a crazy idea. Mazur wrote to Van Vleck. He had met Bloembergen in 1973 when the latter had been Lorentz guest professor in Leiden. Mazur was interested in optics, something he had worked on in Leiden at the beginning of his career (Fig. 13.8). Mazur had also written to MIT and to his surprise received offers from both Harvard and MIT. He chose Bloembergen [69]. In 1984 he became Associate Professor and in 1990 he was appointed to a permanent position. He has contributed to the fields of spectroscopy, light scattering, ultra-short laser pulses in matter, and nanophotonics [70]. He has also enriched the teaching of physics with a number of new methods [71].

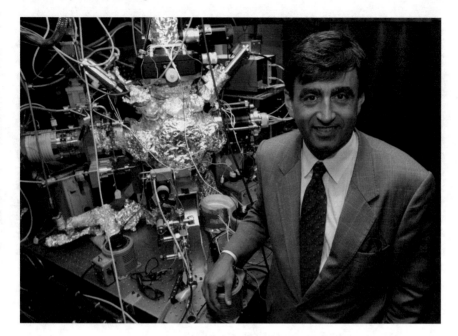

Fig. 13.8 Eric Mazur. *Source* medschool.vanderbilt.edu

Further Research in Nonlinear Optics: Multi-photon Processes (Continued)
In 1984, the overlap between multi-photon research and chemistry, investigating
all kinds of interactions between electromagnetic radiation and molecules, was confirmed in a theoretical article by the chemist Ahmed Zewail at the California Institute
of Technology in Pasadena, California, where Bloembergen had been as a visiting
scholar [72]. Zewail received the Nobel Prize in Chemistry in 1999 *for his studies of
the transition states of chemical reactions using femtosecond spectroscopy.*

Interactions Between the Laser and Matter Bloembergen referred to interactions
in a narrower sense, such as the damage that can be caused by lasers, the breakdown of
matter, melting, evaporation, plasma formation,[10] etc., and laser-plasma interactions.
These types of interactions are now widely used and were mentioned in the previous
chapter on laser applications.

If a material absorbs the laser beam, there is absorption of one photon. This
absorption can be described by the imaginary part of the linear susceptibility.

Eli Yablonovitch and Bloembergen published a study of such phenomena in transparent media, such as kitchen salt crystals (NaCl) [73]. Important studies in the 1980s
were those that set out to determine the duration of a number of processes. These
were performed using pulses with a duration of picoseconds (trillionths of a second)

---

[10]In the case of a plasma, the gaseous particles are (partly) ionized, that is to say, stripped of
electrons.

and later femtoseconds (quadrillionths of a second) in semiconductors and metals. These more application-oriented studies were mainly published in Applied Physics Letters [74].

Mazur continued this work, initially with Bloembergen, but later with his own group, focusing on femtosecond processes [75].

Frans Spaepen was from the Catholic University in Louvain, Belgium, and graduated as a metallurgical engineer. He joined David Turnbull (1915–2007), Gordon McKay Professor of Applied Physics at Harvard, and there got to know Bloembergen. He graduated in 1975. Thereafter he became Assistant Professor and in 1983 and obtained his permanent appointment in the Division of Applied Physics and Engineering where he became a colleague of Bloembergen's. Collaboration between Spaepen and Bloembergen dates from 1979 to 1980. They worked together on a project in which picosecond laser pulses were used to study the behavior of layers of semiconductors and metals. According to Spaepen, Bloembergen realized immediately after developing the fast laser pulses that applications were possible in the interaction with surfaces of different materials. This opened a new research area [76].

13.4   Applications of Non-linear Optics

Some of the major applications of nonlinear optics can be found in the way it opened up other research areas [77], including optical chaos and bi-stability, as exemplified by the ring resonator, used as a switching element at Gigahertz frequencies [78]. Furthermore, the possibility of developing optical computers based on non-linear optics has been investigated. Another area is stimulated Raman scattering, which has been mentioned in previous chapters. Non-linear optics is an important tool for determining the properties of matter. This has led to applications in physics, chemistry, and biology [79], exemplified by a number of applications in biochemistry, such as the use of multiphoton fluorescence for analyzing chemical processes in the skin [80]. In analytical chemistry, non-linear optics is used to analyze crystals, and also for biological applications [81]. The applications described are all nondestructive, i.e., the matter under investigation remains intact.

In fact, the applications of lasers described in Chap. 12 are applications of nonlinear optics. As Bloembergen noted earlier, Maiman's first working laser was itself based on non-linear optics [82]. Bloembergen recalls that as early as 1962 the possibilities for using lasers as weapons were considered. There was a study group on the use of lasers at the Institute for Defense Analysis, where it was stated that 1 Megajoule of energy would be needed to cause significant damage to an aircraft or rocket using light. According to Bloembergen, in order to reach that energy level, one would have to build a neodymium glass laser with a volume of 1 m^3! Moreover, the neodymium glass laser had just been developed in 1962. People did not realize how many problems would have to be overcome to achieve such a thing.

These simple ideas about possible laser applications generated a lot of research.

It was not until decades later that an energy of several hundred kilojoules was reached with the Shiva laser system at the Lawrence Livermore Radiation Laboratory, designed to generate a plasma for nuclear fusion. However, this pulse could only be fired once a day [83].

An important field of application of nonlinear optics is photonics. In the same way that electron information is stored and manipulated in electronics, photons are used in photonics. It is assumed that photons can behave both as particle and wave. Photonics includes the generation, emission (electromagnetic radiation), transmission (telecommunications), modulation, signal processing, amplification (optical amplifier), and detection (photomultiplier) of light. Most applications can be found in visible light. Photonics as a field began with the first working laser in 1960. In the 1970s there followed the laser diode, optical glass fibers for the transmission of information, and the erbium-doped fiber amplifier [84]. The rapid growth of non-linear optics was stimulated by the need for new coherent light sources for optoelectronic devices in the telecom industry [85]. These inventions formed, among other things, the basis for the infrastructure of the Internet. Examples are fiber optics for communications and the converter that converts optical frequencies into microwave frequencies (used with satellite dishes). This also includes the optical computer mentioned earlier [86].

In optical glass fibers, light is transmitted along fibers of optically highly transparent glass in order to transport signals reliably over long distances.

References

1. Armstrong JA. Nico Bloembergen as Mentor in the Golden Age of University Research. In: Levenson MD, Mazur E. Pershan PS and Shen YR. Resonances. A column in honor of Nicolaas Bloembergen. World Scientific, Singapore (1990)
2. Interview Joan Bromberg J and Paul L Kelley met Nicolaas Bloembergen, 27 June 1983. Niels Bohr Library & Archives, American Institute of Physics, College Park, MD USA, www.aip.org/history/ohilist/4511.html
3. Bloembergen N. Nonlinear Optics. World Scientific, Singapore (1996) Fourth Edition
4. Bloembergen N and Damon RW. Relaxation Effects in Ferromagnetic Resonance. Phys Rev 85:699 (1952)
5. Interview Joan Bromberg J and Paul L Kelley met Nicolaas Bloembergen, 27 June 1983. Niels Bohr Library & Archives, American Institute of Physics, College Park, MD USA, www.aip.org/history/ohilist/4511.html
6. Damon RW Relaxation Effects in the Ferromagnetic Resonance. Rev Mod Phys 25:239–245 (1953)
7. Bloembergen N and Wang S. Relaxation Effects in Para- and Ferromagnetic Resonance. Phys Rev 93:72–83 (1954)
8. Dehmelt HG. Modulation of a Light Beam by Precessing Absorbing Atoms. Phys Rev 105:1924–1925 (1957)
9. Bloembergen N. Encounters in nonlinear optics. World Scientific, Singapore (1996)
10. Bloembergen N, Pershan PS and Wilcox LR. Microwave Modulation of Light in Paramagnetic Crystals. Phys Rev 120:2014–2023 (1960)
11. Shen YR. Faraday Rotation of Rare-Earth Ions. I. Theory. Phys Rev 133:A511–515 (1964)
12. Shen YR and Bloembergen N. Faraday Rotation of Rare-Earth Ions in CaF_2. II. Experiments. Phys Rev 133:A515–520 (1964)

13. Shen YR. Faraday Rotations of Divalent Rare-Earth Ions in Fluorides. III. Phys Rev 134:A661–665 (1964)
14. Bloembergen N. Microwave Modulation of Optical Radiation in a Waveguide. United States Patent Office 3.239.670 (1966)
15. Franken PA, Hill AE, Peters CW and Weinrich G. Generation of Optical Harmonics. Phys Rev Lett 7:118–119 (1961)
16. Kaiser W and Barrett CBG. Two-Photon Excitation in CaF_2: Eu^{2+}. Phys Rev Lett 7:229–231 (1961)
17. Lagendijk A. In: Calmthout M van. Al anderhalve eeuw de beste vier vergelijkingen ter wereld. (For over a century and a half the best four equations in the world) De Volkskrant 26 March 2011
18. nl.wikipedia.org
19. Bloembergen N. Encounters in nonlinear optics. World Scientific, Singapore (1996)
20. Bloembergen N Nonlinear Optics and Spectroscopy. Reimpression de Les Prix Nobel en 1981. The Nobel Foundation, Stockholm (1982)
21. Bloembergen N. The Birth of Nonlinear Optics. Nonlinear Optics Conference, Marriot Kauai Beach Resort, Hawaii, 17–22 July 2011
22. Armstrong JA. Nico Bloembergen as Mentor in the Golden Age of University Research. In: Levenson MD, Mazur E. Pershan PS and Shen YR. Resonances. A column in honor of Nicolaas Bloembergen. World Scientific, Singapore (1990)
23. Joan Bromberg interview with JA Armstrong. 17 October 1984. Niels Bohr Library & Archives, American Institute of Physics, College Park, MD USA
24. Bloembergen N. Encounters in nonlinear optics. World Scientific, Singapore (1996)
25. Interview Rob Herber with Pershan, 18 September 2007
26. Armstrong JA, Bloembergen N, Ducuing J and Pershan PS. Interactions Between Light Waves in a Nonlinear Dielectric. Phys Rev 127:1918–1939 (1962)
27. Interview Rob Herber with P Pershan, 18 September 2007
28. Rosenfeld L. Theory of Electrons. Amsterdam (1951)
29. Bloembergen N. Encounters in nonlinear optics. World Scientific, Singapore (1996)
30. en.wikipedia.org
31. Bloembergen N. Encounters in nonlinear optics. World Scientific, Singapore (1996)k
32. Armstrong JA, Bloembergen N, Ducuing J and Pershan PS. Interactions Between Light Waves in a Nonlinear Dielectric. Phys Rev 127:1918–1939 (1962)
33. Interview Frank Elstner with N Bloembergen. Die Stillen Stars. Nobelpreisträger privat gesehen. Heute: Prof. Nicolaas Bloembergen. ZDF-documentary (1987)
34. Bloembergen N. Light Waves Interact. Current Contents 22: Feb 11 (1991)
35. Bloembergen N. Encounters in nonlinear optics. World Scientific, Singapore (1996)
36. nl.wikipedia.org
37. Bloembergen N. Encounters in nonlinear optics. World Scientific, Singapore (1996)
38. Bloembergen N. Light Waves Interact. Current Contents 22: Feb 11 (1991)
39. Born M and Wolf E. Principle of Optics. Pergamon Press, London (1959) First Edition
40. Bloembergen N and Pershan PS. Light Waves on the Boundary of Nonlinear Media. Phys Rev 128:606–622 (1962)
41. Bloembergen N. Encounters in nonlinear optics. World Scientific, Singapore (1996)
42. Bloembergen N and Shen YR. Quantum-Theoretical Comparison of Nonlinear Susceptibilities in Parametric Media, Lasers, and Raman Lasers. Phys Rev 133:A37-A49 (1964)
43. Woodbury EJ and Ng WK. Ruby Laser Operation in the Near IR. Proc IRE 50:2367 (1962)
44. Interview Joan Bromberg with Michael Bass. 29 May 1985, Niels Bohr Library & Archives, American Institute of Physics, College Park, MD USA, www.aip.org/history/ohilist/4496.html
45. Bloembergen N. Encounters in nonlinear optics. World Scientific, Singapore (1996)
46. Interview Rob Herber with N Bloembergen, 7 December 2006
47. Bloembergen N. Encounters in nonlinear optics. World Scientific, Singapore (1996)
48. Bloembergen N. Encounters in nonlinear optics. World Scientific, Singapore (1996)

49. Interview Jeff Hecht with N Bloembergen. 4 November 1984, Niels Bohr Library & Archives, American Institute of Physics, College Park, MD USA

50. Burns WK and Bloembergen N. Third-Harmonic Generation in Absorbing Media of Cubic or Isotropic Symmetry. Phys Rev B4:3437–3450 (1971)

51. Shih H and Bloembergen N. Conical Refraction in Second-Harmonic Generation. Phys Rev 184:895–904 (1969)

52. Schell AJ and Bloembergen N. Laser Studies of Internal Conical Diffraction. III. Second-Harmonic Conical Refraction in α-Iodic Afid. Phys Rev A18 2592–2602 (1979)

53. Bloembergen N and Shen YR. The Coupling Between Vibrations and Light Waves in Raman Laser Media. Phys Lett 12:504–507 (1964)

54. Levenson MD, Flytzanis C and Bloembergen N. Interference of Resonant and Nonresonant Three-Wave Mixing in Diamond. Phys Rev B6:3962–3965 (1972)

55. Kramer SD, Parsons FG and Bloembergen N. Interference of Third-Order Light Mixing and Second-Harmonic Exciton Generation in CuCl. Phys Rev B1853–1856 (1974)

56. Lotem H, Lynch RT Jr and Bloembergen N. Interference Between Raman Resonances in Four-Wave Difference Mixing. Phys Rev B14:1748–1754 (1976)

57. Göppert-Mayer M. Über Elementarakte mit zwei Quantensprungen. Annal Phys 401:273–294 (1931)

58. en.wikipedia.org

59. Levenson MD and Bloembergen N. Observation of Two-Photon Absorption without Doppler Broadening on the $3S$-$5S$ Transition in Sodium Vapor. Phys Rev Lett 32:645–648 (1974)

60. Bloembergen N, Levensson MD and Salour MM. Zeeman Effect in the Two-Photon $3S$-$5S$ Transition in Sodium Vapor. Phys Rev Lett 32:867–869 (1974)

61. Bloembergen N. The Solved Puzzle of Two-Photon Rare Earth Metals Spectra in Solids. J Luminescence 31 and 32:23–28 (1984)

62. webphysics.davidson.edu

63. Bloembergen N. Solid State Quantum Counters. Phys Rev Lett 2:84–85 (1959)

64. Bloembergen N. Comments on the Dissociation of Polyatomic Molecules by Intense 10.6 μm Radiation. Opt Comm 15:416–418 (1975)

65. Bloembergen N. Encounters in nonlinear optics. World Scientific, Singapore (1996)

66. Black J, Yablonovitch E, Bloembergen N and Mukamel S. Collisionless Multiphoton Dissociation of SF_6: A Statistical Thermodynamic Process. Phys Rev Lett 38:1131–1143 (1975)

67. Mazur E, Bural I and Bloembergen N. Collisionless Vibrational Energy Distribution between Infrared and Raman Acive Modes in SF_6. Chem Phys Lett 105:258–262 (1984)

68. Beenakker C en Saarloos W van. In Memoriam: Peter Mazur, 1922–2001. www.lorentz.leiden.nl

69. Interview Rob Herber with E Mazur, 18 September 2007

70. mazur.harvard.edu

71. Mazur E. Peer Instruction. A User's Manual. Prentice Hall, New jersey (1997)

72. Bloembergen N and Zewail AH. Energy Distribution in Isolated Molecules and the Question of Mode-Selective Chemistry Revisted. J Phys Chem 88:5459–5465 (1984)

73. Yablonovitch E and Bloembergen N. Avalanche Ionization and the Limiting Diameter of Filaments Induced by Light Pulses in Transparent Media. Phys Rev Lett 29:907–910 (1972)

74. Liu PL, Bloembergen N and Hodgson RT. Picosecond Laser-Induced Melting and Resolidification Morphology on Si. Appl Phys Lett 34:864-866 (1979)

75. Saeta P, Wang J-K, Bloembergen N and Mazur E. Ultrafast Electronic Disordering during Femtosecond Laser Melting of GaAs. Phys Rev Lett 67:1023–1026 (1991)

76. Interview Rob Herber with F Spaepen, 13 June 2009

77. Miragliotta JA. Analytical and Device-Related Applications of Nonlinear Optics. Johns Hopkins APL Technical Digest 16:348–357 (1995)

78. en.wikipedia.nl

79. Miragliotta JA. Analytical and Device-Related Applications of Nonlinear Optics. Johns Hopkins APL Technical Digest 16:348–357 (1995)

80. Yakovlev VV (ed). Biochemical Applications of Nonlinear Optical Spectroscopy. CRC Press, Boca Raton, Fl (2009)
81. Kissick DJ, Wanapun D and Simpson GJ. Second-Order Nonlinear Optical Imaging of Chiral Crystals. Anal Rev Anal Chem 4:419–437 (2011)
82. Bloembergen N. The Birth of Nonlinear Optics. Nonlinear Optics Conference, Marriot Kauai Beach Resort, Hawaii, 17–22 July 2011
83. Bloembergen N. Encounters in nonlinear optics. World Scientific, Singapore (1996)
84. en.wikipedia.org
85. Miragliotta JA. Analytical and Device-Related Applications of Nonlinear Optics. Johns Hopkins APL Technical Digest 16:348–357 (1995)
86. en.wikipedia.org

Chapter 14
Crowning Glory: The Nobel Prize

"He is of course extremely bright, but he is also very modest in his behavior. He has a child of a moral compass, which keeps him being himself without self-inflation. He is a very generous person," says science historian Gerald Holton about Bloembergen [1].

14.1 Career from the Beginning of the 1960s

1957–1990

In 1957 Bloembergen was appointed Gordon McKay Professor of Applied Physics and in 1960 he became a member of the National Academy of Sciences in Washington (see Chap. 9). He held the McKay chair until 1980. In that period he was also Rumford Professor of Physics for six years. In 1980 he was promoted to Gerhard Gade University Professor, succeeding his former teacher Purcell. The Gerhard Gade professorship was established in 1960 out of the more than $1,000,000 that Gade had bequeathed to Harvard. The instructions for spending this money were as follows: *"to be used as it sees fit. The fund in question […] be used to establish a Chair in memory of my family 'Gade', or to establish scholarships, or for a combination of both purposes."* The President and Fellows of Harvard decided to make $500,000 available for the professorship and put the rest in a Gerhard Gade Fund [2, 3].

In 1990, Bloembergen was granted emeritus status in Harvard.

Ph.D. Students and Postdocs

In 1951, Bloembergen got his first doctoral candidate, Ted Rowland, in his first semester as professor. Rowland obtained his doctoral degree in 1954. Bloembergen's first postdoc was Shyh Wang, who started in 1952. Around 1955 he had six Ph.D. students and two research fellows. In 1965 his group reached its maximum size:

© Springer Nature Switzerland AG 2019
R. Herber, *Nico Bloembergen*, Springer Biographies,
https://doi.org/10.1007/978-3-030-25737-8_14

twelve Ph.D. students and five or six postdocs. From 1983, when he was 63 years old, Bloembergen stopped supervising Ph.D. students. Deli Bloembergen and others felt this was wrong, but Bloembergen claimed he had enough other things to deal with. His last postdoc was Chang-Zai Lu, who came to Harvard in 1987 [4].

Foreign Stays

In 1957 Bloembergen became a guest lecturer at the École Normale Supérieur in Paris.

From 1964 to 1965 he was guest professor at the University of California in Berkeley. There he wrote a number of papers with his former student Ron Shen, who was by then Assistant Professor at Berkeley.

In 1973 Bloembergen was Lorentz Guest Professor at Leiden University and a guest collaborator at the Philips Physics Laboratory in Waalre [5]: "*First we* [Flowers and Deli, RH] *lived six weeks in Oegstgeest; it was a pretty decent apartment and then we lived with Philips in Waalre, and then once a week I went to Leiden for a day to give lectures [...] The atmosphere in Holland was terrible. One third students, one third technical staff, and one third professors. The professors had other things to do, so they never spoke.*" Of course, he also saw Gorter. The latter was already sick but still came to Bloembergen's lectures (Fig. 14.1). Bloembergen recounts that Gorter asked him afterwards: "*Can you show me the way home? It was really pitiful. But he still came to my lecture and we invited him to dinner with Mazur* [the father of Eric Mazur, RH] *and Lila.* Gorter had Alzheimer's disease and when Nico visited him later in the Netherlands, he was already in a nursing home. Gorter's Norwegian wife, Lila, had since withdrawn from the world to live in a monastery. In 1973 their 26-year cooperation and friendship ended [6]. Cor Gorter died in 1980.

During Bloembergen's visit to Leiden he also met Eric Mazur. He was then a sophomore student and went to a lecture by Bloembergen, but didn't understand anything in it [7].

In the autumn of 1979, Bloembergen was Raman Visiting Professor in Bangalore, and at the beginning of 1980, Von Humboldt Guest Lecturer at the Institut für Quantumoptik in Garching and guest professor at the Collège de France in Paris.

Bloembergen was invited to visit the former Soviet Union twice and the People's Republic of China three times. Bloembergen recalls the visits to the Soviet Union in 1967 and 1971 in an exchange program of the Academy of Sciences of the Soviet Union in Moscow and the National Academy of Sciences in Washington. Bloembergen ended up there at the Lebedev Institute (where Prokhorov and Basov were working) and was welcomed there as a V.I.P., long before he had such a reception in Western Europe or the United States [8]. At the time of writing [1996, RH] Bloembergen was probably referring in this remark about Western Europe and the United States to the period after he got the Nobel Prize when he would have been received with the necessary respect everywhere.

In 1967 he also visited the Moscow State University and was received there by Rem V. Khokhlov and Sergei A. Akhmanov who had just received the Lenin Prize for their research. In the hall of the university there was a large mural, where Khohklov and Akhmanow were depicted as young horsemen who triumphantly overcame the fields

Fig. 14.1   Bloembergen trying a laser gun during a 'Science Fair' in Shanghai, China, 1975. *Source* Nico Bloembergen

of non-linear optics. Their research was in the same field as the work done by Bloembergen's group. Khohklov and Bloembergen had met at conferences and privately on many occasions. Bloembergen had invited Khokhlov to his home in Lexington after they had both attended a Gordon Research Conference in New Hampshire. In 1971 Bloembergen and Deli were guests in the Moscow apartment of Khokhlov and his wife Lena. Khokhlov was appointed rector of the Moscow State University, the highest educational position in the Soviet Union. Bloembergen also had a good relationship with Akhmanov. Akhmanov and his wife invited the Bloembergens to their dacha outside Moscow in 1971. Akhmanov also came to the Bloembergens' house after participating in a Gordon Research Conference.

In 1975, the E. Fermi Course 64 on Nonlinear Spectroscopy was held in Varenna, Italy. Bloembergen was course leader and editor of the course reports, the *Proceedings*. Akhmanov was one of the most important lecturers and brought vodka and caviar from Moscow for the evening party. Bloembergen states that these encounters were not only enjoyable, but also an intense intellectual challenge. After the course, Bloembergen was exhausted, so much so that he lacked the inspiration to give the closing speech. Deli made a number of good suggestions, but none really inspired him, so he asked her if she would give the speech, and she agreed. She gave the speech at the banquet: the participants were surprised and she got an ovation when she came to her conclusion. Akhmanov met Bloembergen twice more, in the summer of 1989 in Bretton Woods, New Hampshire, and in 1990 in Irvine, California.

Akhmanov died in 1991. In 1993, a part of the series *Frontiers in Nonlinear Optics* was dedicated to his memory. The Bloembergens honored the many contacts they had made in France, the Netherlands, and Germany.

In 1987 Bloembergen took his last sabbatical, returning to the Institut für Quantumoptik in Garching, which he had visited in 1980. Since his first visit, the institute had moved to a new location [9].

During his stays abroad, for conferences and visits to research institutes, Bloembergen always managed to combine business with pleasure, in the sense that for him the conferences and visits were themselves also part of what was pleasant.

14.2   Research Leader and Teacher

In 1996 Bloembergen wrote an extensive piece about *Teaching in Harvard*.

In this he made the following remark: *Teaching and research are the main duties of a professor. It is often asserted that at famous universities, teaching is considered less important. This statement overlooks the fact that guiding graduate students and even young postdocs in how to carry out research is an advanced form of teaching* [10].

He refers here to what in the Netherlands and Belgium is called research-related education. In the Netherlands there was a rigid separation between education and research. Ph.D. students were no longer given lectures and research-related education was simply not recognized as education by university boards. In the United States, Ph.D. students are called 'graduate students' and the idea of research-related education is rarely ignored.

According to Mazur, it is not the professors at Harvard who are the most important thing for the university, but the students. They are the capital, do the most research, and come up with the ideas [11].

A number of former Ph.D. students and postdocs have described their experience of working with Bloembergen either in written form through interviews. As mentioned earlier, Bloembergen had Ph.D. students from 1951 to the mid-1980s.

John Armstrong, one of his Ph.D. students, tells us the following about the NMR laser period:

I was here as grad student and Research Fellow from '58 to early '63 [...] One trait that we all saw in Nico was his passion for physics. Like all great physicists, he takes physics very seriously. Outside the context of physics, in his personal dealings with students and postdocs, he was quite approachable, one might even say, relaxed. When he stepped into the world of physics, he could be pretty tough on us. Everything about his behavior was suffused with the absolute requirement of high quality. And that's one of the great advantages of working in the group of a great man. You learn what dedication is, what single-mindedness is, and what it can accomplish.

Nico had quite a cast of characters as students. It was a highly international group. There were both theorists, and experimentalists. There were students who have gone on to be corporate managers and executives, there were students who went on to be university faculty, … and there were students who went on to become, as far as I know, dropouts from physics. Some of Nico's students have achievements beyond science and are involved in the international affairs of foreign countries. He was a magnet for a set of students who have not only shown themselves to be very adept at physics, but very versatile in many other areas. To me, this is clear evidence of Nico's talent for leadership.

He, Nico, managed us all; he was the boss, no doubt about that, but I don't believe he derived much ego gratification from exercising his authority. He did not try to embarrass his students with his higher knowledge and ability. Here, again, I see now, he was a valuable role model.

Armstrong also gives a nice description of how things worked between Bloembergen and his Ph.D. students. In the late 1950s and early 1960s there was adequate funding for research. That does not mean that Bloembergen threw money around. For Ph.D. students who worked through the summer, an increase in the fee by $50 to a royal $500 per month was reported to be affordable, and many professors at Harvard did indeed increase the fees. Bloembergen did not. A delegation of the Ph.D. students argued in favor of an increase in Bloembergen's office: he listened attentively and rejected their proposal in a friendly but decisive manner. The Ph.D. students stayed and after Harvard all got impressive careers in industry. Armstrong concludes that they learned to say *No* with Bloembergen, and that that was one of the good sides of working with him [12].

Gordon MacDonald, a geophysicist who had attended Bloembergen's physics classes in the early 1950s, recalls:

At Harvard I initially planned to major in chemistry. I had a disastrous experience, both in the elementary chemistry course and the elementary physics course at Harvard. Both of them were courses that were designed to capitalize on memory as opposed to understanding, and I was turned off. As a matter of fact, during my entire career, I've had half a year in physics. That's all of my formal education in physics. It was so bad, I vowed never to return […] So following that, I took principally geology and math courses, and then later in graduate work sat in on good physicists' lectures - Schwinger, Purcell, and Bloembergen [13].

Speaking about the period in the early 1960s, when Bloembergen went from NMR and masers to nonlinear optics, Peter Pershan recalls:

"Whatever field he was within, he was the master. The older students, that have worked with him, said that he was distant. I never felt that. I always dealt with him in a very good way." [14]

During this period Bloembergen had a postdoc, Chaim Lotem, who had just become a father. His son Zohar had to be circumcised, but there was no family who could be present to act as a godfather. Bloembergen was asked by Lotem and agreed. Just before the event, Bloembergen was questioned by a rabbi about his family background. Unfortunately, his Protestant upbringing turned out to be an insurmountable obstacle and, to the consternation of the Lotems, Bloembergen was not allowed to attend the circumcision. Luckily Eli Yablonovitch, a Ph.D. student of Bloembergen's, was on hand, and he was able to stand in for Bloembergen.

James Wynne, another Ph.D. student of Bloembergen's, stayed on in Harvard in the summer of 1969 after obtaining his Ph.D., and worked with Yablonovitch, then a new Ph.D. student being supervised by Bloembergen. After that, Wynne started working on non-linear spectroscopy at IBM, but was not sure how to deal with certain practical matters and asked Yablonovitch, who was in the middle of his doctorate, for advice. He gave him good advice and the experiments succeeded. However, Wynne could not handle the theory and he again asked Yablonovitch for advice. But Yablonovitch just told him to read *his own* thesis! There he found a passage that perfectly reflected the experiment and the theory behind it:

Now, since I did not remember writing this passage, I must not have thought about it too deeply. It must have been something I put into my thesis because Professor Bloembergen told me to write it. So, even after I had left Harvard and embarked on my own research career in nonlinear spectroscopy, I was being guided by an invisible hand, and the hand belonged to Bloembergen [15].

In January 1982, Eric Mazur came to work as a Research Fellow in Harvard on a one-year appointment with the possibility of extending for another year. Prior to that, he had just been presented with the newly equipped Huygens Laboratory in Leiden with all the latest equipment:

"I remember that just before my PhD, Nico visited Holland. He had arranged then that I would come to work with him and I had invited him to visit the lab. He looked at my setup like this: I had a superconducting magnet in a beautiful lab and he was extremely impressed. He said, "Well, I don't believe I can offer you these kinds of facilities in America. I thought he was joking. I couldn't imagine that."

Mazur had rather naive preconceptions about the United States:

"I had never been there and thought the US was the most modern and efficient country in the world. I arrived at the airport on Saturday and on Monday we went early in the morning to Pierce Hall, an old building. I thought, boy, is this Harvard University? Nico's office was Spartan. I got a small office and he sent me some papers. I tried to read them, but I did not get much further than the second paragraph. I simply lacked too much knowledge in the field of spectroscopy and optics. Then I went to the lab and I couldn't believe my eyes. The table was just a piece of metal on wood. It was a mess, a mess with equipment from the fifties! Nico did not believe in purchasing new equipment. Everything had to be reused, simple, cheap. What struck me was that the PhD students here knew an incredible amount more than in the Netherlands. In the Netherlands, if you obtained your PhD, you did not have to attend any lectures at all. That was the major disadvantage of the study program in the Netherlands: you could never make the right connection between the lectures and research. Only inductive thinking, not deductive thinking, and then research. Here [Harvard, RH] people were able to make clear connections between theory and their research and between theory and other disciplines. So I felt terribly handicapped. Here I am, doctor Mazur and the students know more than I do. Fortunately, I soon noticed that I had more experience in research and that I could write better. After six months, I started to like the research and I just got results. Nico said, 'If I were you, I'd go looking for a permanent job.' I thought: is he dissatisfied? Did I not do it right? You never knew if he was angry. It was very difficult to judge whether he was happy or

not. I was shocked. I went to the lab, where the students were not surprised, because they had had exactly the same kind of relationship as me. Those students didn't see Nico at all, because he had just won the Nobel Prize and travelled all the time. His relationship with the students was not warm. They were very much in awe of him and were afraid to speak. It really was the master and the students. The only people he spoke to were the postdocs, of whom there were only three. Probably Nico was not sure that he could keep me in service. He had to write a research proposal and although that was then no more than two or three paragraphs, Nico very much hated doing that. Fortunately, there was still money coming in and I could, I thought, stay another year and then go back to the Netherlands."

In the end, Mazur liked the work so much that he stayed. Bloembergen was gradually pulling back and was no longer working on spectroscopy. He was also working on the Star Wars project in 1984 (see Chap. 15). Mazur comments:

"I was in the Physics Department, because I was far too afraid I would stay in Nico's shadow if I stayed in Applied Physics. As a result, my interaction with him was greatly reduced. Too bad, in hindsight, but for my own career it was best. Only much later, when I got a permanent position, did I meet him again regularly. He is a great man. I like Nico very much. An incredibly intelligent man. He is sometimes grumpy, but in an outward way. It seems as if I'm contradicting myself, but he's an extremely open, friendly man, a very warm person. He does have a kind of façade and way of doing things that looks rather grumpy.

Non-linear optics started with Armstrong, Pershan, and Bloembergen. That is Bloembergen's main contribution. His name is immediately associated with nonlinear optics. Nico's most creative period was the early period, the sixties. Compared to the early period, the seventies and eighties were not as creative.

Perhaps his biggest handicap was that he couldn't properly assess your level. He was not the best communicator." [16]

Peter Franken, who was the first to demonstrate a non-linear effect in optics (see Chap. 13), was another who did not exactly have a positive opinion about Bloembergen's didactic abilities: *"He also did the so-called first book on nonlinear optics, which I think is a poor book because Nico gave lecture notes, and just transcribed them, but he didn't work hard. That was only a 20-page review paper John* [John Ward, PhD of Oxford, RH] *and I did, and we worked all summer on it, because a review paper is supposed to teach, and not claim, it's supposed to teach. So, I really felt it was a deficiency in Bloembergen's group, they were lousy teachers, but Nico's contributions on the theoretical side of the science were outstanding and very important."* [17]

Gerald Holton (born 1922) is a physicist and science historian at Harvard. He obtained his doctorate in 1948 and was soon given a permanent appointment. He was therefore a long-standing colleague of Bloembergen's. He praised the research climate for physicists in Harvard. He stated that a certain critical mass must be present for successful research, and that physics in Harvard was extremely successful, with 10 Nobel Prize winners. According to him, this was because Harvard distinguished itself from other institutions. The researchers felt valued and had a good relationship with each other: *"Nico had a very good relationship with Purcell and Pound."*

According to Holton, other institutions attracted people who had practiced nothing but their profession throughout their lives, at the expense of their social lives: *"They were busy with differential equations while others went to the movies. They are really loners for a long time. And then the chairman of the department says: 'Now, work together'. That is the first time they heard this, and they don't know how to do it."*

About Bloembergen says Holton: *"He is of course extremely bright, but he is also very modest in his behavior. He has a kind of a moral compass, which keeps him being himself without self-inflation. He is a very generous person."* [18]

Frans Spaepen came from the Catholic University of Leuven and graduated as a metallurgical engineer. He ended up with David Turnbull (1915–2007), Gordon McKay Professor of Applied Physics in Harvard, and got to know Bloembergen there. He obtained his Ph.D. in 1975. After that Spaepen became Assistant Professor and in 1983 he got his permanent appointment in the School of Applied Physics and Engineering and became a colleague of Bloembergen's. The contact between Spaepen and Bloembergen dated from 1979 to 1980. They collaborated in a project in which picosecond laser pulses were used to study the behavior of layers of semi-conductors and metal surfaces. According to Spaepen, Bloembergen had realized immediately after developing the fast laser pulses that applications were possible in the interaction with surfaces of different materials [19]. This opened a new research area.

In Harvard, Bloembergen not only taught Ph.D. students by supervising their research, but also taught so-called 'undergraduate courses' at the rate of half a course per semester. This involved giving three lectures per week. Bloembergen wrote about this in 1996: *I have, as have most of my colleagues, taken this obligation seriously, and have taught both undergraduate and graduate courses. The best way to really learn a subject well is to have to teach it to a critical audience of students. In 1951 a new undergraduate course, "Introduction to Solid State Physics", was called for. I volunteered to teach this course and spent a large fraction of my time on it during my first semester on the faculty. There was no suitable textbook available at that time.* He then states that he had to compile everything for the lecture from various books describing partial aspects of the subject and that he had benefited greatly from this 'course' for his own research on NMR in solid matter. Later courses such as alternating current circuits and thermodynamics at the end of his career gave him a lot of insight into the subjects he was dealing with. He did state, however, that the first time he gave a certain subject, it required a lot of work and was hard for the students. The second time he discussed a subject, the errors and problems would have been taken out of the subject matter, and the third time, everything would run smoothly. He would then start to get bored and ask for another course.

Bloembergen stated that he takes his teaching obligations very seriously and tried to avoid absences as far as possible due to his frequent travels. If there was no other way, he would provide a substitute. During his forty years of teaching at Harvard, he had the good fortune not to be forced to miss a single time because of illness. He assessed exams and theses himself, even if forty or fifty students had to be assessed at the same time.

Being involved in teaching helps Bloembergen to get a clearer view of research presented at conferences and colloquia [20].

Experimental or Theoretical Physicist?

Is Bloembergen now considered as an experimental or a theoretical physicist?

He himself said the following about this in 2006:

"My strength is: I'm not a topper with experiments and I'm not a topper in theory, but the combination is pretty good." [21]

In an interview with Bloembergen by Bromberg and Kelley, the latter made the following remark: *"It's interesting, (perhaps Nico would like to say something about this), my observation is that quantum electronics is an experimentalist's field, and that a lot of the good theory that has been done has been done by the experimentalists because they want to solve practical problems related to their experiments; and there has been really not that much fundamental that's come out of the theoretical work that's been done by theorists in the field."*

Bromberg stated: *"Now, your work is both experimental and theoretical."*

Bloembergen responded: *"But I consider myself an experimentalist. But you know, the theorists didn't carry it enough. And didn't foresee all the implications."* [22]

Pershan (who was first a Ph.D. student and later became a colleague of Bloembergen's) gave the following opinion in an interview: *"Clearly, from the time I knew him I never saw him touch a piece of equipment. His thesis was different. But if you look at his students, all went to careers that went on to experimental physics. In that sense, he was an experimental theorist. Clearly, he contributed to theoretical ideas, what was very important."* [23]

Gerald Holton felt that Bloembergen moved in both areas [24].

The fact is that Bloembergen received the Lorentz Medal for researchers in the field of theoretical physics in 1978.

Administrative Obligations

Bloembergen was pleased that the administrative obligations associated with working as a researcher were kept to a minimum in the 1950s and 1960s [25]. The interview with Mazur already showed that Bloembergen hated writing research proposals [26].

The Bloembergen laboratory occupied half the second floor of the Gordon McKay building. The approximately 470 m^2 that he had were sufficient for his research and he never needed extra space over time, so that never led to any problems. Research funding was no problem either. Bloembergen preferred experimental research on a small scale that did not require large investments. Moreover, at the beginning, everything was 'stuck together with sticky tape' and he only renewed his equipment when his colleagues insisted [27]. This was abundantly clear in the interview with Mazur mentioned previously [28].

He started out in 1949 as a participant in the US Joint Services Electronics Program in Harvard under the management of E. L. Chaffee. This programme was financed by the Ministry of Defense. The research on magnetic resonance was gladly accepted by the program leader. Bloembergen never actually had to write research proposals, only progress reports.

When, five years later, in 1954, Bloembergen began researching the maser, the liberal research policy of the Ministry of Defense had hardly changed. In the early 1960s, Bloembergen started his research on nonlinear optics; again he did not have to submit a formal research proposal. The results of the study were simply presented at the annual meeting of the Joint Services Electronics Program (JSEP).

Bloembergen would also receive additional research funding from the DARPA Materials Research Program at a certain point in time.[1]

Research funding changed in the 1970s. In the future, Bloembergen would have to make research proposals that did not yet have much scope, but would also receive smaller grants. In addition to JSEP, NASA was now subsidizing his research. He was able to do the research he wanted to do after the results of preliminary research of the previous year had been submitted in lieu of a research proposal. The subsidy obtained was then used by Bloembergen to conduct new research. This new research would subsequently be used as a research proposal for the following year. Even later Bloembergen only hired Ph.D. students and postdocs if they brought their own money.

Bloembergen was director of the Joint Services Electronics Program from 1967 to 1983, so his administrative burden increased proportionally during that period. In November 1983 he received a medal because he had been employed for sixteen years [29].

Bloembergen financed all his research from external funds. The first source of money was the university itself. In Harvard this meant that the university paid part of the professor's salary for teaching (in summer there were no students and therefore no payment). Harvard also paid for the building. So, Harvard did not pay anything for the research. Up until the 1970s, Bloembergen financed this entirely from the aforementioned grants, mainly from the Ministry of Defense. According to Bloembergen, these grants were not limited by Congressional supervision, but were awarded after evaluation by highly qualified experts with a broad perspective. In this context, Bloembergen referred to an example from the 1960s. While the Cold War made it impossible to hire a postdoc through the National Science Foundation (NSF), the Office of Naval Research succeeded. They did say, according to Bloembergen: *"Don't advertise it at the front page."* [30]

Bloembergen never had a subsidy from the NSF.

It is clear that Bloembergen's research was mainly financed by the Ministry of Defense. Didn't that cause any problems with Harvard? After all the university had a strict policy that no secret research should be carried out (Fig. 14.2). Everything had to remain public. That meant that in Bloembergen's case everything had to be published without the intervention of the Pentagon. Bloembergen stated that, although it was clear that research into magnetic resonance and lasers still had important military significance, he never felt uncomfortable about this or had the idea that he should

---

[1]DARPA is the Defense Advanced Research Projects Agency. This institution was set up in 1958 *"to prevent strategic surprise from negatively impacting U.S. national security and create strategic surprise for U.S. adversaries by maintaining the technological superiority of the U.S. military."* (website www.arpa.mil).

Fig. 14.2 Deli and Nico in 1979. Deli graduated from Harvard Extension School. *Source* Nico Bloembergen

apologize for it. He pointed out that there had also been important civilian applications of his research (described previously in separate chapters).

Bloembergen did the usual administrative work, standing on committees involved in appointments, promotions, applications, etc. Harvard's directors realized that Bloembergen did not want to do more administrative work than was strictly necessary and took this into account. Bloembergen was on the faculty library committee, a job that didn't take much time.

When the student revolts also reach Harvard in 1967, Bloembergen played an active role in the faculty's negotiations with the students about their grievances. A faculty council was established and Bloembergen was a member of this council for one and a half years. Some of the students' demands were met, but others, such as the demand for representation on the board and appointments committee, were rejected. Academic freedom of expression was enshrined and distinguished from academically intolerable behavior. Student unrest at Harvard was short-lived compared with many other institutes. According to Bloembergen, improvements to the curriculum were introduced as a result of the students' requirements and student contacts with the faculties were improved [31].

14.3   Intermezzo: Scientific Intuition

In Chap. 2 it was asked whether the high cognitive skills observed at primary school and in secondary education are a question of aptitude or upbringing. It was further stated that Bloembergen had a high intelligence or talent, but also that, while these matters are indicative of further development, they can at most be considered a potential for achievement in new areas.

It should be clear that Bloembergen was indeed an achiever. His achievements include experimental and theoretical physics, or more generally, scientific achievements in a natural science. How does such a performance come about? Are there general principles that underlie it, such as diligently working towards an aim with a considerable amount of hard work? Or is it more a question of having a creative approach, which then raises the question of whether there is such a thing as scientific intuition?

There are many misunderstandings about intuition. Myers discusses this in his book: *Intuition. Its Power and Perils*, asking: *Should we follow the example of Star War's Luke Skywalker, by turning off our computers and trusting the Force?* [32]

Intuition is important, but it can also turn out to be completely wrong. Well-known examples are the visual illusions. Two line segments of the same length seem to be different if the arrows point in opposite directions. Our intuition is refuted by the truth. And incidentally, a physicist will never fall for this illusion: he/she will always measure!

A recognized master in this kind of perceptual deception was the graphic artist Maurits Escher (1898–1972). Escher's impossible cuboid is a well-known example of his work.

There are countless examples of the refutation of intuition in the history of science. The fact that the Sun describes an arc in the sky every day could indicate that the Earth is standing still and the Sun is circling around it, or that the Sun is standing still and the Earth is circling around it. According to Myers, intuition preferred the first option. It was only with Aristarchos in the fifteenth century, Copernicus in the sixteenth century, and Galileo Galilei in 1610 that it became clear that the Earth revolved around the Sun.

Myers gives a nice example of intuitive expertise and creativity in mathematics. Andrew Wiles is a British mathematician who proved Fermat's last theorem (already conjectured in the seventeenth century!). Wiles had been turning around the proof for thirty years, until he suddenly had an insight: "*It was so incredible beautiful; it was so simple and elegant. I couldn't understand I missed it and I just stared at it in disbelief for twenty minutes.*" [33] Of course, Wiles had developed a lot of knowledge about the subject in those thirty years. Incidentally, his insight still had to be elaborated into a proof of more than 100 pages [34]. That too is typical: creativity alone is not enough; we also have to do hard work.

According to Robert Sternberg and Todd Lubart, quoted in Myers' book, creativity is made up of a number of different elements. Louis Pasteur said: *La chance ne sourit qu'aux esprits bien préparés* (Luck will favor those who are well prepared). Malcom

Gladwell formulated what he called the *10,000-hour rule*. Gladwell was discussing the career of William Nelson (Bill) Joy, co-founder of the computer company *Sun Microsystems*[2] in 1982. Joy worked hard on computer programs during his studies in Michigan and Berkeley: "*So, so maybe... ten thousand hours? That's about right.*" [35]

Bloembergen had a similar opinion. Asked what is important for students who want to study science, Bloembergen answered: "*Pursue what interests you have and work very hard on it.*" [36]

Hard work, that's what Bloembergen was doing in 1946, when he was carrying out experiments in his laboratory. He often stayed there until midnight and the laboratory became his second home [37]. Even when he fell ill with flu in the Christmas holidays of 1946, he went on working on the theory.

The ten thousand hours of Joy and Gladwell turn out to be a kind of gold standard. The more 'building blocks' are collected, the greater the chance that something will fit.

A second part of creativity, according to Sternberg and Lubart, is the ability to imagine something that is not yet a reality. Once a problem is understood, it can be redefined or a new solution can be found by looking at it in a different way. Bloembergen found his three-level pumping system in this way. First, he thoroughly mastered the practical aspects of masers and then the solution came out of the blue, as it were, from an original thought.

The third part of creativity is an enterprising personality, ready to take risks, tolerate ambiguity, chew away at difficulties, and never give up after failure.

Bloembergen had tried to become a general physicist in 1950–1951. It was a waste of time, but he was eventually able to work on NMR again at Harvard.

The fourth part of creativity is basic motivation. People feel more motivated when challenged by work because it is interesting and satisfying and can give great joy when a problem is solved, as with Wiles who proved Fermat's last theorem. Bloembergen also pointed to motivation as a key factor [38].

The fifth element conducive to creativity is a creative environment. At the end of the 1940s and 1950s, Bloembergen was surrounded by a number of Nobel Prizewinners and future Nobel Prizewinners. In his NMR and maser period, these were Purcell, his teacher and Nobel Prizewinner (1952), Pound, the electronics wizard, Ramsey (1989), Van Vleck (1977), Schwinger (1965), all in the NMR/maser period, with Nobel Prizes in the pipeline. In addition, from outside Rabi (1944) and Bloch (1952) were regular visitors. In his period working with masers and non-linear optics, the Nobel Prizewinners Townes (1964) and Schawlow (1981) were also present, along with many others. Altogether, this made for a highly stimulating environment. Bloembergen also pointed to the competition with others as an important factor [39].

---

[2] Sun Microsystems is now known for the Java programming language, an open language, which can be found on practically all Windows and Apple computers, as well as on mobile applications such as Android from Google.

Myers concludes: *Intuition is nothing more and nothing less than recognition* [40]. Nelissen builds on that: *Successful intuitions arise from organized knowledge and rich experience* [41].

Before the Second World War, Bloembergen had already decided that he wanted to do his doctorate in the United Kingdom, France, or Germany. In the summer of 1945, his brother Evert advised him to seek refuge in the United States, because everything in Europe had been destroyed. In 1945 Bloembergen wrote to three American universities on the basis of physics journals he had used before May 1940 in Utrecht. He chose some subjects 'at random' and didn't really care which university it would be, provided good research could be done. There was little or no scientific intuition there: the only criterion for Bloembergen was good research prospects, and not the actual subject.

Upon his arrival at Harvard University in February 1946, Bloembergen was given the choice of what research he wanted to do, and he only had 10 days to make that choice. He chose nuclear magnetic resonance. He knew something about nuclear magnetism, but more importantly, Purcell had discovered nuclear magnetic resonance in solid matter just six weeks earlier. Within the possibilities available to him, Bloembergen faultlessly selected the research that offered the most prospects. I attribute this to his scientific intuition, which, despite the little knowledge he had of the field at the time, did not let him down.

Bloembergen worked himself in with phenomenal speed. After just two weeks at Harvard he was already able to do experiments with Purcell and Pound and he was working on groundbreaking research.

On 17 May 1956, Bloembergen attended a lecture about a new amplifier for radio frequencies and microwaves. He became interested and understood the importance of a low noise amplifier. Based on his research on nuclear magnetic resonance and knowledge of quantum mechanics in relation to inverse populations, Bloembergen was able to see what such an amplifier should look like. He published the three-level mechanism for microwave frequencies, but he also foresaw that the system could be used for higher frequencies. In the end, this was what made the development of a usable laser possible. Here we see scientific intuition at work: preliminary work on general knowledge and his own research, quickly converted to another area for a completely new idea, and the vision to take the idea far beyond the beaten path.

Another example of Bloembergen's scientific intuition relates to Franken's lecture on non-linear optical effects in 1961. Bloembergen developed the theory of nonlinear optics in a few months and presented the results in a magisterial article. He then entered a field that was completely new to him, namely optics, working at phenomenal speed, taking material from Utrecht as a basis, and going on to describe completely new phenomena.

His post-doc Richard Damon made the following remarks about Bloembergen's insights:

"1-9-50 Talked to Nico Bloembergen; he knows all the answers. This entry in my notebook summarizes the almost mystical insight that Nico had for physical phenomena." [42]

In an interview, Eric Mazur tells us: *"He often asked very good questions. He knew a lot, he could make connections between different areas."* [43]

14.4 The Icing on the Cake: The Nobel Prize

As mentioned earlier, Bloembergen was surrounded by Nobel laureates since his first arrival at Harvard. This was not only important as a creative environment, but was a rare privilege that would be difficult to find in other universities in Europe and also in the United States, except in a very limited number of institutes. This meant that achieving the 'prize to beat all prizes' was not a matter of course, but something that could come within reach.

Bloembergen's work immediately attracts attention with the BPP publication, but he was only a 'graduate student', and although he was mentioned as the first author, the honor rightly went to his teacher Purcell. The latter had been the first to demonstrate nuclear magnetic resonance in solids just six weeks before Bloembergen's arrival in Harvard. Purcell received the shared prize with Bloch in 1952 *for their development of new methods for nuclear magnetic precision measurements and discoveries in connection therewith.* Purcell stated in his speech at dinner: *The Nobel Prize, so long regarded in our science as the highest reward a man's work can earn, must bring to its recipient a most solemn sense of his debt to his fellow scientists and those of the past.* In his speech he mentioned how important it was to re-establish former links between laboratories and also mentioned the Kamerlingh Onnes Laboratory in Leiden, of course, because of Gorter and Bloembergen [44]. The latter believed that Purcell received the prize partly on the basis of the BPP publication. Because of this, some of the honor also shines on Bloembergen. Purcell did wonder whether Bloembergen might have felt missed out, but Bloembergen replied that his permanent appointment at Harvard in 1951 was reward enough [45].

In 1958 Bloembergen received his first important prize: the Oliver E. Buckley Condensed Matter Physics Prize. This is awarded annually by the American Physical Society: *To recognize and encourage outstanding theoretical or experimental contributions to condensed matter physics* [46].

In 1959 Bloembergen received the Morris N Liebermann Memorial Award, this time together with Townes: *For fundamental and original contributions to the maser.* This award is presented by the Institute of Electrical and Electronical Engineers (IEEE) [47]. Significantly, Townes and Bloembergen were awarded an equal share by this organization.

In 1961 Bloembergen received the Stuart Ballantine Medal of the Franklin Institute for the solid-state laser from the Philadelphia science institute of the same name. The medal has been awarded since 1824: *Among science's highest honors, The Franklin Institute Awards identify individuals whose great innovation has benefited humanity, advanced science, launched new fields of inquiry, and deepened our understanding of the universe* [48].

In 1964, the Nobel Prize in Physics was awarded to Townes, Basov, and Prokhorov. In Chap. 11 about the maser and Chap. 12 about the laser, much has already been said about the resulting controversies. It was argued there that Bloembergen was certainly eligible for the prize then.

In 2001, a book was published on the attribution of the Nobel Prizes in physics and chemistry, written by *Richard Friedman: The Politics of Excellence: Behind the Nobel Prize in Science.* As the title suggests, it argues that the award of the prize is also a political matter: science politics. Controversies between the University of Uppsala, the Royal Academy in Stockholm, and the University of Stockholm (all involved in the award ceremony) play a role alongside nationalism (now it's the turn of a Swede), provincialism (we don't believe in the theory of relativity), arrogance, sexism, hostility, inconsistencies, rivalries, and many of the darker sides of human nature. The titles of some chapters are illustrative: *Permanent Battles Will Surely Be Waged for Every Prize, Einstein Must Never Get a Nobel Prize!, Sympathy for an Area Closely Connected with My Own Specialty,* and *Completely Lacking an Unambiguous, Objective Standard* [49].

In 1974 Bloembergen received an important prize: the National Medal of Science. It is awarded by the President of the United States and was established by the House of Representatives in 1959. Each year, twelve to fifteen medals are awarded, until 1979 only in physics, chemistry, biology, mathematics, and engineering, but subsequently also in the social sciences. The National Science Foundation receives nominations and sends them to a committee for review. Bloembergen got the medal of President Gerald Ford: *For pioneering applications of magnetic resonance to the study of condensed matter and for subsequent scientific investigations and inventions concerning the interaction of matter with coherent radiation* [50].

In 1977, a asteroid was named after Bloembergen.

In 1978 Bloembergen received the Lorentz Medal from the Royal Academy of Sciences and Arts[3] in Amsterdam. The medal is awarded every four years to researchers in theoretical physics and is the highest distinction awarded in Dutch physics [51]. Bloembergen was only the fourth Dutchman or physicist of Dutch descent to receive the medal. In 1935 the medal was awarded to Peter Debye (then in Germany), in 1947 to Hendrik Kramers, and in 1970 to George Uhlenbeck (then in the United States). John van Vleck received the medal in 1974.

On 26 September 1978, biologist J. Lever, chairman of the Department of Physics, welcomed Bloembergen, his wife Deli, and their daughter Juliana. Joan van der Waals, professor of experimental physics at the University of Leiden, gave an explanation of the award: *"What is particularly pleasing to the jury is that there is such a clear path from the fundamental work carried out by Lorentz to your own contributions to physics. A path you are clearly aware of and which passes through*

---

[3]The Royal Netherlands Academy of Arts and Sciences (in Dutch KNAW) was founded in 1808 as an advisory body to the Dutch Government—a role that it continues to play today. The Academy derives its authority from the quality of its members, who represent the full spectrum of scientific and scholarly endeavor and are selected on the basis of their achievements. It is also responsible for fifteen internationally renowned institutes whose research and collections put them in the vanguard of Dutch science and scholarship. www.knaw.nl.

the work of previous laureates, such as Debye, Kramers, and Van Vleck." Van der Waals went on to say that what is called 'luck' in physics is in fact the reward for the few who combine detective thinking with real craftsmanship: *"Not only with your fortunate start in the United States, but also later on, you have repeatedly shown that you know where to find new, rich veins. [...] In an astonishingly short time after the discovery of nuclear resonance, you established yourself as a physicist in the classic article with Purcell and Pound: 'Relaxation effects in nuclear magnetic resonance absorption'."* With regard to the development of masers and lasers, Van der Waals noted that, just six weeks after the publication of Bloembergen's article on the solid-state maser, a maser application appeared on the scene (from Bell). And just a year after Bloembergen's publication, the ruby laser was already being used in a radio-telescope in Harvard and Maiman was developing the ruby laser: *"In both cases there is a practical application of your brilliant suggestion."* Van der Waals continued with the observation that the development of nonlinear optics *"to which Dutch researchers have unfortunately hardly contributed"* was an example of concrete physics in which researchers could no longer be divided into theorists and experimenters: *"But the systematic way in which you managed to weave the various threads into a single fabric, and thus provided nonlinear spectroscopy with a consistent basis, is a major achievement in a field of physics where you were preceded by Maxwell, Lorentz, and Kramers."* [52]

After the award ceremony at Kloveniersburgwal, there was a dinner in a restaurant on the Leidseplein. Bloembergen, Deli, and their daughter Juliana were present, as were Eric Mazur (corresponding member of the RAS) and Dirk Polder (professor at the Delft University of Technology, and member of RAS), but their wives have to eat somewhere else, to Bloembergen's annoyance! *Four years later, the RAS made further savings when my French colleague Anatole Abragam noticed that the medal awarded to him was too light for 14 carat gold. The medal had in fact been devalued to gilded silver* [53].

In 1979 Bloembergen became Frederic Ives Medalist. This award is issued every two years and is granted *for distinguished work in optics* [54].

In 1980 Bloembergen received the Humboldt Research Award. The aim of the award, which is given to a maximum of a hundred people a year, is as follows: *The award is granted in recognition of a researcher's entire achievements to date and to academics whose fundamental discoveries, new theories, or insights have had a significant impact on their own discipline and who are expected to continue producing cutting-edge achievements in the future* [55]. In 1987, Bloembergen received the award again.

In 1981 Bloembergen's time finally came and he was awarded the Nobel Prize.

The press release of the Swedish Royal Academy of Sciences went as follows: *The Royal Swedish Academy of Sciences has decided to award the 1981 Nobel Prize in Physics by one half jointly to Professor Nicolaas Bloembergen, Harvard University, USA, and Professor Arthur L. Schawlow, Stanford University, USA, for their contribution to the development of laser spectroscopy, and by the other half to Professor Kai M. Siegbahn, Uppsala University, Sweden, for his contribution to the development of high-resolution electronic spectroscopy.* The press release went on to say the following about Bloembergen: *A(nother) class of non-linear, optical methods of spectroscopy is based on the mixing of two or more light waves. This type of phenomenon was demonstrated shortly after the laser was introduced, and the theory for it was comprehensively explored around the same time by Nicolaas Bloembergen and his co-workers. Of particular interest is four-wave mixing where three coherent light waves act together in generating a fourth light wave. By this method it is possible to generate laser light far outside the visible range, in both the infrared and the ultraviolet directions. The method has thus drastically extended the range of wavelengths accessible to laser spectroscopy studies* [56] (Fig. 14.3).

Bloembergen had the following opinion about the granting of prices: *I think I was awarded the 1981 Nobel Prize for laser spectroscopy, but that was partly for the work on the maser, which they should have appreciated as early as 1964, but by then they already had three people: Townes and Basov and Prokhorov. You cannot give more than three. If they had given four in 1964, I would have received it in 1964* [57].

Fig. 14.3 King Carl Gustav of Sweden awarded the Nobel Prize to Bloembergen in 1981. *Source* Nico Bloembergen

In Stockholm, not only Deli, but also the children Antonia, Brink, and Juliana were present. Nico's brothers, Herbert and Auke Reitze, were also present, led by Evert.

On 8 December, Bloembergen gave the *Nobel Lecture*, focusing on his scientific work.

In his lecture he considered nonlinear precursors in magnetic resonance, nonlinear optics, nonlinear spectroscopy, mixing of four wavelengths of laser beams, nonlinearity of higher orders, and transparency in optics [58].

On 10 December, the day of the award ceremony, Ingvar Lindgren of the Swedish Royal Academy of Sciences delivered the following reflection on the work of Bloembergen:

Laser light is sometimes so intense that, when it is shone on to matter, the response of the system could not be described by existing theories. Bloembergen and his collaborators have formulated a more general theory to describe these effects and founded a new field of science, we now call non-linear optics. Several laser spectroscopy methods are based upon this phenomenon, particularly such methods where two or more beams of laser light are mixed in order to produce laser light of a different wave length. Such methods can be applied in many fields, for instance, for studying combustion processes. Furthermore, it has been possible in this way to generate laser light of shorter as well as longer wave lengths, which has extended the field of application for laser spectroscopy quite appreciably [59].

Bloembergen gave a speech at the banquet on behalf of the three prizewinners. He referred to the contribution of the laser to improvements in communication technology (Fig. 14.4). He also pointed out that communication would become increasingly important and increasingly influence people's lives. He mentioned some applications of the laser and ended with the fact that science, like Nobel's dynamite, could be used both for good and for evil [60].

Asked what he thought was his most important contribution to the award of the Nobel Prize:

Oh, my own opinion is that the pumping principle is very important. And I think I have received part of the Nobel Prize for the work of 1956, although they do not mention it. I received the Nobel Prize shared with Schawlow for laser spectroscopy. I did mention nonlinear optics and spectroscopy in my Nobel Prize lecture, but I think they also wanted to reward me in part for my work in 1956, and that is well described there. The third person, Kai Siegbahn, who got half the prize, that was for a completely different area. There was clearly a compromise in the selection committee; some wanted the Swede and electron spectroscopy and others optical laser spectroscopy. This split the subject into two parts, two different parts; half goes to Siegbahn, Kai Siegbahn, and Art Schawlow and I only got a quarter each [61].

In 1983 Bloembergen received the IEEE Medal of Honor. This is the highest distinction awarded by this institute, presented annually to one person. IEEE stands for Institute of Electrical and Electronics Engineers, a huge organization with more than 400,000 members. The medal is awarded *for an exceptional contribution or an extraordinary career in the IEEE fields of interest*. Bloembergen got the prize *for*

Fig. 14.4 Family gathering in Stockholm during the time of the Nobel Prize awards. From left to right: Daughter Antonia, Nico, son Brink, Deli, and daughter Juliana. *Source* Michael Salour

pioneering contributions to Quantum Electronics including the invention of the three-level maser [62]. So, after 27 years there was finally recognition for Bloembergen's pump principle!

Also, in 1983, Bloembergen became Commander in the Order of Orange-Nassau of the Kingdom of the Netherlands. In the classification of this order, Commander takes third place in a hierarchy of six. An appointment in the Order is granted for special contributions to society.

The year 1983 was a good one for Bloembergen as far as prizes were concerned. He also got the Dirac Medal. This prize is awarded annually by the *University of New South Wales, Sydney, Australia, jointly with the Australian Institute of Physics on the occasion of the public Dirac Lecture* [63].

In 1989 Bloembergen was awarded the Alexander von Humboldt Medal.

In 2001 Utrecht University named a building after Bloembergen, one in which nuclear magnetic resonance (NMR) is applied to fundamental research in chemistry.

The Russell Varian Prize in 2005 would be the last: *The Russell Varian prize honors the memory of the pioneer behind the first commercial Nuclear Magnetic Resonance spectrometers and co-founder of Varian Associates. The prize is awarded to a researcher based on a single innovative contribution (a single paper, patent, lecture, or piece of hardware) that has proven of high and broad impact on state-of-the-art NMR technology.* This prize has only been awarded annually since 2002, and so far only to professors with an emeritus degree. Bloembergen received it for his dissertation and the BPP publication on nuclear magnetic resonance: *The awarded paper proposed a semi-quantitative prediction for Bloch's relaxation times T1 and T2, based on an appropriate adaptation of transition probability theory (as originally presented by Weisskopf and Wigner) combined with the assumption that relaxation is dominated by the effects of molecular Brownian motion on a "fluctuating local field" acting on each spin. The paper introduced the notion of "motional narrowing" and established NMR as an essential tool for the experimental study of molecular motion, a situation that still persists today* [64].

References

1. Interview Rob Herber with G Holton, 21 September 2007
2. Bentinck-Smith W and Stouffer E. History of Named Chairs. Harvard University, Cambridge (1991)
3. Harvard Class of 1921. Twenty-fifth Anniversary Report. Harvard University, Cambridge (1946)
4. Bloembergen N. (ed) Encounters in Magnetic Resonances. World Scientific, Singapore (1996)
5. Bloembergen N. Encounters in nonlinear optics. World Scientific, Singapore (1996)
6. Interview Rob Herber with N Bloembergen, 7 December 2006
7. Interview Rob Herber with E Mazur, 18 September 2007
8. Bloembergen N. Encounters in nonlinear optics. World Scientific, Singapore (1996)
9. Bloembergen N. Encounters in nonlinear optics. World Scientific, Singapore (1996)
10. Bloembergen N. (ed) Encounters in Magnetic Resonances. World Scientific, Singapore (1996)
11. Interview Rob Herber with E Mazur, 18 September 2007
12. Armstrong JA. Nico Bloembergen as Mentor in the Golden Age of University Research. In: Levenson MD, Mazur E, Pershan PS and Shen YR (eds) Resonances. World Scientific, Singapore (1990)
13. Interview Finn Aaserud with G MacDonald, 16 April 1986, Niels Bohr Library & Archives, American Institute of Physics, College Park, MD USA, www.aip.org/history/ohilist/4754
14. Interview Rob Herber with P Pershan, 18 September 2007
15. Wynne JJ. The Invisible Hand of Bloembergen. In: Levenson MD, Mazur E, Pershan PS and Shen YR (eds) Resonances. World Scientific, Singapore (1990)
16. Interview Rob Herber with E Mazur, 18 September 2007
17. Interview Joan Bromberg with P Franken, 8 March 1985. Niels Bohr Library & Archives, American Institute of Physics, College Park, MD USA, www.aip.org/history/ohilist/4612.html
18. Interview Rob Herber with G Holton, 21 September 2007
19. Interview Rob Herber with F Spaepen, 13 June 2009
20. Bloembergen N. (ed) Encounters in Magnetic Resonances. World Scientific, Singapore (1996)
21. Interview Rob Herber with N Bloembergen, 6 December 2006
22. Interview Joan Bromberg J and Paul L Kelly with N Bloembergen, 27 June 1983. Niels Bohr Library & Archives, American Institute of Physics, College Park, MD USA, www.aip.org/history/ohilist/ 4511.html
23. Interview Rob Herber with P Pershan, 18 September 2007
24. Interview Rob Herber with G Holton, 21 September 2007
25. Bloembergen N. (ed) Encounters in Magnetic Resonances. World Scientific, Singapore (1996)
26. Interview Rob Herber with E Mazur, 18 September 2007

27. Bloembergen N. (ed) Encounters in Magnetic Resonances. World Scientific, Singapore (1996)
28. Interview Rob Herber with E Mazur, 18 September 2007
29. Bloembergen N. (ed) Encounters in Magnetic Resonances. World Scientific, Singapore (1996)
30. Interview Rob Herber with N Bloembergen, 6 December 2006
31. Bloembergen N. (ed) Encounters in Magnetic Resonances. World Scientific, Singapore (1996)
32. Myers DG. Intuition. Its Powers and Perils. Yale University Press, New Haven (2002)
33. Myers DG. Intuition. Its Powers and Perils. Yale University Press, New Haven (2002)
34. nl.wikipedia.org
35. Gladwell M. Outliers. The Story of Success. Little, Brown and Company, New York (2008)
36. Interview Edward Goldwyn with N Bloembergen, 2005. Op vega.org.uk/video/programme/156
37. Mattson J, Simon M. Pioneers of NMR and Magnetic Resonance in Medicine. The Story of
    MRI. Dean Books, Jericho, 1996
38. Interview Edward Goldwyn with N Bloembergen, 2005. Op vega.org.uk/video/programme/156
39. Interview Rob Herber with N Bloembergen, 6 December 2006
40. Myers DG. Intuition. Its Powers and Perils. Yale University Press, New Haven (2002)
41. Nelissen JMC. Intuïtie: over intuïtie en probleemoplossen (Intuition: about intuition and prob-
    lem solving). Panama-post 27:313 (2008)
42. Damon R. To Nico Bloembergen. In: Levenson MD, Mazur E, Pershan PS and Shen YR (eds)
    Resonances. World Scientific, Singapore (1990)
43. Interview Rob Herber with E Mazur, 18 September 2007
44. Purcell   EM.   Banquet   Speech".   Nobelprize.org.   29   July   2011
    nobelprize.org/nobel_prizes/physics/laureates/1952/purcell-speech.html
45. Interview Rob Herber with N Bloembergen, 7 December 2006
46. http://www.aps.org/programs/honors/prizes/buckley.cfm
47. http://www.ieee.org
48. http://www.fi.edu/franklinawards/
49. Friedman RF. The Politics of Excellence: Behind the Nobel Prize in Science. New York (2001)
50. http://www.nsf.gov/od/nms/recip_details.cfm?recip_id=47
51. KNAW. Professor Bloembergen met de Lorentz medaille bekroond (RAS. Professor Bloem-
    bergen awarded the Lorentz medal). Press release, September 1978
52. KNAW. Bijzonder bijeenkomst der Afdeling Natuurkunde, Amsterdam (RAS. Special meeting
    of the Department of Physics, Amsterdam, 30 September 1978)
53. E-mail from N Bloembergen to R Herber, 5 December 2011
54. Frederic Ives Medalist for 1979. J Opt Soc Am 12:1423-1428 (1980)
55. http://www.humboldt-foundation.de/web/humboldt-award.html
56. Press Release: The 1981 Nobel Prize in Physics. Nobelprize.org. 2 Aug 2011
    http://nobelprize.org/nobel_prizes/physics/laureates/1981/press.html
57. Interview Rob Herber with N Bloembergen, 6 December 2006
58. Bloembergen N. Nonlinear Optics and Spectroscopy. Réimpression des prix Nobel en 1981
    (Reprint of the 1981 Nobel Prizes). The Nobel Foundation, Stockholm (1982)
59. Lindgren I. Award Ceremony Speech. Nobelprize.org. Nobelprize.org. 2 Aug 2011
    http://nobelprize.org/nobel_prizes/physics/laureates/1981/presentation-speech.html
60. Bloembergen N. Banquet Speech. Nobel Banquet, 10 December 1981. Odelberg W. Les Prix
    Nobel, Stockholm (1982)
61. Interview Rob Herber with N Bloembergen, 6 December 2006
62. http://www.ieee.org/documents/moh_rl.pdf
63. http://newt.phys.unsw.edu.au/phys_news/Dirac2011.htm
64. http://www.euromar.org/york/laudatio2005.html

Part VI
Star Wars

Chapter 15
Strategic Defense Initiative (SDI)

List of Abbreviations

APS   American Physical Society
DEW   Direct Energy Weapons
KEW   Kinetic Energy Weapons
SDI   Strategic Defense Initiative
SDIO   Strategic Defense Initiative Organization
UCS   Union of Concerned Scientists

> *"It was so remarkable. Most scientists did not believe in Reagan's Star Wars idea. But Reagan said: 'We're going to do that'."* Bloembergen in an interview [1].

Speech of Reagan

On 23 March 1983, the then President of the United States, Ronald Reagan (1911–2004), gave a speech that later became known as the *'Star Wars speech.'*

Reagan began his speech by noting that the defense budget was more than a series of figures. The defense budget had been increased for two years, but in his view it needed to be increased further. Since the beginning of the atomic age, attempts had been made to reduce the risk of war through arms control. This was based on mutual deterrence, i.e., between the United States and the (then) Soviet Union. Reagan stated that the Soviet Union was arming itself at a much faster rate than the United States, both in terms of conventional weapons and nuclear weapons. Moreover, according to Reagan, the weapons developed by the Soviet Union were technologically equivalent to those of the United States. The Soviets, Reagan claimed, were building up an offensive military force. Instead of mutual deterrence, his idea was that the US needed a system based on preventing an attack rather than avenging victims.

© Springer Nature Switzerland AG 2019
R. Herber, *Nico Bloembergen*, Springer Biographies,
https://doi.org/10.1007/978-3-030-25737-8_15

What if free people could live secure in the knowledge that their security did not rest upon the threat of instant U.S. retaliation to deter a Soviet attack; that we could intercept and destroy strategic ballistic missiles before they reached our own soil or that of our allies? I know this is a formidable technical task, one that may not be accomplished before the end of this century. Yet, current technology has attained a level of sophistication where it is reasonable for us to begin this effort. It will take years, probably decades, of effort on many fronts. There will be failures and setbacks just as there will be successes and breakthroughs. And as we proceed, we must remain constant in preserving the nuclear deterrent and maintaining a solid capability for flexible response. But isn't it worth every investment necessary to free the world from the threat of nuclear war? We know it is! [...] Proceeding boldly with these new technologies, we can significantly reduce any incentive that the Soviet Union may have to threaten attack against the United States or its allies [...]

As we pursue our goal of defensive technologies, we recognize that our allies rely upon our strategic offensive power to deter attacks against them. Their vital interests and ours are inextricably linked - their safety and ours are one. And no change in technology can or will alter that reality. We must and shall continue to honor our commitments. I clearly recognize that defensive systems have limitations and raise certain problems and ambiguities. If paired with offensive systems, they can be viewed as fostering an aggressive policy and no one wants that.

But with these considerations firmly in mind, I call upon the scientific community in our country, those who gave us nuclear weapons, to turn their great talents now to the cause of mankind and world peace: to give us the means of rendering these nuclear weapons impotent and obsolete [2].

Maybe Reagan's idea came from a 1940 film, *Murder in the Air,* in which he himself had played the main role. The film featured a secret weapon called the *Inertia Projector* that could remotely interrupt the power supply to a device [3].

Reagan's initiative was soon nicknamed '*Star Wars*' by reference to the popular series of films by George Lucas with that name [4], and was abbreviated to SDI.

15.1   Developments Prior to Reagan's Speech

Killing Rays and Their Inventors

Although Reagan's speech did not refer to any particular system, it was clear to everyone that they mainly assumed lasers in one form or another.

The use of 'death rays' has always appealed to the imagination. In countless comic books it has become a *fictional or theoretical weapon often utilized by mad scientists, super villains and alien species in fiction - it is generally portrayed as a large and powerful laser that can be utilized for highly destructive means* [5]. The idea behind Reagan's film *Murder in the Air* is not dissimilar to this, and the killing ray also features in James Bond films like *Goldfinger* from 1964 [6], the TV animated films *The Simpsons* from 1989 [7], and the aforementioned *Star Wars* series.

But there are examples in physics that indicate that physicists and those who finance their projects also had a wake-up call when the killing ray came on the scene.

Harry Grindell Matthews (1880–1941) was a Welsh engineer who had many inventions to his name, such as a transmitter/receiver for communicating with aircraft, proposed in 1911. Around 1924 Matthews came up with a killing jet, which could stop the engines of cars and planes from a great distance. When the British Government asked for a demonstration, he refused. But he did demonstrate his invention to journalists by remotely igniting a charge of gunpowder and hit the headlines by doing so [8].

One newspaper article spoke of *destroying any forces or airplanes attacking a city*. Matthews put it like this: *Let me recall to you the air attacks on London during the war* [the First World War, RH]. *What happened? Searchlights picked up the German raiders and illuminated them while the guns fired. Hitting some but more often missing them. But supposing that instead of a searchlight you direct my ray. So soon as it touches the plane, this bursts into flame and crashes to the earth* [9].

In 1924, all the ingredients listed in 1983 were already present, that is, striking out an aerial attack, and the science and technology that should make this happen (Fig. 15.1). Just one point was not addressed in 1983 but was in 1924, and that was the possibility of failure. Matthews could not demonstration that his invention would work in practice.

In 1934, the *The New York Times* carried a report on Nicola Tesla (1856–1943), inventor, electrical engineer, and physicist, with the following headlines: *NEW "DEATH-BEAM" Invention Powerful Enough to Destroy 10,000 Planes 250 Miles Away, He Asserts*. According to Tesla this was *the most important of the 700 inventions made by him so far* [10]. This invention, too, was surrounded by great secrecy, but never had any practical value and was never even seen by anyone else.

In Chap. 12 it was mentioned that Gould already dreamt in 1959 of intercepting atomic missiles with powerful lasers, even though no such lasers were available at that time. However, military scientists were impressed and that was what Gould was all about.

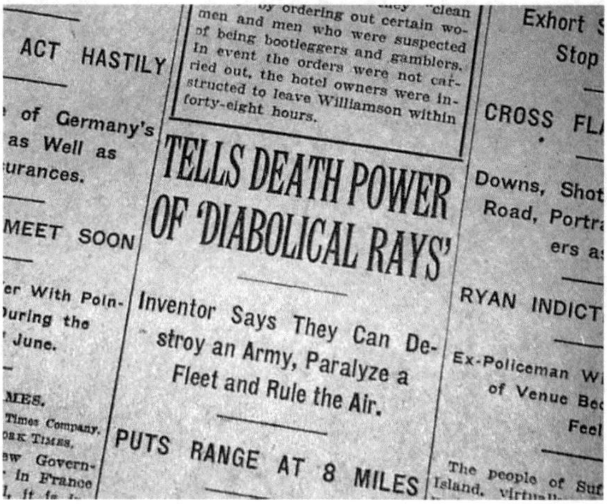

Fig. 15.1 New York Times from 21 May 1923 mentioning 'Death Rays'. *Source* Carradice P. The man who invented the death ray. www.bbc.co.uk

In 1960, Hughes Research Labs organized a press conference on the occasion of Maiman's first working laser. Hughes referred therein to *atomic radio light*. The press jumped on this with headlines like: *LA man discovers science fiction death ray* and *Death ray possibilities by* scientists, inspired by the fact that Maiman could not deny that the laser might one day be used as a weapon.

In the same year, Bell held a press conference about their laser, following the publication by Schawlow et al., and again the expression '*death rays*' came up.

In 1962, the Pentagon considered the possibility of using lasers as a weapon (Chap. 13). A study group on the use of lasers was set up at the Institute for Defense Analysis. This group claimed that 1 MJ of energy would be needed to cause significant damage to an aircraft or rocket using light alone. According to Bloembergen, in order to reach that energy level, a neodymium glass laser of 1 m^3 would have to be built. The neodymium glass laser was only developed in 1962.

Yet the idea of the killing ray continued to circulate both in the Pentagon and in the laboratories of the Ministry of Defence. Los Alamos, the laboratory where the atomic bomb was born, was the main breeding ground for ideas to develop such a scheme. And the most important man in Las Alamos was Edward Teller.

Edward Teller (1908–2003) [11]

Teller was born in Budapest from Jewish parents and had left Hungary in 1926 to study chemical engineering in Karlsruhe, Germany. His father had warned him that there was no place for Jews in an increasingly anti-Semitic Hungary. In 1928, he lost his right foot in Munich when his bicycle collided with a tram. In the hospital, one of his visitors was Hans Bethe (1906–2005), who would later become a physicist and astronomer. Bethe's mother was Jewish and he fled Germany for the United States in 1933. Teller had come into contact with Arnold Sommerfeld and others in Karlsruhe and decided to do his doctorate with Werner Heisenberg in Leipzig. There Teller became friends with the Soviet Russians George Gamow and Lev Landau. Felix Bloch (mentioned earlier in connection with NMR) also worked in that group. Isodor Rabi (also connected with NMR) sometimes came to visit. After Teller's doctoral graduation in 1930 (when he was only 21!) Heisenberg asked if Teller would like to stay on as an assistant, and indeed he did. After his doctorate in Leipzig he started working in Göttingen. Due to the Nazi takeover in 1933, Teller fled Germany in 1934 and went to the United Kingdom. He was assisted by the British Academic Assistance Council. From the period in Hungary and Germany, Teller was left with a deep aversion to both fascism and communism. After a short period in the United Kingdom, Teller went work with Niels Bohr in Denmark and then back to the United Kingdom. In 1935 he was invited by Gamow to become professor of theoretical physics at George Washington University in Washington, DC. He remained there until 1941.

In 1942, he became one of the countless physicists who worked on the Manhattan atomic bomb project, after he and Bethe had been asked as theoretical physicists by Robert Oppenheimer. That project was initially part of Chicago University, but moved to the Los Alamos National Laboratory in New Mexico in 1943.

Differences between Oppenheimer and Teller were already coming to the fore in 1943. This would have major consequences for the development of physics in the coming decades.

At 8.15 am local time on 6 August 1945, the first atomic bomb was dropped. It fell on Hiroshima in Japan. The number of people who died as a result of the bomb will never be known, but estimates range from 78,000 to 200,000. After the report that the bomb had fallen, Oppenheimer called the entire staff together and applauded them as if a very important scientific result had been achieved.

On 9 August, the second atomic bomb was dropped. This time Nagasaki was the target.

The hangover about what had been done would come soon after. Oppenheimer said he was *a little scared of what I have made*. Most scientists, technicians, and others who had worked on the bomb like army staff wanted to go home. Bethe commented: *We all felt like soldiers, we were doing our duty*. Oppenheimer went back to Berkeley and Bethe to Cornell. Teller could have returned to Chicago, but continued to work on his super project, the hydrogen bomb. He was both intimidated by Oppenheimer and jealous of him and his success in Los Alamos. He was deeply disappointed when Oppenheimer came on 16 August, the day after the capitulation of Japan, to tell him that the war was over and that there was no longer any need for the hydrogen bomb. However, Oppenheimer's successor, Norris Bradbury, agreed that Teller could continue his research in a small group. But in the meantime, the future of Los Alamos had grown uncertain due to the prospect of international supervision under the auspices of the International Atomic Agency in the context of arms control. In February 1946, Teller left for Chicago.

Politically, the situation changed rapidly in 1946. Secretary-General Stalin of the Communist Party of the Soviet Union gave a speech in February, in which he referred to capitalism as the source of all crises and armed conflicts. Washington understood this as a declaration of the Third World War. Former UK Prime Minister Churchill gave a speech in March in the presence of President Truman in which he stated that an iron curtain had been lowered across the continent from the Polish city of Stettin in the Baltic Sea to Triest on the Adriatic Sea. The Cold War was thus a fact and was the cause of ensuing developments like the arms race and SDI. On 18 April 1946, a conference was held on the Superbomb in Los Alamos. Goodchild states: *Those who had left to return to academia or elsewhere were now fast realizing that the genie they had helped to escape was not to be returned easily in the bottle*. After the conference, nothing seemed to happen. Los Alamos was preparing a series of experiments on the island of Bikini.

In February 1948, the communists in Czechoslovakia carried out an unexpected coup. In April 1948 the Soviet Union blocked train and road traffic to Berlin. These events led to an intensification of the Cold War and created a favorable climate for a further arms race. In 1949 the Soviet Union detonated its first nuclear bomb and President Truman announced a rapid development of the hydrogen bomb.

Los Alamos needed Teller again. Together with the Polish mathematician Stanislaw Ulam, Teller devised the principle of the hydrogen bomb, a combination of nuclear fission and nuclear fusion. However, he did not become the head of the project group that was to develop the bomb.

In 1952, Teller left Los Alamos to work at the new Livermore Laboratory at the University of California near San Francisco. After the detonation of the hydrogen bomb on 1 November 1952, Teller was called the 'father of the hydrogen bomb'.

In 1954, Teller testified before a committee of the United States Atomic Energy Commission that Oppenheimer would be a safety risk. He was the only one who claimed that. The result was that he was rejected by the scientific world.

He began to move more and more in military and political circles and became the darling of conservative politicians. He worked tirelessly for strong defense and challenged the international treaty banning above-ground nuclear testing, signed in 1963.

Teller and SDI

In the Livermore Laboratory, a group of researchers led by Lowell Wood had been designing the X-ray laser since 1974. The X-ray laser (wavelength below 10 nm) was expected to make three-dimensional structures visible in cells, for example, just as lasers in the optical wavelength range (400–700 nm) can be used to produce three-dimensional holograms. The Livermore Laboratory, which should actually be called the Lawrence Livermore National Laboratory, has a military purpose: *Our mission is to ensure the safety and security of the nation through applied science and technology* [12]. In 1978 a group at Livermore was working on a project to use the X-ray laser as a powerful beam to eliminate enemy missiles. The developers involved were Peter Hagelstein and George Chaplin. The problem in designing such a laser was the pump (see the discussion of the optical pump in Chap. 11). This pump had to be extremely powerful to generate the necessary energy. The idea arose of using a small atomic bomb as a pump. A major problem was that the nuclear explosion would destroy the laser construction itself shortly after the laser had delivered its pulse (which would move at the speed of light). A nuclear explosion was obviously not feasible on Earth, but some did see this as a possibility in space. According to Teller, these were 'third generation weapons'. The first generation was the atomic bomb and the second the hydrogen bomb. In 1979 Hagelstein's design was to be tested and in 1980 this design was to be tested in conjunction with Chaplin's X-ray laser.

In November 1980 Reagan was elected president. Reagan had already established contacts with right-wing Republican movements before his election as president and he had been convinced that, in the event of an attack by the Soviets, sitting still and waiting or hitting back with missiles were not serious options.

In that same month of November, Hagelstein and Chaplin's experiment succeeded, and Lowell Wood (a protégé of Teller and leader of the secret O-group) immediately informed Teller of this. A new group was formed under the leadership of Tom Weaver, and soon a hundred people were working on the X-ray laser. The project was called *Excalibur*, after the magical sword in the Welsh legend of King Arthur. Now they had something to show, Teller and Wood started lobbying at Congress in February 1981. Prior to this, the project should normally have been assessed at the Ministry of Defense, but Teller managed to interest a group of Reagan's entourage who were

Fig. 15.2 Edward Teller pleads with President Ronald Reagan for Star Wars. *Source* en.wikipedia.org

strong supporters of such a system (Fig. 15.2). Reagan abolished the *President's Science Advisory Committee*. Over the years, this committee had become a channel for expressing progressive scientific opinions to the White House. Teller's candidate George Keyworth became scientific advisor to the president. Keyworth later commented: *Bluntly, the reason I was in that office was because Edward first proposed me, and the President very much admired Edward.*

It was no big surprise that, when the new *White House Science Council* was appointed, Teller was one of the fifteen members. The first formal meeting of this council was held in July 1981. In October 1981, Edwin Meese, Reagan's advisor and a member of both the *National Security Council* and the cabinet, received a letter from Karl Benedetsen. The latter had known Teller for years and had set up a non-profit organization with him under the name '*High Frontier*'. Benedetsen told Meese that 50 million dollars would have to be put on the table for the *Excalibur* project. Although that project had shown that there were in principle some opportunities, Benedetsen gave the impression that the system was already fully operational and could be used as a weapon as early as 1986. According to him, with the injection of 50 million dollars, the X-ray lasers could be ready for use even earlier, in 1984. *Excalibur* had to compete with the *Global Ballistic Missile System*, brainchild of Daniel Graham, a retired general who was also a member of *High Frontier*. The two camps, Teller's and Graham's, hurled insults at each other. In response to Graham's objection that lasers in space were vulnerable because they could not defend themselves, Teller came up with the idea of launching them into space from submarines. This meant that another technological feat had to be accomplished: the submarine launch of a spaceborne X-ray laser equipped with a nuclear bomb, not to mention actually preparing it for action once in place, and all this in an extremely short space of time. In November 1981, the *Excalibur* report by *High Frontier* reached the President and Graham's system was not even mentioned, so Teller could celebrate his victory. On 7 January 1982 the group met Reagan and then all went quiet.

In February, it was announced that a panel had been set up by the *White House Science Council* under the chairmanship of the physicist Edward Frieman, who would assess the project. One of the panel members was Charles Townes.

Remarkably, Teller was not a member of this panel. The panel's report could not be expected before the end of 1982, by which time Teller was becoming increasingly impatient. In June 1982 he sought television publicity with conservative commentator William Buckley and openly asked for a meeting with Reagan. The President had seen the broadcast and granted Teller's request. On 15 September 1982, Teller met the president for a half-hour meeting. Reagan seemed to accept what Teller said, but William Clark, Minister of the Interior, had serious doubts. In June there had been a meeting of the Frieman panel with the people from *Excalibur*. Livermore's group leader, Weaver, stated that USD150–200 million per year would be needed to continue the development of *Excalibur*. So, many times the amount of the one-off grant requested by Benedetsen in October 1981 would eventually be required. Work on the project had already slowed right down and there was no question of an acceleration. Then in the autumn of 1982 the draft report of the Frieman panel was published. The X-ray laser was unrealistic and unworkable technology; the Graham project was also downgraded. When Teller heard of this draft report, he was furious. He threatened to cancel his membership of the *White House Science Council* if the section of the report on the X-ray laser was not revised.

Teller asked his former colleague Bethe to give an opinion on *Excalibur*. Despite the rivalry between them, he attached great importance to Bethe's judgment. After a visit to Livermore in February 1983, Bethe was full of praise for Hagelstein's presentation of his model. However, Bethe's assessment of the whole project was negative [13]. According to him, even if one were "*as optimistic as possible within the limits of the laws of physics and geometry*" the system would not work [14].

In the same month of February 1983, the second report of the Frieman panel was published. This was even more negative than the first report, and *Excalibur* seemed to be doomed. However, there had been another development. The White House and the Pentagon would soon make the project unstoppable.

Admiral James Watkins, operational commander of the fleet, had heard of Teller's plans and asked him to come and explain them. On 20 January 1983 Teller visited Watkins and defended his plans with great enthusiasm. Watkins was convinced and on 5 February 1983, he explained the idea at a meeting of the *Joint Chiefs of Staff*. To his own surprise, everyone supported his proposal to break the standstill on the deterrence strategy with the SDI program At a meeting of the *Joints Chief of Staff* on 11 February, General Vessey read Watkins' proposal, which contained the statement that the task of defense was *to protect the American people, not just to avenge them*.

Reagan greatly appreciated this sentence. All concerned were informed, the Chiefs of Staff by Watkins, and Reagan and his assistants by Teller himself.

Reagan then told his astonished assistants that he was ready to announce the SDI publicly. The soldiers and Robert MacFarlane, an assistant to William Clark, wanted Watkins' proposal to be worked out properly by the Chiefs of Staff, to put some flesh on the bones, as it were. Teller and many others were invited to attend Reagan's announcement without knowing what to expect. Reagan's television speech began on 23 March 1983 at 18:00 Eastern Standard Time.

Press, Film, and Comic Strip

As mentioned earlier, there was lively press coverage of *Star Wars*. This is confirmed by some headlines in *Time Magazine* from 1980 to 1982, i.e., before Reagan's SDI was announced: *Science: A Technology to Transform War* [15], *Space: Battlestar Columbia?* [16], and *In Sight: Killer Lasers* [17]. The last article even gave a glimpse of how the battle would proceed, according to Richard De Lauer, head of the research department of the Pentagon: *Lethal warships in space. Laser beams fired from orbiting rockets. Satellites zapped out of the sky. It sounds like the script of Star Wars, but according to testimony last week, the Pentagon's top weapons man believes it is perilously close to becoming a reality.* This shows not only that De Lauer had been watching too many science fiction films and read too many comics, but also that the term *Star Wars* was already in use before Reagan announced the SDI program. In May 1982, the New York Times headline was: *STAR WARS: PENTAGON LUNACY; WASHINGTON—Space, according to the experts, is the next battlefield. It will be an arena for laser death rays, orbiting battle stations, and heroic acts. Yet the vision holds a flaw, one carefully brushed aside by those in the defense industry and the military who are pushing for exotic new arsenals. The weapons could be knocked out of action by a single nuclear blast in outer space* [18]. The last sentence shows that criticism of space weapons was not only being expressed in the seclusion of the defense world, but had actually been made public.

The two-hour-long mega production that was always referred to—*Star Wars*—was released in 1977 under the direction of George Lucas, who wrote the story himself. In fact, it wasn't much of a story (a princess has to be saved from the clutches of a kind of science fiction devil), but the special effects were breathtaking as things went in those days, and appealed to the imagination of the public and the press. Laser swords, battles between hundreds of spaceships, killing beams, huge space stations, it couldn't get better. Before that, there had been two films by Stanley Kubrick that had also made a great impression. In 1968, he made *2001, A Space Odyssey* [19]. The core of the film is the intelligent computer HAL, which controls humans instead of the other way around. An old fear, which also recurred in the discussions about the SDI process.[1] In 1964, Kubrick made the film *Dr. Strangelove or: How I Learned to Stop Worrying and Love the Bomb* [20]. In this film, an insane general triggers a process that will lead to a nuclear war. The president asks for advice from Dr. Strangelove

---

[1]The film was so well made that there are still people who believe that NASA had approached Kubrick to 'fabricate' the moon landing in the studio.

(also known as 'Merkwürdigeliebe'), a former Nazi scholar and strategy expert who calls the president 'Mein President' or even 'Mein Führer'. Strangelove explains the principle of the doomsday machine, easy to understand, credible, and convincing.

The previously cited book by Goodchild about Teller has the subtitle 'the real Dr. Strangelove', thus suggesting that Teller might be precisely that person. That's nonsense. Strangelove is a Nazi; while Teller was Jewish and, in fact, fled the Nazis in 1933. Strangelove is more like Dr. Wernher von Braun, a Nazi scientist (he was a member of the NSDAP and SS) who designed the V2 ballistic missile, produced using slave labor. In 1945 Von Braun was secretly transferred to the United States, where he later became the spiritual father of the Apollo moon project. The satirical singer-songwriter Tom Lehrer sang in 1961 about Wernher von Braun: *Once the rockets are up, who cares where they come down? 'That's not my department', says Wernher von Braun [...] 'In German and English I know to count down und I'm learning Chinese', says Wernher von Braun* [21].

Mary Shelley's novel *Frankenstein, or, the Modern Prometheus* [22] from 1818 describes the prototype of the mad scientist and has been made into numerous films [23].

Comic strips also contributed to the ongoing public debate. A first example from the countless comic books is *The Hulk*, a superhero from 1962, involuntarily transformed by the physicist Bruce Banner, who causes a gamma explosion [24]. It is striking that the stereotype killing rays are connected with a 'crazy scholar'. The 'crazy scholar' can almost always be traced back to Johann Wilhelm Goethe's classic work *Faust*. Heinrich Faust was modeled on Johann Georg Faust (circa 1480–1538), a researcher and a lecturer by name. Faust looks back on his life and comes to the conclusion that he has failed both as a scientist and as a person. As a scientist, he lacks deeper insight and as a human being he cannot fully enjoy life. Deeply depressed and exhausted, he promises the devil Mephisto his soul, when Mephisto succeeds in freeing him from his dissatisfaction and restlessness [25].[2]

---

[2]The previously quoted book about Teller by Goodchild has the subtitle 'The Real Dr. Strangelove', thus suggesting that Teller can be identified with Strangelove. This is nonsence, because Strangelove is a Nazi, while Teller was Jewish and, in fact, fled from the Nazis in 1933. Strangelove is more like Dr. Wernher von Braun, the Nazi scientist (member of the NSDAP and SS) and designer of the ballistic missile V2, produced using slave labor. In 1945 Von Braun was secretly transferred to the United States, where he later became the spiritual father of the Apollo Moon project. The satirical singer-songwriter Tom Lehrer clearly refers to Wernher von Braun in his song from 1961, "*Once the rockets are up, who cares where they come down? "That's not my department", says Wernher von Braun [...] In German and English I know to count down and I'm learning Chinese, says Wernher von Braun.*" (www.youtube.com/watch?v=kTKn1aSOyOs). Comic strips also took part in the background of the public debate. Two examples from the numerous comic books are: The Hulk, a superhero from 1962, involuntarily transformed by the physicist Dr. Bruce Banner, who caused a gamma explosion (*The Incredible Hulk*, No 1 (1964)). In *The Avengers*, in 1966, the scientist Arthur Parks produced hand-held laser weapons. It is striking that the stereotypical killing rays are combined with a 'crazy scholar'. The latter can almost always be traced back to Johann Wilhelm Goethe's classic work Faust. Heinrich Faust was modeled on Johann Georg Faust (circa 1480–1538) a researcher and a teacher. Faust looks back over his life and comes to the conclusion that he has failed both as a scientist and as an individual. As a scientist, he lacks deeper insight and as a human being he cannot fully enjoy life. Deeply depressed and lacking vitality, he promises his

15.2   Emergence of the Strategic Defense Initiative (SDI) Up Until 1987

In March 1983, three days after Reagan's speech, the second X-ray laser test was carried out in Livermore. The underground detonation failed. Teller appeared two weeks later before the committee of inquiry of the *House Armed Service Committee*. He stressed what he saw as the threat posed by the Soviet Union, referring to the activities of the three Nobel Prizewinners in 1964, awarded for the laser. Teller compared the American winner with the Soviets: *Charlie Townes publishes beautiful papers about what is happening in the center of our galaxy. Basov and Prokhorov are working with the Soviet military on lasers. Your guess who is currently ahead.*

Caspar Weinberger, the Minister of Defense, set up three new committees to assess SDI. Although the main committee (*Defense Technologies Study Team*), under the chairmanship of former NASA director James Fletcher, hammered the SDI in its report, it paradoxically recommended providing USD 26 billion for the entire SDI. The amount was recommended for the period 1983 to 1989, with 1 billion for the X-ray laser. The cause of this paradox lay in the growing tension between the United States and the Soviet Union, mainly caused by the aggressive attitude of the Reagan government. SDI also played an important role in this: if SDI were to function, the Soviet Union would no longer be a threat to the United States, while the United States would still potentially pose a threat to the Soviet Union. The *balance of powers* would then be disturbed.

George Shultz, Minister of Foreign Affairs from 1982 to 1989, had heard from the Soviets in the course of 1983 how concerned they were about SDI. Reagan wanted to be re-elected in 1984 and moderated his tone. In the light of the critical Fletcher report and also the criticism of colleagues at Livermore, Teller needed a positive experiment. On 16 December 1983, the fourth test finally succeeded and Wood immediately informed Teller. Teller was extremely pleased with the first results supporting the possibility of the X-ray laser. In the meantime, none of the funding of 1 billion for the X-ray laser promised in the Fletcher report had yet materialized. Teller asked for additional funding: *"We are now entering the engineering phase of X-lasers"*. However, Weaver of the research group had stated a year earlier that at least five more years' research were needed, so Teller's remark showed a drastic overestimate of what was really possible. Teller had not sent copies of his letter to the scientists at the Livermore Laboratory and when research leader Woodruff heard about this, he was furious and went to Teller to find out what was going on. He told him that the construction of the X-ray laser was still years away and that even after the last test there were still many uncertainties. However, Teller refused to send a correction, saying *my reputation would be ruined.* In February 1984, there was growing criticism of the results of the successful trial from within Livermore and from other defense laboratories [26].

---

soul to the devil Mephisto, when Mephisto succeeds in releasing him from his dissatisfaction and restlessness. (de.wikipedia.org/wiki/Faust._Eine_Tragödie).

E. Walbridge of the Argonne National Laboratory in Chicago, wrote in July 1984: *Nuclear-pumped X-ray laser weapons are constrained to have an output beam divergence that is at least as great as a certain minimum value. Because this value is large, such weapons used for ballistic missile defense will require disturbingly high megatonnages in space: for example, a total of greater than or equal to 73 megatons within range of 1,000 attacking missiles, with individual warheads up to 3.7 megatons or higher* [27].

Wood, head of the O-group, but not Woodruff's boss, approached Walbridge in an aggressive and hostile way after the publication. But Wood could not refute the article either, because the investigation was secret. This was typical of what would happen in the years after 1984: a bitter battle in the 'Science War' (Fig. 15.3). One group, the *Union of Concerned Scientists* (UCS), including Richard Garwin and Hans Bethe, reported that it was an illusion to build a perfect or near-perfect defense system. A fully operating system would have to include 2400 space stations of 100 tons each in space. Each station would cost 1 billion dollars, making a total of 2.4 trillion dollars. But Livermore and Las Alamos would not let that pass. They found a calculation error which would reduce the number of stations needed to 300. Robert Jastrow, a physicist involved with Livermore and Teller, accused the UCS of "*shoddy work*" that had been performed with the sole intent of "*making the President's plan seem impractical, costly, and ineffective.*" Garwin reacted with his own calculation and came to 2263 space stations.

Fig. 15.3 Lowell Wood (left) in 1997. He was called 'Dr. Evil'. *Source* home.earthlink.net

Garwin in turn accused Jastrow of having "*a career of hyena-like behavior*".
There were also sharp contradictions about SDI in Washington. In November
1984 Reagan was re-elected for four years. The Minister of Defense Weinberger
had changed from a critical follower of SDI into a supporter and the Pentagon set up
the *Strategic Defense Initiative Organization* (SDIO) [28]. General James Abramson
became head of this organization, which had been established on the recommendation
of the Fletcher Commission. SDIO had to deal with all matters relating to *ballistic
missile defense,* which included both the *Directed Energy Weapons* (DEW) and the
Kinetic Energy Weapons (KEW) [29].

Meanwhile, however, the Foreign Minister Shultz was building up ever better
relations with the then Secretary-General of the Communist Party of the Soviet
Union, Konstantin Chernenko. Shultz realized that SDI would promote a significant
re-armament of the Soviet Union and stand in the way of a new arms control treaty.

Livermore already had a new plan. Instead of a laser that was "*a thousand times
brighter than the brightness of the atomic bomb explosion*", there had to be a laser that
would be as much as "*a billion times brighter*". This project was named *Excalibur
Plus* or *Super Excalibur.* The laser would be permanently positioned on a geosta-
tionary orbit 35,786 km above the equator. The inventors stated that, even at this
distance, a rocket could be taken out at launch.

However, the people running Livermore, like Teller and Wood, were running far
ahead of the foot soldiers working on the science there. Teller was talking about the
tremendous prospects that could be realized within a short time if enough money
was put into it, while the scientists still saw years of experiments ahead of them and
considered even a simple application impossible within the coming five years [30].

**The American Physical Society (APS)—Bloembergen Co-Chairs the Study
Group**
This association of physicists had over 53,000 members in 2017, publishes more
than twelve journals, and organizes about twenty conferences a year [31]. Within the
APS there was more and more discussion about SDI. It was felt that too little was
published about the technological possibilities and limitations of SDI, or that these
cases were dealt with in secret documents. On 20 November 1983, the Council of
Representatives commissioned a study of the scientific and technological merits of
DEW. In November 1984, an APS Study Group was established with scientists and
engineers from government laboratories, industry, and universities. The Study Group
had to set out the results in a non-secret report [32]. Bloembergen was asked to chair
the Study Group. This put him right in the middle of a political and social debate
characterized by fierce polemics between the interested parties. As may be expected
after his totally apolitical past, Bloembergen limited his efforts to science. The following quotation is significant because Bloembergen reduced everything to science
and apparently did not realize that, although one might agree on purely scientific
merits, the problems already begin with the application. He was astonished by the
fact that scientists of different political colors could have totally different opinions
about SDI: "*At least one should agree on science and technological facts.*" [33]

In an interview with Jeff Hecht about his work, Bloembergen commented on laser applications: *And of course there is the big thing of the high-power lasers, defense laser weapons.* Hecht then remarks: *I was a little surprised to see you getting involved with that field. They twisted your arm?* to which Bloembergen responded: *They twisted my arm, I guess. I don't relish it, but I think it is a job that has to be done.* [34]

Later Bloembergen stated that he had been asked to become chairman three times and that he had declined three times before finally agreeing, because he wanted to know why scientists were divided into two camps and totally polarized [35].

In an interview for this biography, Bloembergen said the following about the procedure to set up the Study Group: *"It was so remarkable. Most scientists did not believe in Reagan's Star Wars idea. But Reagan said: 'We're going to do that'. And even the National Academy didn't dare to question it. They should have done so. So then the American Physical Society started to investigate the problem. They wanted me to be chairman, but my condition was that everyone from the Study Group must have access to the secret documents. If we didn't have access to those documents, that report wouldn't impress anyone. They would all say, 'What they claim is all the same. If you knew what I know, it would make a big difference.' So we got access."* Bloembergen continued: *"Everybody had to have highest class security, top secret, in fact a little more than top secret for one small thing, which concerned five of us, and that was to discuss the X-ray laser that Teller had proposed. [...] And the second requirement was that none of these people had gone on public record politically, beforehand. And I had never written anything with political implications. But, you had to be expert, either in lasers or accelerators in space. So they asked me to chair it and I said I don't want to do this. I don't have secret facilities at Harvard, and so on, and it is a lot of administration. And then they said... well, would you do it with us, and I said I'd like to select a co-chairman, and that was Kumar Patel from Bell Labs, where he was a vice chairman, and he had a lot of secretarial help, and all the rest. He did all the organizing, typing of the proceedings of all the meetings. That took up half of my time for two years."* [36]

The investigation by the Study Group was limited to the Direct Energy Weapons, so Kinetic Energy Weapons were excluded from the investigation. Moreover, the research was limited to the physical basis of high-intensity lasers, high-intensity particle beams, radiation control, and radiation spread (Fig. 15.4). Other issues studied were the determination, localisation, and identification of the target, discrimination, the interaction of radiation and matter, the degree of lethality, power sources, and survival rates.

A *Council Review Committee* was set by the APS, presided by Georg Pake of Xerox in Palo Alto, an old acquaintance of Bloembergen's from his NMR days (see Chap. 7). Charles Townes and Ard Schawlow were also members of this committee. Schawlow was interviewed about this in 1985 for the magazine *Laser and Applications* and made the following remarks about his work on that committee (which was still working hard in December 1985):

Fig. 15.4 Chandra Kumar N Patel in 2014. *Source* omicsgroup.com

Question from *Laser and Applications*: *There is a lot of talk in the laser community these days about the Strategic Defense Initiative. How realistic is SDI? Is that a reasonable thing to be done with lasers?*

Answer Schawlow: [...] *I'm on the American Physical Society's directed energy weapon study group review committee. But my only function there is to make sure that the report makes sense to a person like myself who is not involved in the SDI program. I think, from all that I've read, though, that no matter how much money you put in now, you couldn't do it. It's questionable whether you're ever going to do anything satisfactory* [37].

Wood and Teller did not sit still while the APS committee was working. In January 1985 Wood sent a memo entitled *Re: CHARGE!* in which he trumpeted that Roger Batzel, director of Livermore, had provisionally agreed to bring *Super Excalibur* to the attention of the Congressional Committee on Finance during one of their hearings. Outside Livermore, there was outright pessimism about *Super Excalibur*, especially at Los Alamos. After a failed experiment, confidence in the entire X-ray laser project had been lost there. The director of Los Alamos, Donald Kerr, told Batzel at a dinner party that Batzel, Wood, and Teller should be dismissed because they did not represent Livermore at all. Kerr was afraid that Wood and Teller would damage not only the reputation of Livermore, but also the reputations of all the national laboratories.

Woodruff told the *SDI Working Group* in mid-February 1985 that *Excalibur* should function in 1992 and *Super Excalibur* perhaps in 1995.

In March 1985 Wood and Teller attended an experiment that caused euphoria in both of them, as good as it had gone in their eyes. According to both, this experiment, which succeeded in focusing a beam, was the proof that *Super Excalibur* was possible. On 3 April, Teller even told an audience at the University of California campus that the X-ray laser was not paper, but had become reality. In doing so, he put his own credibility on the line.

In June, Teller met Reagan and managed to make the president so enthusiastic that the latter promised him an extra 100 million dollars on the spot. Teller had also contacted McFarlane, but he was a lot less positive. McFarlane on Teller: "*My instinctive reaction is that he was not an impartial analyst. He was an attorney.*" However, Teller's merchandise and the positive results of the experiment in March ensured that both Livermore and the X-ray laser would receive extra funding. Then came the setback. Los Alamos had calculated that the results showed that the beryllium reflectors were probably responsible for radiation, which had little or nothing to do with the X-rays generated. This led to frustration and anger among the Livermore employees [38].

On 30 October 1985, a hearing of a subcommittee of the Senate took place, the '*Subcommittee on Strategic and Theater Nuclear Forces*'. The Secretary of State Fred Ickle expressed the familiar position: *As I have stressed here before, Mr. Chairman, the strategic defense initiative is not an optional program at the margin of our defense effort. It is a key element in our long-term policy of enhancing United States and allied security and of reducing the risk of nuclear war.* General Abrahamson of the SDIO compared the requested subsidies for SDI with those of the Manhattan and Apollo projects and found that SDI cost less than those other projects. Richard Lau, from the Office of Naval Research, stated that the software to be used in the SDI should be examined critically. Frederick Brooks, computer specialist at the University of North Carolina, said: *I see no reason why we could not build the kind of software system that SDI requires with the software engineering techniques that we have today.* [39]

In the Washington Post of 30 October, however, the following had been written: *President Reagan's dream of an effective 'Star Wars' system of antimissile defenses is almost certainly doomed to failure, according to a growing number of top computer programming experts who say there is no conceivable way to write and test the software that would be needed to operate it with adequate reliability* [40].

A conference of national laboratories on the various nuclear fission programs was held at the end of October. It was an explosive gathering, figuratively speaking, as the Livermore people first heard of the disastrous results from Los Alamos. Many Livermore employees suddenly realized that they might have been working for nothing for four years. Woodruff resigned from Livermore the day after the conference, tired of the years of struggle against Wood and Teller.

In November, the first meeting between President Reagan and the new Secretary-General of the Soviet Union, Michael Gorbachev, took place. The meeting, at which SDI was also discussed, yielded nothing concrete in Geneva, but it did create a relaxed atmosphere. McFarlane later wrote that Gorbachev *had to conclude one of two things; either Reagan was being cynical with all his preaching about eliminating nuclear weapons, and his real intention was to bankrupt the Soviet Union; or he was incredibly ignorant* [41].

On 3 December 1985 there was a hearing of the *'Subcommittee on Strategic and Theater Nuclear Forces'*. During this meeting, technical issues about the software were discussed. In this session, the various software specialists contradicted each other. Solomon Buchsbaum of Bell Labs: *"Can such a large, robust, and resilient system be designed - and not only be designed, but built, tested, deployed, operated, and further evolved and improved? I believe the answer is yes."* Danny Cohen of the University of Southern Carolina: *"It is my own judgment [...] that the computing requirements for the battle management of the SDI can be met."* David Parnas, of the University of Victoria, British Columbia, Canada, was against: *"SDIO is, in my opinion, dangerous to the country to undertake."* He further argued that software is never perfect, and because it could only be used once in this case, it would not be possible to learn from mistakes [42].

At the end of December another experiment was done. It cost 30 million dollars and the results were disastrous. Hagelstein left Livermore [43].

On 11 and 12 October 1986, Reagan and Gorbachev met in Reykjavik. The Soviets were prepared to abolish all long-range missiles, as long as SDI stayed within the laboratory, but Reagan refused [44].

November 1986 saw the harbingers of the storm that would arise after publication of the SPS report. The government of the United States announced that parts of the report to be published were secret and could not be published. Bloembergen then commented diplomatically: *"There are certain questions on items that the [Strategic Defense Initiative Office (SDIO) at the Department of Defense] feels should be classified."* [45]

In view of the difficulties surrounding *Excalibur*, Canavan, a member of the group around Teller came up with a new idea: a non-nuclear system of countless small satellites that could fire mini-missiles at incoming projectiles: this was *'Brilliant Pebbles'* [46].

15.3   SDI After 1987: Devastating Criticism

On 25 September 1986 [it actually reads 1987, which is incorrect, RH] the report was sent to SDIO to assess any security risks [47].

The plan was to publish the report in the month of January 1987 in *Reviews of Modern Physics*. However, the government blocked the publication because it wanted to keep parts of the report secret. Thomas Marshall, a member of the Study Group, comments: *Obviously, the report will have some political impact, but it's to the advantage of the government to publish it because it makes the problems with SDI publicly available.* A meeting with SDIO was convened in December 1986 to discuss the problems. Havens of APS: *The purpose of the meeting was to end up with an unclassified report and I think we've done that.* According to members of the Study Group, many small technical changes were made. The report was sent to Washington on 15 January [48].

A week later, the Defense Secretary Weinberger raised objections and the Ministry tried to stop publication of the report. Some prominent physicists accused Weinberger of *delaying for political reasons, the release of their report that reaches negative conclusions about some key components in the Strategic Defense Initiative.* The secretariat of the Ministry found some parts of the report *problematic and unpublishable* [49].

On 20 March 1987, the APS report was released for internal circulation. On 30 March it was ready for the Review Committee, and on 8 April the Review Committee sent it back with a positive opinion. On 10 April, the SDIO's approval for publication was received [50] and on 17 April the report went to Washington under embargo. On 21 April there was a press conference in the *National Press Club* in Washington. The press release was issued on 23 April [51]. The press statement had the neutral title: *FEASIBILITY OF DIRECTED ENERGY WEAPONS FOR SDI IS EXAMINED*, but the first few lines were already devastating for SDI: *The development of an effective ballistic missile defense utilizing directed energy weapons (DEW) would require performance levels that vastly exceed current capabilities, according to a panel of experts on directed energy technology convened by the American Physical Society. In a report released today entitled "THE SCIENCE AND TECHNOLOGY OF DIRECTED ENERGY WEAPONS," the study group concludes that insufficient information exists to decide whether the required performance levels can ever be achieved.* Chairman Pake of the *APS Review Committee* stated that the *security review resulted in small but significant deletions from the Report. In particular, some of the countermeasures available to the offense could not be discussed in detail. They believe, however, that the Report 'performs a significant service to APS members and to the general public by providing an authoritative technical background and relevant technical assessments'* [52].

The APS report was 201 pages long and covered 26 subjects. Bloembergen gives an insight into the state of affairs within the committee: "[...] *we came out with a unanimous report, no minority opinions or appendices. Very interesting. When we discussed an issue, sometimes for hours, I said 'Why can't we agree?' And then either the subject matter was not scientific or technical, but had political things in it, and we said: that won't do in our report, we won't discuss this issue; or it was scientific or technical, but there was a dispute about the wording, and some people said: 'That's not politically neutral.' Then we worked it out, and we succeeded, and we had very like thinking on that committee. Probably on purpose. They were not officially labeled, but all these people could agree on the purely scientific aspects using neutral language, and that's why we got a unanimous report. Fascinating!"* [53]

It would be impossible to summarize the report in a few paragraphs. It only cites publicly available literature for reasons of confidentiality. We find the following note on the X-ray laser: *Nuclear-explosion-pumped X-ray lasers require validation of many of the physical concepts before their application to strategic defense can be evaluated.* According to the report, the physics behind the X-ray laser and its structural development still had to be studied. Furthermore, it had not been demonstrated that it was even possible to develop an X-ray laser that could be used for military purposes. Below an altitude of 80 km, the use of this type of laser actually becomes impossible due to atmospheric perturbations. With gamma lasers, which have even shorter wavelengths and therefore higher energy content than the X-ray laser, it is stated that these lasers *represent an extremely high risk and long term approach to strategic defense.*

The main overall conclusion was:

Although substantial progress has been made in many technologies of DEWs over the last two decades, the Study Group finds significant gaps in the scientific and engineering understanding of many issues associated with the development of these technologies. Successful resolution of these issues is critical for the extrapolation to performance levels that would be required in an effective ballistic missile defense system. At present, there is insufficient information to decide whether the required extrapolations can or cannot be achieved. Most crucial elements required for a DEW system need improvements of several orders of magnitude. Because the elements are inter-related, the improvements must be achieved in a mutually consistent manner. We estimate that even in the best of circumstances, a decade or more of intensive research would be required to provide the technical knowledge needed for an informed decision about the potential effectiveness and survivability of directed energy weapon systems.

In addition, the important issues of overall system integration and effectiveness depend critically upon information that, to our knowledge, does not yet exist.

This general conclusion is further explained by, among other things, an estimate that the existing candidates for *Direct Energy Weapons* must have at least a factor of 10 higher energy output and beam quality to be considered seriously for use against a ballistic missile system [54].

Such a devastating report required a reaction from the stakeholders in the SDIO, Livermore, and Teller himself. A complicating factor was that on 24 April, so one day after the report was revealed, the members' council of the *APS* issued the following position on SDI: *In view of the large gap between current technology and the advanced levels required for an effective missile defense, the SDI program should not be a controlling factor in the U.S. security planning and the process of arms control. It is the judgment of the Council of the American Physical Society that there should be no early commitment of the development of SDI.* This position was further explained by the following statements:

1. *Even a very small percentage of nuclear weapons penetrating a defensive system would cause human suffering and death far beyond that ever before seen on this planet.*
2. *It is likely to be decades, if ever, before an effective, reliable, and survivable defensive system could be deployed.*
3. *Development of prototypes or deployment of SDI components in a state of technological uncertainty risks enormous waste of financial and human resources* [55]. This statement went a lot further than the APS report, because there was a clear political position against SDI, whereas the APS report only raised scientific objections.

SDIO responded to the report on 28 April. They already knew the report because they had screened it in connection with the confidentiality procedure: *We find the conclusions to be subjective and unduly pessimistic about our capability to bring to fruition the specific technologies needed for a full-scale development decision in the 1990s* [56].

In *Science* of 1 May 1987 Colin Norman wrote: *President Reagan's Strategic Defense Initiative (SDI) has come in for a lot of criticisms since it was launched 4 years ago. But none of it is likely to be as damaging as a report published last week by the American Physical Society, which was not even intended as criticism [...] It underscores the enormity of what must be accomplished before any of these technologies are ready to become workable parts of a missile defense system* [57].

Luis Marquet of the SDIO reacted to Norman's article. Marquet first posed: *We are pleased with the overall content of that report, in that it accurately describes many of the basic technical hurdles that must be overcome and, at the same time, acknowledges the considerable progress made to date.* But he disagreed completely with the summary: *However, it is important to differentiate between the main body, which is factual, and the executive summary, which goes beyond the facts and becomes subjective in its speculation regarding potential rate of progress* [58].

Wood from Livermore and Canavan from Los Alamos then took the initiative and claimed that the report contained a calculation error of a factor of 100. Frederick Seitz, former president of the APS, found that there were clear mistakes in the report.

Bloembergen had the following to say about the remarks by Seitz: *"There was a group by Seitz, Fred Seitz, he had been President of the National Academy, but he was very conservative. For a long time, he also published many reports denying global warming. He went on for years. On our report, he said: it is fatally flawed, no proof of it. And he said, it was politically motivated and he compared us with the Lysenko affair*[3] *in the Soviet Union [...] He went completely off his rocket, and*

---

[3]Trofim Lysenko (1898–1976) was a biologist and agricultural scientist from the Soviet Union. He was a fierce opponent of genetics, as it was practiced in the West, and characterized it as a

that is in the Congressional Record. And what they did is that went to a Republican Subcommittee, who wanted to listen to them and wanted to debunk this report." [59]

As a result, a number of conservative republican congressmen (including the well-known Newt Gingrich) took the initiative to send a letter with these remarks to the secretary of the APS, William Havens. In this letter, the authors considered it a major problem that none of the rapporteurs had to deal with SDI-related technology over the previous 25 years. They also claimed that the report did not meet the usual criteria of peer reviewing [60]. William F. Buckley, founder of the conservative magazine *National* Review, wrote in a UPS column that a *scientific team responsible for a regular feature in National Review [...] has examined the American Physical Society's findings. [...] The APS' projections are based on false estimates.* Buckley asserted without a trace of evidence: *And these false estimates are creatures of ideology, not science* [61].

Warren Strobel of *The Washington Times,* a daily paper founded by the Unification Church of Sun Myung Moon, known from the Moonies sect, also wrote a negative piece, quoting Wood and Seitz. However, co-chairman Patel is quoted: *Those individuals who have decided to attack the report have chosen to misrepresent what is presented in those 423 pages,* but the article ends with a quote from Allan Mense, SDIO chief scientist: *the thing that is most lacking in the APS report is peer review. How they can let this document get published in Modern Physics without having a peer review process is beyond me.* The document was published on 11 June [62].

The APS Study Group responded on 18 June with a letter to Val Fitch, chairman of the APS. The latter sent the letter from the Study Group on 19 June to Weldon, leader (not chairman) of the small group of Republicans that responded to the APS report on 8 June. The Study Group mentions the criticisms of Wood, Canavan, and Seitz. First, the Study Group admitted that there had been a *clerical error in accommodating changes to SDIO's classification policy.* However, the Study Group firmly concludes that: *The large errors claimed by Wood and Canavan simply do not exist. These claims turn out to be based both upon Wood and Canavan's errors in physics and upon their extravagant assumptions of the performance of unproven technologies. SDIO officials have agreed that the numbers in the APS Report are substantially correct. This has been publicly stated by Dr. Louis Marquet of SDIO.* The latter is quoted above. In his letter to Weldon, Fitch referred to the following assertions made by Seitz: *That statement attempted to draw a comparison between the publication of the APS report and the decline of German science under the Nazis. Such a bizarre attack has no place in a technical discussion. I bring it up only to emphasize that, for some, the debate has become highly charged and emotional, making it difficult for them to maintain objectivity* [63].

However, the people of Livermore, Los Alamos, and the SDIO were fighting for their jobs and would not take this lying down. On 8 July Canavan wrote a reply to

---

reactionary and pseudoscience. He contrasted with his own "vernalization", which assumed that hereditary qualities could be changed under the influence of the environment. The Communist Party decreed this as the official doctrine and said that deviations from it would not be tolerated. en.wikipedia.org.

Fitch, in which he took him to task: *That the APS is satisfied with a review by a group with limited involvement in strategic defense, of which two thirds had taken public positions opposing strategic defense in advance, clarifies the APS's attitude towards peer review.* Canavan complains about the fact that *the editor of the APS's Physics Today declined to print any of these analyses of the APS Report, even though they were fully and correctly presented, and the editor of the APS's Reviews of Modern Physics refused to accept any comments on the APS Report from outside the Study* [64].

Seitz published his statement to the outside world in *The Scientist*: *In my view, however, this report is not worthy of serious consideration [...] There are major errors, by factors of as much as 100-fold. Furthermore, these errors and inaccuracies are, as far as we can tell, always in one direction - such as to make the plan for defending the American people - indeed the people of the free world - against a Soviet nuclear attack seem more difficult than it really is... I know of no precedent, in my long association with the American Physical Society, for the issuance of so seriously flawed a document as this, under the aegis of that council* [65]. Seitz thereby accused not only the Study Group, but also the APS Council.

The APS (not the Study Group) reacted in *The Scientist* on 7 September, through their Director of Information Robert Park: *The comments by Frederick Seitz, excerpted in THE SCIENTIST (July 13, 1987, p. 13) were based on the arguments of Wood and Canavan. They were not, as THE SCIENTIST indicated, made in testimony before a House committee, but rather before a coalition of Republican congressmen who support SDI* [Weldon's group mentioned above, RH]. Park also discussed Seitz' background: *Seitz is identified in THE SCIENTIST only as president emeritus of Rockefeller University. That is much too modest [...] Currently, he serves as chairman of the board of the George C. Marshall Institute and chairman of the Science and Engineering Committee for a Secure World, and is associated with the Center for Peace and Freedom - all organizations that exist to promote SDI. He has also in recent years served the International Conference for the Unity of Sciences, an organization affiliated with the Unification Church.* In the newspaper distributed by that church, *The Washington Times*, an article critical of the APS report had previously been published. Park continued his remarks by saying: *The ubiquitous Seitz is also the chairman of the SDI Advisory Committee of the Department of Defense. That group has been urging the creation of a Strategic Defense Initiative Institute to provide SDIO with technical advice* [66].

On 15 September, Bloembergen and Patel appeared at a hearing of a committee of the House of Representatives. Wood and a number of others would also appear at this hearing. Bloembergen was the first to be given the floor. He had prepared a statement in which he listed everything once again. He concluded with a summary of the report: *At this point in time, uncertainty prevails, if or when directed energy weapons would become deployable in space.* Bloembergen mentioned the criticism by Seitz, Canavan, and Wood and the response of Fitch and the Study Group. He also discussed the criticisms of Wood and Canavan, who found that there was an error of a factor of 100 in the report. Bloembergen refuted this claim: *Thus, the discrepancy of a factor 100 is eliminated. [...] other allegations by Wood and Canavan similarly lack a basis in fact."* Patel then made the following statement: *"It is abundantly clear that by changing the required conditions to unrealistic values, lower capabilities for various technologies can be deduced.*

John Spratt (Democrat, South Carolina) was acting as chairman and asked who had elected the members of the Study Group. Bloembergen replied that the Council of the APS had done so. Spratt asked him this because it had been claimed that the members of the Study Group had no experience with the subject, although in fact each member of the Study Group did indeed have such experience. Bloembergen and Patel explained this and Patel stated: *So I think what we have here is, in fact, a collection of individuals who are highly knowledgeable in the field.* Spratt then asked: *"All 17 of these members came to the same basic conclusion, they supported your result and finding without finding it necessary to file a dissenting minority report?* Bloembergen confirmed: *No dissenting opinion in any topics, they all subscribe to the complete report.*

Wood then came up to speak. He reiterated his views: *The APS Response's points are not difficult to address individually; taken together they demonstrate that the APS Report's analysis and scaling are flawed and should be redone, removed, or ignored. [...] The APS Study on Directed Energy Weapons will not be one of the brighter chapters of the Society's attempts to participate in the political life of this country.*

Dave McMurdy (Democrat, Oklahoma) wanted to know what the reaction was to Seitz's claim that academic institutions and associations should not engage in politics. McMurdy, however, stated: *Scientists in open countries should treasure the freedoms they have and devote their reports to scientific affairs [...] I don't find any conclusion that says that the Society is against SDI.* Patel and Bloembergen confirmed that, and regarding McMurdy's request: *Is this a political document or is it an engineering report,* Bloembergen replied: *Strictly engineering and science.* Patel also added: *We cannot comment on what the council does. That is a different issue. The two of us are only responsible for the report that is published.* Bloembergen went on to discuss the confusion that arose because the Study Group and the Council were seen as one. Bloembergen emphasized that the Council was not present. Wood now stated that, five members of the senior review group that was to assess the report had spoken out publicly against SDI. Jim Courter (Republican, New Jersey) asked about a technical issue that was assessed totally differently by Wood and the Study Group. Wood responded that in the Study Group there were twelve academics doing research, three working in government laboratories, and one working in a defense laboratory [P. Avizonis, RH]. Wood then accused Avizonis of not having developed anything useful for SDI. Wood further stated that, because there were no engineers in the Study Group, they could have no idea what the real costs of the systems would be: *Now, an engineer would know...* Patel replied that he (Patel) would not know what was meant by a good scientist or engineer: *I think attacks against physicist engineers are totally unwarranted, either based upon facts or otherwise.* Bloembergen added: *Now, it is also true that anything that can possibly be misread will be misread by certain people with preconceived ideas about physics and engineering.* [67]

Teller subsequently joined the fray in an interview that was by published in *The Scientist* on 21 September 2011. As usual, he made a number of rather categorical statements: *The APS report is really very peculiar.* [...] *For instance, a number of technical people working on SDI have submitted to the Physical Society their arguments, to be published together with the original report. They were refused. And I believe that it is really a sad situation where, in questions of such great importance, free speech does not prevail in science, where free speech should be most honored* [68]. Bloembergen reacted in *The Scientist* of 30 November 1987 with: *The technical people referred to by Teller have always been free to submit manuscripts for publication. Any manuscript submitted to the Reviews of Modern Physics goes through established review procedures. Anyone can submit manuscripts for publication to any journal of the American Physical Society, which has never impeded free speech, as Teller claims. The critique of the APS study report alluded to by Teller was first published in The Congressional Record and The Wall Street Journal. Since the nature of the critique has nothing to do with the basic issues of physics, it was deemed more appropriate to provide a forum for the critique in Physics Today. Gregory Canavan has prepared the critique in a form suitable for that journal, and C.K.N. Patel and I have prepared rebuttals to the points raised* [69].

On 16 November, Seitz published in *The Scientist* a response to the September article by Park of the APS: *Perhaps my greatest concern about the APS report lies in the fact that the council of the society has used the report as an instrument for achieving political purposes* [70].

In the same month of November, the debate between Canavan, Bloembergen, and Patel, announced by Bloembergen in *The Scientist*, was published in *Physics Today*. Canavan criticized the APS report on ten points; Bloembergen and Patel reacted to each. The important thing was that SDI supporters (SDIO, Los Alamos, and Livermore) were over-optimistic in their estimates of, for example, the assets that would be needed. The APS Study Group, consisting of sixteen people from very different political backgrounds, had listened to the numerous presentations on SDI-related research. Bloembergen and Patel thus concluded: *It is remarkable that a unanimous publication emerged from this effort. This was possible only because the group eliminated all politically motivated statements and narrowed the choice of scientific parameters to a reasonable range [...] We reject Canavan's statement that the "analysis and scaling projections [in the APS report] are sufficiently flawed that they should be redone or removed". We reject in stronger terms the additional, gratuitous statement, "Even in a tutorial one does not wish to use calculations that are off by factors of 2^n, where n ranges up to 20." Here Canavan endorses and lends scientific credibility to a letter that appeared in The Wall Street Journal. We had until now declined public comment on this letter, as it is political rather than scientific in character. However, an allegation in PHYSICS TODAY of an error of 2^{20} cannot be ignored* [71].

Finally, Wood sent a letter to Spratt, chairman of the committee of the House of Representatives, where Wood was heard in September. He concluded about the power requirements: *The APS Report misleadingly attributed to the SDIO its extraordinarily high estimates of housekeeping electric power needs of defensive space platforms, that the SDIO had provided the Report's authors with far lower estimates of these housekeeping power requirements, and that the Report's authors apparently were not knowledgeable of the basics of contemporary space electric power supply technology* [72].

However, this letter no longer played any role since it was sent after the formal procedure of an oral hearing. It was a rearguard action. The APS report had been devastating: SDI appeared to be based solely on wild speculation.

In addition, there were developments in the Soviet Union that rendered the introduction of SDI superfluous.

With the arrival of Gorbachev, major changes were made in the Soviet Union that would eventually lead to the implosion of that state. In February 1987 the dissident atomic physicist Andrei Sakharov spoke at a forum organized by Gorbachev. Sakharov had been responsible in the Soviet Union for the development of the hydrogen bomb, and thus was Teller's Soviet counterpart. He stated at a closed session that SDI would be a kind of Maginot line in space, expensive and vulnerable to counter-attacks. SDI could never act as a shield to protect the population.

Gorbachev adopted that position at the end of February and broke the deadlock with the proposal to ban all medium range rockets such as the Pershing.[4] The Commission agreed to take all appropriate measures to prohibit the use of such devices. The Reagan administration succeeded in obtaining concessions from the Soviet Union using SDI as a stick and on 8 December 1987 a treaty was signed between the US and the Soviet Union which banned medium range rockets. In the spring of 1988 Sakharov gave a lecture in Boston, at which Teller was also present. Sakharov once again condemned SDI and stated that if the system was ever implemented, there was a good chance that a nuclear war would break out. The audience were left anxious and shocked in Boston. After Sakharov's departure, Teller defended his case again with passion. In 1988, Teller, Wood, and Canavan attempted to promote *Brilliant Pebbles*. Reagan was very interested, but could not do much more because his term of office ended in 1989. His successor, George Bush senior, was caught between two fires regarding SDI. Vice President Dan Quayle was a strong supporter, but departing SDIO head General Abrahamson submitted a report in the spring of 1989, in which he proposed a radical change of the SDI program with *Brilliant Pebbles*. X-ray lasers, *Excalibur*, and *Super Excalibur* were eliminated, and with *Brilliant Pebbles* a total of 25 billion dollars in cost reduction could be achieved. Bush agreed to adopt the program *Brilliant Pebbles* against an attack by the Soviet Union and its successor in 1991, the Commonwealth of Independent States [73]. The goal was changed into a limited defense against limited attacks.

In 1989 Bloembergen wrote that the APS report only covered part of SDI. The DEW budget represented only 20–25% of the total SDI budget. He further concluded that since 1987 *nothing has changed that view*, referring to the conclusions laid out in the APS report. At the end of the article the following note was added: *This manuscript was submitted in September 1988. Subsequent developments reaffirm the validity of the conclusions of the APS DEW study. At this time DEW deployment*

---

[4]The MGM-31A Pershing was the missile used in the Pershing 1 and Pershing 1a field artillery missile systems. It was a solid-fueled two-stage ballistic missile designed and built by Martin Marietta. en.wikipedia.org.

is not a serious option. Political interest in and support for SDI is on the wane in 1989 [74]. Bloembergen referred to hearings of the House of Representatives in 1988, in which it was stated: *No matter how much peacetime testing were done, there would be no guarantee that the system* **would not fail catastrophically** *during battle as a result of a software error* [75].

It is quite remarkable that Goodchild in his book about Teller does not say a word about the APS report, because it turned out to be fatal for SDI.

On 2 August 1990 Iraq attacked neighboring Kuwait and occupied it in just two days. The United Nations subsequently adopted a resolution giving Iraq until 15 January 1991 to evacuate Kuwait. Iraq refused to do so, and on 17 January a coalition of 34 mainly Western and Arab countries invaded Iraq. Iraq had previously threatened to attack Israel if it were attacked. In the six weeks that the war would last, Iraq launched 42 *Al Hussein* Rockets on Israel and 48 on Saudi Arabia. The *Al Hussein* missiles were Soviet *Scud* missiles with extended range [76]. They carried conventional warheads and flew at an altitude of 150 km and a speed of 5400 km/h. Because of the great distance the rockets had to travel from Iraq to Israel, accuracy was poor. The coalition reacted to the *Scud* attacks by setting up a missile shield using *Patriot,* an anti-missile missile that was fired from mobile installations [77]. This was an excellent opportunity to test at least one rocket shield. *Patriot* was used altogether forty times against the *Scud,* but the results remain controversial to this day. President George Bush senior claimed a success rate of 97%, but the U.S. Army estimated only 80%, and this was later lowered to 70%. Theodore Postol of MIT and Reuven Pedatzur of Tel Aviv University calculated in 1992 that fewer than 10% of the *Scuds* were knocked out and maybe not even one [78]. In any case, it seems likely that only a few *Scuds* were brought down. In January 2001 Defense Secretary William S Cohen joined the doubters: *The Patriot didn't work,* he said. We see here the same story as with SDI. The institutes involved were the Lincoln National Laboratory, the Aerospace Corporation, a military industrial research company, and the Lawrence Livermore National Laboratory. Again, terms like *false and unsupported* (Postol) were used. This time, MIT was also involved [79].

This demonstrated the failure of a missile shield that was only a faint reflection of the space shield.

In 1992, Pentagon scientist Aldric Saucier was given whistleblower protection after he was fired and complained about *wasteful spending on research and development* at the SDI [80]. The Army lifted the security clearance on Saucier on 13 April, after trying to fire him since 14 February [81].

In 1993, President Bill Clinton's government withdrew the SDIO and further restricted the program by using only ballistic missiles launched from the Earth. It was renamed Ballistic Missile Defense Organization or BMDO [82]. This brought the SDI to a definitive end as a spaceborne defense program.

In 2003, the Second Gulf War took place with the invasion of mainly Western armies in Iraq. Iraq reacted again with the deployment of a reduced version of the *Scud,* the *Al Samud-2.* Once again, *Patriot* missiles were deployed. The *Al Samud-2* was only used for short distances (less than 150 km) [83]. In this case, a new version of the *Patriot* did succeed in intercepting a number of Iraqi missiles, but only four *Patriots* were launched [84].

Bloembergen said in an interview in 2006 that the ballistic defense that was revived by President George Bush junior in 1998 was still not successful at that time. Republican politicians continued to dream of a missile shield, at least a shield erected on land [85]. In 2008 Bush junior wanted to install a rocket shield in Poland and the Czech Republic. However, this met with strong opposition from Russia.

Democrat president Barack Obama argued in 2009 that a new system was needed. On 17 September 2009, the Obama administration announced that it would shelve the Bush administration's European missile defense system and replace it with an entirely new missile defense architecture. This decision to stop the deployment of 10 interceptors in Poland and an X-band radar in the Czech Republic had two extremely positive results: it scrapped a technically flawed missile defense system that could never produce a useful level of defense for Europe, and it averted a potentially disastrous foreign policy confrontation with Russia. Again, Postol was among the critics [86].

This system would not be directed against Russia, but against Iran. The system would be based on the *Aegis* system [87], carried by warships. Because the system was not directed against Russia, the Russians were prepared to cooperate with the United States and NATO [88]. According to the Pentagon, the SM-3 rocket used by this system intercepted 84% of the targets in tests. However, Theodore Postol (MIT) and George Lewis (Cornell) stated that only 10–20% were actually hit: *The system is highly fragile and brittle and will intercept warheads only by accident, if ever.* [89]

The *Aegis* system would become operational in 2011, but could only be used against slow Iranian missiles.

Almost thirty years later, therefore, it appears that slow rockets or fast rockets launched up close by an anti-missile shield are capable of destroying a small fraction of ballistic missiles. This is a long way from the original goal of making an impermeable shield in space that would stop all intercontinental rockets carrying atomic warheads. This proves once again that the authors of the APS report were right. Nevertheless, the *Star Wars* project shows some of the features of an ideology that changes with changing circumstances. In 2013, the missile shield in the Czech Republic and Poland was put on the back burner and replaced by President Barack Obama with a plan called *Missile Defense.* This system was to be located in Alaska and serve as a defense against a possible North Korean attack [90].

In 2016, the US activated a land-based missile defense station in Romania, which would form part of a larger and controversial European shield. Russia saw this as a security threat, but this was denied by NATO [91].

In 2017, the administration of President Donald Trump in its turn launched the idea of protecting South Korea against attacks from North Korea by installing the THAAD anti-missile system [92]. Heavy protests from North Korea [93] and China [94] followed.

References

1. Interview Rob Herber with N Bloembergen, 6 December 2006
2. http://pierretristam.com/Bobst/library/wf-241.htm
3. www.tcm.com/tcmdb/title/921/Murder-in-the-Air/
4. www.imdb.com/title/tt0076759/?ref_=sr_2
5. http://villains.wikia.com/wiki/Death-Ray
6. www.imdb.com/title/tt0058150/
7. www.imdb.com/title/tt0096697/?ref_=sr_1
8. Carradice P. The man who invented the death ray. On: www.bbc.co.uk/blogs/waleshistory/2011/02/harry_grindell_matthews_death_ray_inventor.html
9. TELLS DEATH POWER OF 'DIABOLICAL RAYS'. The New York Times, 21 May 1924
10. NEW "DEATH-BEAM" The New York Times, 11 July 1934
11. Goodchild P. Edward Teller, the real Dr. Strangelove. Harvard University Press, Cambridge, MA, USA
12. https://www.llnl.gov/
13. Goodchild P. Edward Teller, the real Dr. Strangelove. Harvard University Press, Cambridge, MA, USA
14. Bethe H. In: Tomas E. Space: Cover Stories: Roaming the High Frontier. Time Magazine, 26 Nov 1984
15. Demott JS. Science: A Technology to Transform War. Time Magazine, 15 Sept 1980
16. Time Magazine. 27 April 1981
17. Time Magazine, 15 March 1982
18. Broad WJ. STAR WARS: PENTAGON LUNACY. New York Times 13 May 1982
19. www.imdb.com/title/tt0062622/
20. www.imdb.com/title/tt0057012/?ref_=fn_al_tt_1
21. www.youtube.com/watch?v=kTKn1aSOyOs
22. en.wikipedia.org
23. www.imdb.com/title/tt3906082/
24. Brodsky S et al. The Incredible Hulk. Marvel Comics No 1 (1964)
25. de.wikipedia.org
26. Goodchild P. Edward Teller, the real Dr. Strangelove. Harvard University Press, Cambridge, MA, USA
27. Walbridge, E. (1984). Angle Constraint for Nuclear-Pumped X-Ray laser weapons. *Nature, 310,* 180–182
28. Goodchild P. Edward Teller, the real Dr. Strangelove. Harvard University Press, Cambridge, MA, USA
29. Bloembergen, N., & Patel, C. K. N. (1987). Report to The American Physical Society of the study group on science and technology of directed energy weapons. *Reviews of Modern Physics, 59,* S1–S201
30. Goodchild P. Edward Teller, the real Dr. Strangelove. Harvard University Press, Cambridge, MA, USA
31. www.aps.org
32. Bloembergen, N., & Patel, C. K. N. (1987). Report to The American Physical Society of the study group on science and technology of directed energy weapons. *Reviews of Modern Physics, 59,* S1–S201
33. Interview Frank Elstner with N Bloembergen. Die Stillen Stars. Nobelpreisträger privat gesehen. (The Silent Stars. Nobel prizewinners seen privately) Heute: (Now) Prof. Nicolaas Bloembergen. ZDF-documentary (1987)
34. Interview Jeff Hecht with N Bloembergen. 4 November 1984, Niels Bohr Library & Archives, American Institute of Physics, College Park, MD USA
35. Interview Frank Elstner with N Bloembergen. Die Stillen Stars. Nobelpreisträger privat gesehen. (The Silent Stars. Nobel prizewinners seen privately) Heute: (Now) Prof. Nicolaas Bloembergen. ZDF-documentary (1987)

36. Interview Rob Herber with N Bloembergen, 6 December 2006
37. Breck Hitz C. Laser Pioneer Interviews: Arthur L. Schawlow. A Nobel Laureate Reviews the Theoretical Work on Stimulated Emission that Led to the Laser. Laser and Applications. Dec 1985 53–57
38. Goodchild P. Edward Teller, the real Dr. Strangelove. Harvard University Press, Cambridge, MA, USA
39. Strategic Defense Initiative: hearings before the Subcommittee on Strategic and Theater Nuclear Forces of the Committee on Armed Services, United States Senate, Ninety-Ninth Congress, first session, October 30; November 6,21; December 3,5, 1985
40. Rensberger B. Computer Bugs Seen as Fatal Flaws in "Star Wars". The Washington Post 30 Oct 1985
41. Goodchild P. Edward Teller, the real Dr. Strangelove. Harvard University Press, Cambridge, MA, USA
42. Strategic Defense Initiative: hearings before the Subcommittee on Strategic and Theater Nuclear Forces of the Committee on Armed Services, United States Senate, Ninety-Ninth Congress, first session, October 30; November 6,21; December 3,5, 1985
43. Goodchild P. Edward Teller, the real Dr. Strangelove. Harvard University Press, Cambridge, MA, USA
44. en.wikipedia.org
45. No Writer Attributed. Suspicious Secrecy. The Harvard Crimson, 19 November 1987
46. Goodchild P. Edward Teller, the real Dr. Strangelove. Harvard University Press, Cambridge, MA, USA
47. The American Physical Society. FEASIBILITY OF DIRECTED ENERGY WEAPONS FOR SDI IS EXAMINED. Press Release, 23 April 1987
48. Cohen NS. Classified SDI Report Revised for publication. The Harvard Crimson, 10 February 1987
49. Yoo JC. Weinberger's Office Delays SDI Report. The Harvard Crimson, 11 April 1987
50. The American Physical Society. FEASIBILITY OF DIRECTED ENERGY WEAPONS FOR SDI IS EXAMINED. Press Release, 23 April 1987
51. Bloembergen, N., & Patel, C. K. N. (1987). Report to The American Physical Society of the study group on science and technology of directed energy weapons. Reviews of Modern Physics, 59, S1–S201
52. The American Physical Society. FEASIBILITY OF DIRECTED ENERGY WEAPONS FOR SDI IS EXAMINED. Press Release, 23 April 1987
53. Interview Rob Herber with N Bloembergen, 6 December 2006
54. Bloembergen, N., & Patel, C. K. N. (1987). Report to The American Physical Society of the study group on science and technology of directed energy weapons. Reviews of Modern Physics, 59, S1–S201
55. American Institute of Physics. APS STATEMENT URGES NO EARLY COMMITMENT TO SDI DEPLOYMENT. News Release, 24 April 1987
56. Letter from the Strategic Defense Initiative Organization to the APS Council, DEW Study Group Members, 28 April 1987
57. Norman, C. (1987). Doubt Cast on Laser Weapons. Science, 236, 509–510
58. Marquet, L. C. (1987). APS Report on SDI. Science, 236, 1411–1412
59. Interview Rob Herber with N Bloembergen, 6 December 2006
60. Letter from C Weldon to W Havens, 8 June 1987
61. Buckley WF Jr. Scientists Stomping Out Star Wars. Universal Press Syndicate, June 1987
62. Strobel W. Scientists punch holes in report critical of SDI missile defense. The Washington Times, 11 June 1987
63. Letter from CKN Patel to VL Fitch and letter from VL Fitch to C Weldon, 18 and 19 June 1987
64. Letter from GH Canavan to WW Havens, 14 July 1987
65. Seitz F. APS Report Has Numerous Errors. The Scientist 1(17)13 (1987)
66. Park, R. L., & Th, A. P. S. (1987). Report Weathers Its Critics. The. Scientist, 1(20), 12

67. Hearings before the Defense Policy Panel and Research and Development Subcommittee of the Committee on Armed Services. House of Representatives. March 26, July 8, and September 15, 1987
68. Gwynne, P. (1987). Teller on SDI, Competitiveness. The. *Scientist, 1*(21), 14
69. Bloembergen, N. (1987). Letter to the Editor. The. *Scientist, 1*(26), 10
70. Seitz, F., & The, A. P. S. (1987). Report: The Flaws Remain. *The Scientist, 1*(25), 13
71. Canavan, G. H., Bloembergen, N., & Patel, C. K. (1987). Debate on APS Directed-Energy Weapons Study. *Physics Today, 40*, 48
72. Letter from L Wood to JM Straat, Jr. 22 December 1987
73. Goodchild P. Edward Teller, the real Dr. Strangelove. Harvard University Press, Cambridge, MA, USA
74. Bloembergen, N. (1989). The Science and Technology of Directed-Energy Weapons. *Interdisciplinary Science Re, 14*, 362–364
75. US Congress. (1988). *Office of Technology Assessment. SDI: Technology, Survivability, and Software*. Washington, DC: US Government Printing Office
76. en.wikipedia.org
77. science.howstuffworks.com/patriot-missile.htm
78. www.nytimes.com/1991/10/31/world/israel-plays-down-effectiveness-of-patriot-missile.html
79. Broad WJ. M.I.T. Studies Accusations of Lies and Cover-Up of Serious Flaws in Antimissile System. New York Times. January 2 (2003)
80. Scientist Said to Assert Fraud in 'Star Wars'. New York Times. 2 March (1992)
81. Lardner G Jr. Critic of Falsifying Credentials. The Washington Post. 14 April (1992)
82. www.federalregister.gov/documents/1994/08/24-20706/organizational-charter-ballistic-missile-defense-organization-bmdo
83. www.globalsecurity.org/wmd/world/iraq/al-samoud_2.htm
84. www.defense-aerospace.com/article-view/feature/18858/patriot-missile-had-mixed-record-in-iraq.html
85. www.armscontrol.org/act/2010_05/Lewis-Postol
86. Goodchild P. Edward Teller, the real Dr. Strangelove. Harvard University Press, Cambridge, MA, USA
87. en.wikipedia.org
88. Shear MD, and Scott Tyson A. Obama Shifts Focus of Missile Shield. The Washington Post. 18 September 2009
89. Broad WJ and Sanger DE. Review Cites Flaws in U.S. Antimissile Program. New York Times. 17 May 2010
90. www.ucsusa.org/nuclear-weapons/missile-defense/how-gmd-missile-defense-works
91. www.bbc.com/news/world-europe-36272686
92. www.cnbc.com/2017/09/11/south-korea-missile-defense-thaad-system-cant-do-the-job-alone.html
93. www.express.co.uk/news/world/853865/WW3-North-Korea-warning-US-anti-missile-THAAD-system-Kim-Jong-Un
94. www.express.co.uk/news/world/856869/World-War-3-China-North-Korea-news-US-anti-missile-system-South-Korea-THAAD

Part VII
Taking Things Easy

Chapter 16
Emeritus, But Not Retired

Bloembergen used ruby lasers for his research on nonlinear optics. Ruby also has a more familiar variant as an ornamental stone. The beauty of the gem only shines in all its splendor if the stone is given the so-called brilliant cut. There is a main surface, called the table, perpendicular to the viewing direction. Equally important are the many facets that are fashioned, 57 in total [1]. These facets give the ruby its brilliance.

Just as the brilliantly cut ruby has a main surface, so Bloembergen's life was based around physics. However, his brilliance came only with the many other facets of his life. A number of characteristic facets are described in this last chapter.

Emeritus, but not retired

In 1990, Bloembergen was appointed Gerhard Gade University Professor, Emeritus, at Harvard, thereby leaving his faculty. Bloembergen kept his room in the laboratory and went on doing what he was already doing before his retirement: giving lectures, writing review articles, and editing work. In honor of his seventieth birthday and as a farewell to the faculty, a symposium was organized by Bloembergen's former students and staff on Sunday (!) 27 May 1990. About two thirds of those students and staff were present at the symposium. The participants come from all round the world: Germany, France, Italy, Norway, Austria, Brazil, China, India, Israel, Japan, and Kazakhstan. The presentations made during the day were recorded in the book *Resonances. A volume in honor of the 70th birthday of Nicolaas Bloembergen*. The editors were Marc Levenson, Eric Mazur, Peter Pershan, and Ron Shen [2]. John Armstrong recalls in his contribution, entitled *Nico Bloembergen as Mentor in the Golden Age of University Research*, that after winning prizes in the 1980s Bloembergen was anything but inactive: *A lesser man might might have rested on his laurels; Nico continued to do competitive research, even competing with his former students.*

However, Bloembergen had been phasing himself out gradually since the 1980s. The number of Ph.D. students and postdocs he supervised gradually decreased after 1981, until he completely ceased to be a supervisor in 1989. Retirement was not therefore a major transition for him (Fig. 16.1).

© Springer Nature Switzerland AG 2019
R. Herber, *Nico Bloembergen*, Springer Biographies,
https://doi.org/10.1007/978-3-030-25737-8_16

Fig. 16.1 Bloembergen in
1995. *Source* Mary Lee,
Harvard News Office

In 1996 Bloembergen published two books containing his collected work: *Encounters in Magnetic Resonance* (550 pages) and *Encounters in Nonlinear Optics* (622 pages). Both books include a selection of Bloembergen's most important work and feature autobiographical introductions and comments by the author. This was a formidable undertaking, especially for a man of 76 years. Regarding his retirement, Bloembergen said in 1996: *Retirement has not brought sudden change, but it has provided even more freedom than academic life had already permitted. As I reminded my Harvard colleagues at my retirement dinner, "A professor can do as he pleases, but a professor emeritus can do as he damn well pleases" [...] Since I have no teaching obligations and no longer provide guidance for laboratory research, I have time to read about many new developments in nonlinear optics* [3].

Since my retirement from the Harvard faculty in 1990, I enjoy a kind of continuous sabbatical. Invitations to present historical overviews, keynote lectures, or colloquium talks keep coming. These always involve encounters with old friends and with new, younger colleagues [4].

My wife and I pursue our interests in art and science, respectively. We have, of course, more leisure time for reading and sports, and for seeing children and

Fig. 16.2 Michael Salour (former graduate), Deli, and Nico on his birthday in 2012. *Source* Michael Salour

grandchildren. We know we are slowing down, but we keep body and mind in motion and have avoided boredom. We hope our state of health will allow us to remain active for years to come [5].

In March 2000, Bloembergen would be 80 years old. Fifty former Ph.D. students and Ph.D. students gathered in Harvard on Thursday 12 May for a day full of memories, and of course for a symposium. Science took first place, even though Bloembergen was eighty years old. The day was organized by Ron Shen from Berkeley and Eric Mazur from Harvard. Under *Reminiscences from the 50s*, the symposium mentioned contributions from, among others, Benedek, Rowland, and Sorokin, under *Reminiscences from the 60s*, Armstrong, Myers, and Shen, under *Reminiscences from the 70s*, Yablonovitch and Salour, and under *Reminiscences from the 80s*, Mazur and Yang [6].

Harvard University awarded Bloembergen an honorary doctorate in 2000: 'Doctor of Science' (Fig. 16.2).

Deli wanted to leave cold New England. She felt she had lived there long enough and wanted the warmth she had known in her youth in the former Dutch East Indies [7]. Deli and Nico went to Florida to see if they could live there, but there were too many alligators and in summer it was hot and humind. In the end, the place of choice would be Academy Village, an innovative retirement community in Tucson, Arizona. Deli: "*I had to wait 10 years.*" [8] Bloembergen gave the following reasons for moving to Tucson: "*I like Tucson. It's much easier to find tennis partners. There are three reasons we are moving to Tucson. First my wife wants to get away from the New England weather. Second, the continuing care aspects of the community are very attractive. Being able to own a freestanding home is a plus. And third, the optical sciences are there ...*" [9].

Bloembergen was appointed Professor of Optical Sciences at the University of Arizona in 2001 (he was almost 81). In 1996 and 1997, he had been visiting scientist there. James Wyant, director of the Optical Science Center at the time of Bloembergen's appointment explained: "*His mind is extremely sharp. He is very good at finding the heart of the problem and helping to solve it. He is excellent at getting students excited about optics.*" [10].

In 2008 Bloembergen was awarded an honorary 'Doctor of Science' by his employer, the University of Arizona.

On his 90th birthday in March 2010, Bloembergen was offered a symposium by the University of Arizona. A tennis tournament was organized in which Bloembergen took part. On the occasion of the 50th anniversary of the laser in 2010, there was a partnership between the American Physical Society, the Optical Society, the International Society for Optics and Photonics, and the IEEE Photonics Society called *Laserfest*, which organizes many events during the year. Bloembergen was invited to attend the May conference in Los Angeles. Nico travels with Deli to Lindau in the summer of 2010, then they visited Paris and he subsequently spent a few days in Amsterdam. Then the Bloembergen's travelled back to Tucson. According to Bloembergen in a phone interview: "*We were very tired when we were back.*" [11].

In July 2011 Bloembergen travelled from Tucson to Hawaii to attend a congress. He gave a lecture with the title: *The Birth of Nonlinear Optics*, in which he stated that non-linear optics was in fact created by Maiman with the demonstration of the first working laser in 1961, and that non-linear optics was in fact fifty years old.

Activities within Harvard

No Harvard Man
Was Bloembergen a real 'Harvard man' and what would that mean?

In the Harvard student magazine, *The Harvard Crimson*, it is not clear: "*What is a Harvard man? Is he a nerd, a cheat, a cad?*" [12].

Frans Spaepen made the following remarks about the term 'Harvard man' in an interview: "*A Harvard man means someone who has graduated here, from the college actually. This is related to Harvard, the elite university which is difficult to enter and within which it is difficult to establish contacts within the highest circles. It certainly has a strong elitist element. This is especially the case for the large group of rich and influential members, while it is not so for those who have graduated with a Ph.D. there. As a professor, you do of course contribute, but you are not part of that elite.*" [13].

Purcell was a Harvard man. Bloembergen says: "*Purcell, like many others, had a love for Harvard. Of course, I was not a Harvard man.*" Bloembergen had not studied at Harvard [14]. This is a strange matter, because Bloembergen was after all attached to Harvard from 1951 to 1990 and before that also as a graduate student, but apparently the fact that he had not obtained his master's degree there was more important.

Erasmus Lectureship: Dutch Culture in Harvard

From the very beginning, Bloembergen was involved in a cultural circle of Dutch academics in New England. This circle was founded around 1953 after the flood disaster in Zeeland, South Holland, and North Brabant. Bloembergen was one of the initiators of the circle, whose purpose was to give lectures about Dutch culture and to maintain social contacts. The circle has since changed name several times and has now been swallowed up by the Netherland-America Foundation [15]. This was founded in 1921 and, according to the website, has the following purpose: *NAF is the leading bilateral foundation initiating and supporting high-impact exchange between the Netherlands and the United States, including the NAF/Fulbright Fellowships and programs in the arts, business, public policy and historic preservation* [16].

Spaepen comments: *"Within Harvard an attempt was made to establish something academic. Not a chair: people were a bit afraid that someone would end up in one of those chairs and we would hear nothing more about them. What they did, and that was a very good idea, was to found a lectureship, where someone could come for a semester to promote the civilization of the Netherlands."* [17].

The Erasmus Lectureship was founded in 1967. This is a subsidized chair at Harvard, sponsored by Dutch companies, organizations; and individuals. A Dutch professor occupies the chair for one year. This was called *The Netherlands Civilization* Chair. The first professor invited was the art historian Frans van Regteren Altena [18]. Others include (but not exhaustively) Hans Daalder (political scientist), Simon Schama (a British historian who published three books about the Netherlands), Klaas van Berkel (science historian and historian), Joep Leerssen (literary scholar and historian), Frans Brüggen (conductor and founder of the Orchestra of the Eighteenth Century), Nicolas Standaert (sinologist), Carlos Steel (ancient and medieval philosophy), Rob Zuidam (composer), and Arie Gelderblom (literary scholar) [19].

Since 1993 the teaching assignment has been expanded with money from the Flemish government to *Civilization of the Netherlands and Flanders*, at the initiative of Spaepen, among others: *"There is a small committee that selects people and proposes people to do that. It is therefore necessary to have some level of cohesion within a particular department and also to give a number of public presentations. Bloembergen has been one of the founding members of this initiative. He has been involved from the outset, always in the selection committee. That has certainly been a credit to him. He was also very enthusiastic about the expansion to include a Flemish contingent."* [20].

In 1972 Bloembergen was awarded the *Half-Moon Trophy* by *The Netherland Club of New York*. This was given because he was a person of Dutch descent who had made a highly creditable contribution in the United States [21].

Activities outside Harvard

Scientific journals
Bloembergen did the usual work as editor for the following scientific journals:

- Associate Editor of *Physical Review* (1956–1958)
- Associate Editor of *Journal of Chemical Physics* (1960–1962)
- Associate Editor of *Journal of Applied Physics* (1964–1967)
- Associate Editor of *IEEE Journal of Quantum Electronics* (1965–1970)
- Advisory Editor of *Optics Communications* (1969–1989)
- Advisory Editor of *Physics* (Amsterdam, 1975–1990)
- Advisory Editor of *Il Nuovo Cimento* nowadays *EJP Plus* (from 1979)
- Advisory Editor of *International Journal of Nonlinear Optical Physics & Materials* (from 1990, Honorary Member)

Government and companies: Bloembergen's fascination for applications
Bloembergen was an advisor to a number of government organizations and companies. In 1953 he started as an advisor to the Lincoln Laboratory at MIT, also in Cambridge just like Harvard. Benjamin Lax was head of the Solid State Group there. When asked whether Bloembergen had become a maser advisor, Lax answered: *"Well, we picked Bloembergen not because of the maser. We were working on magnetic phenomena, resonance phenomena, which is related to some of those things, and Bloembergen was one of the experts and had done some very interesting experiments with his students, and consequently we chose him as one of our consultants."* Bloembergen turned out to be of great value to MIT; *"He discussed with me, I guess it must have been in 1956, prior, or I think about the time of his publication."* This publication referred to was the one about the three-level model: *"I proposed to Meyer and McWhorter, after Bloembergen talking to me, that we work on a maser. In fact, I ordered one of them to quit what he's doing and to do that, McWhorter. He was working on surface physics, and yes, that I recruited him and Jim Meyer to work on the three level maser, that's correct. This was before he actually put it in print, but Bloembergen outlined to me the idea of the three-level maser prior to publication."* [22]. The Lincoln Laboratory was a laboratory of the Institute for Defense Analysis. And it should be noted that Bloembergen only got American nationality in 1958.

A very important role for Bloembergen was as advisor for the oil company Schlumberger from 1953. In an interview about this, he commented: *"Schlumberger studied the possibility of using nuclear magnetic resonance techniques in bore holes to distinguish between oil and water in geological formations. This consulting job involved travel by train and car to Ridgefield, Connecticut, where the company had a nice modern research facility. In later years, I occasionally had to travel to their headquarters in Houston, Texas. The company was of French origin and I enjoyed the many French touches of the competent leadership. The financial remuneration was an essential supplement to my academic salary and was used to make the down payment on our house in Lexington. I was fascinated by the relationship between scientific research and engineering applications. Unfortunately, the project of NMR logging of oil wells turned out to be impractical at that time. My services were not too useful and were terminated in the early sixties.*" [23]. In an interview in 2006; Bloembergen made a few further remarks: *Schlumberger Wells Survey: that's a company that if they have drilled a borehole for oil, there with tools to check what is in the borehole: water or oil. I earned my first money with that. They had asked Purcell to become a consultant. He didn't feel like it, but said "why don't you ask Bloembergen". That was in 1953. Schlumberger is a very big business, on the New York Stock Exchange, big in oil exploration.*" [24].

Bloembergen sold the maser patent to Bell Telephone and Bell thus had free use of the patent for the United States and Canada: *But everywhere outside North America I sold the rights to Philips and they applied for and received them in twelve countries. The only thing that led to was in England and France, where they used it for microwave reception of signals over the ocean via a satellite. So, ten years later, in 1965, I received a sum of money from Philips and I paid the study costs of my three children.*" [25].

The Perkin-Elmer Company in Norwalk, Connecticut, known for its contributions to the Hubble Space Telescope, bought the other rights for the maser (optical instrumentation) from Bloembergen, with the exception of the rights for the government. In addition, Bloembergen became an advisor to the company in the period 1961–1967: "*I then got options on shares and those shares have gone up sharply.*" [26].

From the beginning of the 1960s, Bloembergen was an advisor to the NASA Research Advisory Committee on Electrophysics.

From 1968 to December 1973, Bloembergen was advisor to United Aircraft Corporation in East Hartford, Connecticut (United Technologies Corporation since 1975). This multinational developed among other things the Pratt & Whitney jet engine and Sikorsky helicopter, also worked on defense assignments. Bloembergen worked on the Science Advisory Board of the company. From 1976 to 1992 he was again advisor to United Technology. An example of the activities of Bloembergen in this company can be found in a report of the January 1984 Research Committee at West Palm Beach in Florida:

The committee urges OATL [Optics and Applied Laboratory, RH] *to follow carefully the feasibility of the weapon systems being proposed.*

Feasibility of:

- *Short-range situations, as Tactical Air Defense, ground or ship based OK*
- *anti-satellite systems ASAT OK*
- *not for ballistic missile systems BMS*

Signed among others by *Nicolaas Bloembergen* [27].

In 1972 Bloembergen became an advisor at Itek Corporation in Lexington, Massachusetts. Itek was a company that made its living from defense orders and specialized in cameras for spy satellites. In addition, the company worked on the first Computer Aided Design software for electronic printing and on an optically readable disc [28]. Bloembergen got the title of 'Director' and, due to the close connections between Itek and the Pentagon, was subject to so-called 'clearance'. His contract fell into the category 'TOP SECRET'. Bloembergen was Chairman of the Scientific Board [29]. According to him: "*Itek was a small conglomeration of optical and other items, but was bought by Litton Industries. I made some money with that, because I saw that they were better in the stock market.*" [30]

Other advisory positions were for the Institute for Defense Analysis (IDA), the Los Alamos National Laboratory, DARPA Materials Research Council, the Board of Directors of the Optical Society of America, and the Energy and Research Development Administration.

Los Alamos National Laboratory was mentioned in the Chap. 14, in the context of 'Star Wars'.

The IDA was MIT's Lincoln Laboratory, also mentioned earlier. Founded in 1956, its aim was to bring together a group of major universities for civil, non-commercial research. In 1958, a department was established to support ARPA [31].

(D)ARPA was discussed in Chap. 11 in the context of the maser. This organization of the Pentagon is described by Erica Fuchs as follows: *[...] it was also in the 1960s that ARPA established its critical organizational infrastructure and management style. Specifically, ARPA decided against doing its own research. Instead, it empowered its program managers-scientists and engineers on loan from academia or industry for three to five years to fund technology developments within the wider research community. Within this environment, there was little to no hierarchy. To fund a project, program managers needed to convince only two people: their office director and the ARPA director* [32]. For Bloembergen, with his preference for the greatest possible freedom of research, this provided an ideal working environment as a consultant.

The Optical Society of America studies theory and application of light and publishes a number of scientific journals.

In 1975, the Energy and Research Development Administration (ERDA) was a partial continuation of the Atomic Energy Commission and was mainly involved in promotion. In 1977, ERDA combined with the Federal Energy Agency to form the US Department of Energy.

American Physical Society

In 1989, Bloembergen becomes vice president of the American Physical Society (APS). For his candidacy, he wrote: *It is of prime importance that APS avoid compromising its scientific character. APS should refrain from issuing politically motivated public pronouncements, he believes, but it should continue to contribute unbiased scientific analyses relevant to political issues, on the model of the APS-sponsored studies of nuclear reactor safety, solar power and, most recently, directed-energy weapons* [33].

In 1990 Bloembergen became president elect, in 1991 president, and in 1992 past president. This four-year period involved a lot of work and was anything but a formality. At that time, the APS had 40,000 members, including 4000 members from other countries, and there are now 53,000 members [34]. The scientific journals of the APS were growing well and becoming favorite venues for the exchange of ideas between physicists around the world. During Bloembergen's four-year period, the APS dealt with geopolitical and social issues including arms control, environmental issues, and the participation of minorities. Together with the Association of Physics Teachers, the APS and the APS launched a campaign to improve the teaching of natural sciences. Bloembergen was asked to chair the Campaign Advisory Group and was still doing so in 1996 [35].

Bloembergen and Townes

Bloembergen and Townes were competitors throughout their careers. In 1947 Bloembergen gave his first lecture for the American Physical Society. Townes was also present in the audience. Townes and Schawlow presented their first working maser in 1955. In 1956 Bloembergen had got the idea of the three-level laser (see Chap. 11), and Townes and Schawlow had shamelessly plagiarized the three-level principle, without mentioning it. They fought each other, not only over the maser and laser, but also with regard to population inversion, but without becoming enemies. Bloembergen remarked: "*Charles and I had some vigorous discussions about physics.*" [36].

Peter Pershan: "*I never noticed anything that I would consider as hostility between them. They were both gentlemen. Clearly they were competitors.*" [37].

Townes was much more prominent than Bloembergen and did not always fight fairly, as we have seen with the hijacking of Bloembergen's three-level principle. Bloembergen was more modest and presented a somewhat dull image in public.

Regarding the difference between Townes and Bloembergen, Eric Mazur remarked: "*It has nothing to do with science but with his [Bloembergen's, RH] character and the character of Townes. It is more the aura that Townes radiates. Nico was never like that. Townes gives great lectures. Nico's lectures are not particularly exciting.*" [38].

Bloembergen and Deli had clear opinions about the controversy, as expressed in an interview about Townes. Bloembergen declared: "*It is clear: the great man in lasers, the greatest man in lasers, is Charles Townes. He is generally recognized and he has received many more prizes than I have. The first name that always comes up is Charles Townes and rightly so, he just has a little more. Everyone knows the highest mountain on earth, Mount Everest. But what is the name of the second highest? Nobody knows that. I am number two. I am not the top man. Of course, without Charles Townes someone else would have done it. He has written an autobiography: 'How the laser happened'. In the end he says: if I hadn't been there, how long would it have taken? My answer is: only a few years, because the Russians were already working independently of him. So, he was completely wrong.*"

Deli: "*And he's not as modest as Nico either, that's also the case.*"

Flowers: "*And ambitious.*"

Deli: "*Yes, very ambitious and he still wants, in his old age, he is now 90, he still wants to prove that he is the first. He doesn't need to prove that at all, because everyone already knows that.*"

Flowers: "*But I never had a quarrel with him. Townes wanted to become a University Professor at Harvard, after leaving MIT.*"

Deli: *"That didn't work out and then he went to Berkeley and became University Professor there."*
Flowers: *"He is very happy there. He left the field of lasers and went into astronomy. But the astronomers there didn't like that: he took over everything from them."* [39].

Contacts with the Dutch
Bloembergen maintained contacts with other Dutch people in the United States, but especially with Dutch people who also had something to do with physics. Uhlenbeck and Goudsmit knew him since January 1946, as discussed in Chap. 6.

Indirect link with Lorentz
Bloembergen was a friend of Lorentz' granddaughter Geertuida Boeke-de Haas from 1955 to 1975 [40]. She was the daughter of Geertruida Luberta de Haas-Lorentz (1885–1973) and Wander Johannes de Haas (1878–1960). Geertruida Lorentz was the eldest daughter of Hendrik Lorentz and Aletta Catharina Kaiser. Boeke-de Haas was also a physicist and obtained her doctorate under her father's supervision. De Haas was professor of experimental physics in Leiden. Boeke was a chemical engineer. The Boeke-de Haas couple lived in Concord, New Hampshire. After the couple moved to North Carolina, their contacts diminished somewhat.

Bram Pais
Bram Pais was born on 19 May 1918 in Amsterdam and died on 28 July 2000 in Copenhagen. Pais grew up in Amsterdam, where he also attended secondary school at Mauritskade (now the Amstellyceum). Pais then studied at the University of Amsterdam, where he graduated in 1938. He went on to Utrecht University, where he graduated with an MSc in 1940. He then became an assistant, but on 23 November 1940 he was struck by the German measure that all Jewish officials were to be dismissed. Despite this setback, in 1941 Pais managed to get his Ph.D. under Léon Rosenfeld in Utrecht [41]. Bloembergen knew Pais from this period [42]. Pais went into hiding in various places in Amsterdam, but was arrested on 15 March 1945 together with his friend Lion Nordheim and held in the prison on the Weteringschans. Tineke Buchter came to his aid. She had persuaded Hendrik Kramers to write a letter in German to Heisenberg, in which Kramers pleaded for the release of the promising young physicist Pais. Heisenberg, however, wrote back to say that he could do nothing. Buchter then called a senior Nazi officer and asked if she could visit him. She showed the officer a copy of Kramer's letter to Heisenberg. The officer picked up the phone: *"Hast du die Jude Pais? Lass ihn gehen"* (Do you have the Jew Pais? Let him go.) Pais was free, but Lion Nordheim had been shot a few days earlier. Pais' parents also survived the war, but his sister Annie was gassed with her husband in Sobibor [43].

Pais wanted to go to the United States after the war, just like Bloembergen. Together with Bloembergen, Pais was put by the Dutch government on the list of promising physicists, who were eligible for further study in the United States. However, Pais did not have a father who could provide the necessary finances and could not leave. That is why Pais went to Copenhagen for a year, to the institute run by Niels Bohr. In 1946 Pais left for the Institute for Advanced Study in Princeton [44]. Pais actually wanted to work for Pauli, but by then he was back in Switzerland. In 1950 Bloembergen met Pais in Princeton: *"Bram Pais had his room right next to Einstein, so he experienced everything during the last five years of Einstein's life. But when I came there as a former funnel, he said: do you want to meet Einstein? I said, well, well. I shook hands with Einstein for two minutes."* [45].

In 1963 Pais left Princeton and went to Rockefeller University in New York, where he stayed until his retirement in 1988. In 1982 he published his biography of Albert Einstein *Subtle is the Lord*, which was a great success [46]. After 1988 Pais spent half of his time in the United States and half in Denmark.

Pieter van Heerden

In Chap. 4 we met Van Heerden as assistant to Milatz. In wartime, Van Heerden did a thesis on solid-state detectors he was developing for nuclear physics. Bloembergen then worked as assistant to Van Heerden, investigating the secondary emission of these crystals [47]. Van Heerden obtained his doctorate in Utrecht on 30 July 1945 [48].

In an interview, Bloembergen recounted: *"Pieter van Heerden had started fighting in Indonesia as a soldier, as a volunteer [...] I don't understand why he had done that, but at least he did, or he had done that and then he wrote me, that he wanted to leave and that he actually wanted to come to America."* [49].

On 14 March 1946 Bloembergen held a colloquium in Harvard about Van Heerden's solid-state detector. This had been discussed when Bloembergen described his work in Utrecht. Bloembergen was so successful with the lecture that Piet van Heerden was asked to become Assistant Professor in Harvard.

Dennis Gabor was the first to find a solution for generating wave fronts and storing them. This was in 1947, while working on improving the electron microscope: *Pieter van Heerden was an Assistant Professor at Harvard University at the time* [1948, RH] *and he remembers rushing into Julian Schwinger's office to proclaim the solution to the optical phase reconstruction problem. Schwinger quickly rebuked Van Heerden, saying that it was well known that such a thing was impossible, and that young Pieter should have read the article more carefully. 'But Julian' Pieter said 'he shows pictures' To which, Schwinger read the article for himself and recognized its importance. The cry of 'Show me the hologram!' has been a continuing theme in the field ever since* [50].

Gabor was working with a mercury lamp, but that did not give good results because this type of light source does not emit monochromatic light. Monochromatic light sources are a prerequisite for producing interference patterns. Only when the helium-neon laser became available did further research bear fruit. Emmeth Leith and Juris Upatnieks succeeded in 1962 in making three-dimensional images on photo plates. Van Heerden published his ideas on 3D-holographic storage in 1963. They were so innovative that General Electric, where Van Heerden worked, pointed to another physicist who had 'demonstrated' with computer simulations that the ideas were nonsense. Van Heerden was asked to resign from General Electric [51].

According to Bloembergen: *"Van Heerden did not obtain a permanent appoint-ment* [at Harvard, RH]. *and so went to General Electric in Connecticut. There he made a very important discovery and there he also got a patent on three-dimensional holography [...] Edwin Land offered him a job at Polaroid and said: "I want you to do this work in the morning, but in the afternoon you are free to do your own things. And what did Pieter van Heerden do in his own time: he made a 'theory of everything'. He finally published a book, at his own expense, with Wistik [52] and he constantly wanted to talk about the fundamental theory of everything [...]*

This haunted him in the last decades of his life but he could never convince anyone. He wanted to write an article and he said "well then we should actually go to philosophy, the history of physics then". That was never accepted and then he wrote back to me: "Can't you do something about it." I said, 'I have done everything I can, they asked me to give my opinion and I said, 'I don't understand it all, but I don't see anything clearly wrong with it.' That's what I wrote: "I don't see anything wrong, so why don't you give him the benefit of the doubt, and accept it." Which they wouldn't. [...]

And his Dutch brother and his best friend in Holland knew he was derailed and said, "Nico, you have to do something about it." I said, "You can't do anything about it, if no one wants it. What can I say then? [53].

In 1996, the Dutch journal De Volkskrant published an article with the title: *Pieter, Cor and the accelerator mafia.* In this book, Van Heerden expressed his dissatisfaction with the way in which Dutch physicists also ignored his ideas and refused to publish his writings; *"Snots they are, the Dutch physicists, shameless rascals, unaware of the grand tradition of scientific practice they are involved in."* Cor, Pieter's brother, now also converted to the insight that the existing theories were incorrect, but according to Pieter, Cor's own theory was nonsensical [54].

Other activities after 1990: Is there more to life than physics?
Was there room in Bloembergen's life for anything other than physics?

The Erasmus Lectureship discussed above shows a certain social commitment, at least to the university world.

Bloembergen answered the question of whether he was socially active: *"Thanks to my wife, I became interested in world population problems and wrote a lot about them."* [55].

That was too modest, however. As a Nobel laureate, he was invited everywhere and had been ensnared by a number of organizations.

The Heidelberg Appeal was published in Rio de Janeiro in 1992. This was a declaration calling for the rejection of irrational ideologies, and in particular *an irra-tional ideology which is opposed to scientific and industrial progress, and impedes economic and social development.* This irrational ideology was mainly associated with anti-technological and pseudo-scientific tendencies within the environmental movement. To date, 4000 scientists have signed the declaration, including 72 Nobel laureates. Bloembergen was one of them. Conservative institutions embrace the state-ment as support for the rejection of the thesis that human activity is warming the Earth. However, this statement is incorrect. The first point of the statement is very clear: *We want to make our full contribution to the preservation of our common heritage, the Earth* [56].

In 1993, Bloembergen worked with a group of fellow scientists for the United States Supreme Court. The case in question concerned Jason Daubert and Eric Sculler, both of whom were born with serious birth defects. The children and their parents were suing Merrell Dow Pharmaceuticals because the complainants claimed that the drug Bendectin was the cause of the anomalies. Bendectin was a combination of vitamin B6 and doxylamine and was prescribed to pregnant women to suppress nausea and vomiting. The group of scientists, including Arno Penzias and Frederick Seitz, called themselves *Letter Amici Curiae* (friends of the court) and were acting for Merrell Dow. The issue at stake in this case was, in fact, the credibility of the experts put forward by both parties: *The court stated that scientific evidence is admissible only if the principle upon which it is based is 'sufficiently established to have general acceptance in the field to which it belongs'. The court concluded that petitioners' evidence did not meet this standard [57].* The judgment would prove to be the standard for whether or not to admit expert testimony, and would henceforth be called the Daubert standard [58]. The main expert for the plaintiffs, William McBride, was later struck off the medical register in Australia because he had forged the results concerning Bendectin's teratogenic effects[1] [59].

However, Bloembergen did more than just sign declarations.

Skull Valley in Utah is the Goshute Indian Reservation. To the east of Skull Valley was a nerve gas storage facility and to the north the Magnesium Corporation, both of which posed a serious environmental threat [60]. Private Fuel Storage wanted to provide 'temporary' storage for highly radioactive waste from eight nuclear power plants over 3 km^2 of Skull Valley [61].

In 2000, Bloembergen became a member of a group called *Scientists for Secure Waste Storage (SSWS)*, which concerned itself with scientific aspects of the radioactive waste problem in the valley; the SSWS was neither an advocate or an opponent, but tried to approach the problem exclusively from the scientific angle. Supporters included Hans Bethe and Norman Ramsey. The U.S. Nuclear Regulatory Commission granted a license to Private Fuel Storage in 2005, but the U.S. Bureau of Indian Affairs and the U.S. Bureau of Land Management have so far refused to grant permission to use the area.

In 2001, a large number of scientists, including Bloembergen, addressed the Senate and the House of Representatives with an urgent request to refuse any funding that was contrary to the Anti-Ballistic Missile (ABM) Treaty [62].

In 2001, at the commemoration of the 100th anniversary of the Nobel Prize, 100 Nobel prizewinners, including Bloembergen, issued a brief but urgent warning about the dangers threatening the world, such as global warming and modern weapons (missiles) [63].

---

[1]Teratogenic effects are the combined consequences of consuming a harmful substance, such as alcohol, on a developing fetus; these may manifest themselves as growth deficiencies and/or mental retardation. Fetal alcohol syndrome is an example. medical-dictionary.thefreedictionary.com/ teratogenic+effect.

Also in 2001, the Association of American Universities wrote a letter to President George Bush junior urging him *to support Federal funding for research using human pluripotent stem cells* [Stem cell research, RH]. *We join with other research institutions and patient groups in our belief that the current National Institutes of Health (NIH) guidelines, which enable scientists to conduct stem cell research within the rigorous constraints of federal oversight and standards, should be permitted to remain in effect. The discovery of human pluripotent stem cells is a significant milestone in medical research. Federal support for the enormous creativity of the US biomedical community is essential to translate this discovery into novel therapies for a range of serious and currently intractable diseases* [64]. Bloembergen was one of the many signatories.

Bloembergen belonged to a large group of sponsors (all scientists) who published a report in 2002: *Journal of the Federation of American Scientists: Public Interest Report*. The report contained the articles *Dirty Bombs: Response to a Threat, Making Sense of Information Restrictions after September 11, The "War on Terror" and the "War on Drugs": A Comparison* [65].

In 2003 Bloembergen signed a declaration against the war in Iraq: *The undersigned oppose a preventive war against Iraq without broad international support. Military operations against Iraq may indeed lead to a relatively swift victory in the short term. But war is characterized by surprise, human loss and unintended consequences. Even with a victory, we believe that the medical, economic, environmental, moral, spiritual, political and legal consequences of an American preventive attack on Iraq would undermine, not protect, US security and standing in the world* [66].

In 2004, Bloembergen was one of the signatories of the Union of Concerned Scientists' declaration Promoting Scientific Integrity: *On February 18, 2004, 62 leading scientists—Nobel laureates, leading medical experts, former federal agency directors, and university chairs and presidents—signed a scientists' statement on scientific integrity in policymaking* [67]. This declaration was preceded by a *SCIENTISTS' STATEMENT: Scientific input to government is rarely a dominant factor in public policy decisions, but this input should always be weighed up from an objective and impartial perspective to avoid perilous consequences. This principle has long been adhered to by presidents and administrations of both parties in forming and implementing policies. The administration of George W. Bush has, however, disregarded this principle.*

According to the journal *ChildRight*, in 2005, Bloembergen and 99 other Nobel prizewinners initiated an international fund for children's rights, based in Amsterdam; *"Children are the future of humanity; they have the right to a dignified existence."* [68]. In 2007, the Nobel prizewinners no longer supported the organization; Nelson Mandela took legal action against it, and in 2010, it turned out that the former director had embezzled money [69].

The Bloembergen Family in Lexington

Work

At the beginning of his career at Harvard, Bloembergen worked day and night. But did he work so hard later on, for example in the evening and at weekends? Pershan says: "*I think we all work like that. Some people are known to be so compulsive, that they leave no room for a social or a non-professional life. I don't think Nico is that way. He enjoyed relaxing.*" [70].

His daughter Juliana said in an interview that her father came home strictly on time at six o'clock every evening [71]. The sports mentioned below also indicate Bloembergen's more relaxed behavior in family life than when he was still single.

Health

Bloembergen recounts: "*When I was 79, I had a cycling accident. I was riding my bike in a large car park, which was part of a rest home where we had acquaintances I wanted to visit; I wanted to look around a bit and I said, well, where does the parking lot lead to? The parking lot was closed with a low hanging chain and I was riding into the setting sun and did not see the chain. I was riding at least 15–20 km per hour down when I came against the chain. I had a helmet on and my head was not hurt, but I fell over on my left shoulder. I was in great pain and thought: what should I do? First, go to the institute and complain about the chain, I thought. I went home, that was only five minutes by bike. Deli said: Jesus, what happened to you? I went to Harvard Clinic and after taking the X-ray of the shoulder—it was Sunday evening and an X-ray laboratory had to come to our home—the doctor declared: 'Nothing is broken'. But the pain didn't stop and after two days I went back to Harvard Clinic. Then, after taking a thorax picture, it appeared that I had broken my shoulder blade and four ribs and that there was fluid in my left lung. According to the doctors, the fluids in the lung would disappear by themselves and the ribs and shoulder blade would heal by themselves. Since then, my left shoulder has been much less mobile, but I still play* [tennis, RH] *with my right arm.*

About my eightieth birthday I thought I had a severe cold or flu. I then took six Advils [ibuprofen, which may affect the stomach, RH] *and got a major stomach hemorrhage. Two days later Werlhof's disease was diagnosed, which I also had as a baby. You then have too few platelets; they call that idiopathic, which means you have a certain clinical picture, that the doctors use if they don't understand it. I also had all those purple spots. I got prednisone and have been swallowing it for five years now. I made too few platelets, but that is now in balance with a reasonable dose of prednisone.*"

Deli remarked: "*He looks so young now. He used to have a thin, wrinkled face and now it's a bit smooth. Now he is not allowed to ski anymore, because there may be a dangerous bleeding in the head. He was complaining a lot about that and then his brother* [Herbert, RH] *said: How can you complain if you are no longer allowed to ski when you are eighty!*"

Bloembergen: "*Yes, again a part of your life that is taken away.*" [72].

Sports

This brings us to skiing and other sports. Bloembergen recalled the time he skied in Cambridge: "*I only went with two of the children, in the Boston area, and we went Saturday and Sunday before dawn, to be there for the opening of the ski lifts, because then there was no queue; it was so cold, 8.30 or 9 o'clock and those lifts went on until four o'clock. Every time you wanted to go on the lift in the middle of the day you had to wait an hour. We went up and down 8 times. That was exhausting."*

"Before the last bell rang, we went up in the half darkness. And then we drove back [...] But New England is not such a good place for skiing, because it is either icy, down to minus ten to minus twenty, or it is nice weather and then it is overcrowded with people from New York City and Washington. But the Rocky Mountains is paradise. Aspen is the main area, with Snowmass Mountain, Buttermilk Mountain, and Aspen Highlands; I skied them all. But Utah, that's even better, cheaper, less popular. We rented a simple apartment in the city for a week, at the edge of the city [Salt Lake City, RH]. *It was not in a resort, but within 40 min you could drive to 10 different ski resorts. Luckily we had a friend, who had a car, who came from Tucson, and we always skied with him. He would come to pick us up from the airport and then drive us to the apartment. Every morning we drove to a different ski area. And for the last 20 years I've been skiing for free, because if you were over 65, you'd get a free pass on the lifts in many of those ski areas. That is no longer the case, because too many people now come. Above 80 it was still possible and the last time I skied in Alta, where you had to be above 80, I skied there for free. I could use 'mileage' from my air travel there. And then I also had a free airline ticket [...] I started tennis when I was 40, I had some time, and where we live now, I also have a tennis court."* [73]

Bloembergen also did sports with colleagues, such as Pershan, who recounts as follows: *"I am a sailor. Once we sailed together here in Boston, but he didn't know the boat and I didn't know the boat. But the most fun was at a conference in California. I was still working with him on nonlinear optics. We worked in the morning and had a break in the afternoon. Some people had rented a boat so we did, too. There were many people from Bell Laboratories and someone said 'OK, everybody will rent a boat. It will be Harvard against Bell Laboratories.' We won, and they were embarrassed. I never tell the story in public. Nico was also a good tennis player and a very good squash player, but he was too good for me. Nico used to play squash with Jim Owens, who was a champion of the Harvard squash team, and they were a match for each other."* [74].

Bloembergen didn't like the typical American sports like basketball, baseball, and ice hockey [75].

Finally, Bloembergen added: *"I am not a good sportsman, but I like to do sport, because it is difficult."* [76].

Bloembergen and the Nobel Prize: did it change him?
Today (2016) the Nobel Prize amounts to 10 million Swedish crowns, which is approximately $ 885.000. Bloembergen about 1981, when he got the prize: *"The Nobel Prize then was $ 180,000 or $ 190,000 and I only got a quarter of it, $ 46,000, a third of which went on to all the family there in Stockholm, so I had $ 30,000 left. That's how it always went in history, of course. In the 1950s it was only a few thousand.*

I was awarded the Nobel Prize at the right age. When you get it young, 30, 40 years, you think the rest of your life: I have to perform more. Then they all take strange steps to do something special.

My own teacher, Purcell, who received the Nobel Prize in 1952, when he was 35, was struggling then. He said: 'I don't know what kind of subject to give you'.

Josephson in England, who got the prize for superconductive tunneling at low temperatures, then started doing psychology. That was terrible. Well, every three years he gives a lecture on psychology. Nobody understands.

You shouldn't get the prize too young. Not too old either, when you are 80, because then you can't enjoy it anymore. Not so much, I have just one colleague, Glauber, and who got the prize last year [that was in 2005, RH]. *He had just turned 80 and that brought him new life, but that was only for a few years.*

If you are about 60, then you still have 20 years for doing fun things." [77].

Deli made the following remarks about Bloembergen after he got the prize: *"He always says it hasn't changed his life. Well, of course that's not true. Of course, it changes your life. It may not change in your own eyes, you are always the same person, and also for your wife, but the outside world treats you differently. He doesn't feel that. I feel it much more than he does. All kinds of people do things for you and you get used to that, that they meet you off the train and meet you off the plane. People do things for you and want to be nice to you. You are invited to give lectures everywhere, even when it doesn't matter. They always want his autograph. He always gets notes from students: can you send me pictures with your signature. Once he received a letter from a primary school teacher. They had talked about Nobel Prizes in the classroom. The children all wrote him small notes, they were children of about ten years old. They wanted Nico to put a signature on a piece of paper for each child. I said to Nico: 'You have to do that. You have to encourage the children, because they took the initiative to write all those notes, there were thirty.' And he didn't want to react to that. I said: 'You have to do it.' So, he did that. I got all the papers and he put his autograph on them."*

Deli added: *"The wife of the president of the Academy of Sciences asked me: 'What is to like be married to a Nobel prizewinner? I said: 'Nobel prize winners are just like all other men. Some are wonderful, most are in the middle, and some are awful. Nico is about in the middle.' They always think you're different from someone else. Of course, many of those Nobel prizewinners are not nice at all."*

Bloembergen: *"I am in the middle because of personal qualities, but I am also in the middle because of scientific qualifications. I am not at the bottom, not at the top... I feel at home with the Nobel prizewinners..."* [78].

Antonia, the eldest daughter, made the following criticism during a family dinner: *"Dad, you never emigrated or immigrated, you just moved from one physics lab to another."* [79].

Lindau, an el dorado for Nobel laureates

Every year, the South German town of Lindau on Lake Constance provides hospitality to the Nobel prizewinners. The Swedish count Lennart Bernadotte van Wisborg was the patron of this conference from 1951. It is held near Mainau Castle. In 1981

Bernadotte's much younger wife, Sonja, took over the task and became patroness. Since 2008, their daughter Bettina has been patroness. Every year at the end of June, around six hundred students from many different countries are invited to attend the week-long conference. They can then talk to about twenty Nobel prizewinners in panel discussions, small group discussions, and over lunch and dinner. There are also presentations by the laureates [80]. Each year, a different discipline is honored: economics, physiology or medicine, physics, and chemistry. In-between there are interdisciplinary meetings. Dutch physicists Gerard 't Hooft, Nico Bloembergen, and Martinus Veltman went regularly to give lectures [81]. Bloembergen and Deli were present at the last interdisciplinary conference in 2010: Bloembergen enjoyed the contact with students and young researchers. Students praise the informal atmosphere that prevails. Bloembergen attended the conferences ten times from 1985. In 2005, Bloembergen commented on the motivation for these meetings (in German) '*Es sind die junge Leuten die es interessant machen*' (*It is the young people who make it interesting*) [82]. In 1997, Bloembergen gave a lecture on *Human Population Growth*. This was really the odd one out, since all the other lectures that year were more or less about physics. This is an example of the 'Halo effect': if someone has received the Nobel Prize in Physics, it is often assumed that he will also be good at other things [83]. In 2004, Bloembergen gave a lecture on *Lasers in Peace and War*, thus returning to physics. The last time Bloembergen gave a lecture in Lindau was in 2008: *From Millisecond to Attosecond Laser Pulses* [84].

Deli and Nico

Deli had a strong personality and that generated some friction. Evert Bloembergen remarked: "*She was not an academic type at all, but she was very decisive. Deli set limits on him, you know.*" [85].

Deli found Harvard to be 'male chauvinistic' [86]. Laurel Thatcher Ulrich describes some striking examples of 'Harvard's Womanless History' at Harvard's Radcliffe Institute for Advanced Study, such as the lack of women in portrait galleries and the selectivity regarding *what the University chooses to celebrate about its past*. Harvard's official web pages contained an 'Introduction' of 1200 words without mentioning a single woman [87].

Deli laments: "*A dean retired and then all the lectures were put on hold; that man was so famous and they never talked about the wives, and that makes me so mad.*

And the women, they always give the dinner parties and sit at home, and those men don't even look after their children. There was a good friend of mine and I said, 'I'm also going to say something at that farewell dinner.' There were six deans at that farewell dinner. And everyone tried to keep me away from it. One of those deans said to Nico: 'Is Deli going to talk to the ladies?' And Nico said: 'No, at the dinner.' And another one said: 'Are you going to say something, Nico', and he answered: 'No my wife will already be talking.' And that's how they tried to keep me out. So I gave my five-minute talk and enjoyed it quite a lot. They always say that behind a famous man stands a woman. I say, she is not behind him at all. How far behind, one step, three steps? Half of the women at that dinner were crying. They all came over to me and all the deans just pretended I didn't exist.

We were in Germany and there was another pompous old windbag. Yes, he says, he had never heard a good speech from a woman. I say: Oh no? Well, I do know a few of them. I've heard a lot of bad speeches from men. Half an hour later, Nico would give a lecture at that international meeting. The man approached me, he wanted to ridicule me. He says: 'Would you like to introduce your husband?' It was a big congress and I thought, will I do that now or not? I had half an hour to write it all down. The story ended with the French revolution. Three men would be beheaded there. The first was a priest, the second was a lawyer, and the third was a scientist. The judge asked the priest: with your head down or with your head up? The priest says: 'Down, I bow before God'. The guillotine falls down and stops just a few centimeters above his neck. The priest is allowed to go. Then comes the lawyer, who also wants to go with his head down: once again the axe stops and the lawyer is allowed to go. Then the scientist comes along and wants to lie with his head up to see what is happening. He says: 'Hay, wait a minute, I know what's wrong with this machine.

I ended by saying: I don't know how smart the next speaker is, you'll have to judge for yourselves".

The German Minister of Education thought it was such a beautiful story that he asked if Deli had it on paper and if he could have a copy. Deli said: "*I sent him a copy and I got a very nice note back from him.*" [88]. This is one of the many guillotine jokes in circulation. There are variations with a philosopher instead of a scientist, or a Dutchman, a Frisian, a Belgian, and so on.

Bloembergen said this about Deli: "*My wife is a good musician and she plays the piano. She is almost a professional. She could have had it easy if she hadn't gone to America with me and had to raise three little children. So yes, my wife has great artistic gifts, in music as well as handicrafts and theatre.*" [89].

Deli played piano in the trio *Philomusica,* which performed at the Arizona Senior Academy. The latter is *a charitable organization that brings together professional retirees in an intellectual setting to stimulate and foster their continued productivity and creativity* [90]. Deli also made quilts, that is, bed covers that consist of at least three layers.

When asked whether Deli ever talked about the period when she was interned in a Japanese camp in the Dutch East Indies, Nico and Deli's youngest daughter said: "*She once showed me a wound she had got in the camp. There was no penicillin and it was not very clean, so in the tropical climate the wound wouldn't heal. The other thing was that she was very good in catching flies, because every day in the camp they had to catch twenty to forty flies. Her view was that she didn't hold the Japanese people responsible. She hated the Japanese army.*"

Concerning the character of her mother, Juliana remarked: "*My mother had a reputation. She was very aggressive and she was very strong, not afraid to give her opinion. I think that was her reputation. She probably had children at a younger age than she wanted. There wasn't much money. So I think it was difficult. My father had his career, of course, and that was difficult for my mother. She wanted her own role. So she cut one out for herself.*

Her influence on me was music. We all went for music lessons. She and I played the most. In school or university, we were never pushed.

I think in the beginning my parents didn't feel accepted, but that changed over the years, certainly in Arizona, where so many different people live. They were never interested in American history, but gradually they developed some interest in the national parks and other things." [91].

Juliana again: *"My father had some time for the children. He was usually back home for dinner at six o'clock. He had to be home for dinner, but he started work early. He was probably the first in the office. After dinner he fell asleep, or watched the 6.30 television news. He was never a big TV watcher, but my mother was. Usually, he would read."* [92].

Pershan on Bloembergen and Deli: *"They have a long standing relationship. How can an outsider even begin to understand? Clearly, they have a very meaningful and important relationship. He is never as aggressive as Deli was. Deli is. She is easy to speak to about things that might be controversial. Deli doesn't worry if she offends anybody. She didn't want to offend anybody, but if she says something that does offend she thinks that is their problem. Nico is more reserved."* [93].

Eric Mazur: *"Nico is greatly overshadowed by Deli. Deli is an extraordinary woman, but incredibly dominant. She often disagrees with Nico in public."* [94].

Spaepen: *"Nico and Deli were totally different."* [95].

Family: Nico Bloembergen's brothers and sisters
Evert, the oldest, was more or less predestined to succeed his father Auke in the fertilizer company (ASF/VCF). During the war and shortly afterwards, there was nothing to do at the company and father Auke advised Evert to look for something else. Evert joined the Military Authority in Hilversum, his home town, immediately after the war.[2] He wrote a brochure: *Who is the boss and why?* Evert then joined the head office of ASF/VCF, later Albatros. Afterwards he was sent to a subsidiary in Quebec and after returning became factory director in Kralingse Veer. After the retirement of his father Auke in 1955 he became co-director. In 1958 Albatros was taken over by Koninklijk Zout-Organon (now Akzo) [96]. Evert thought he would become president and director, but that turned out not to be the case [97]. He was not the man to play a subordinate role and resigned in 1962. In 1963 he divorced from his first wife Tilly Dagevos. From 1963 until 1971 he was a member of the board of the Brederode Construction Company [98]. That company developed from a construction company to a project developer. Large-scale projects were developed, such as Hoog Catharijne in Utrecht, the biggest city mall in the Netherlands [99]. In 1965 Evert married Machteld Koningsberger, daughter of professor Victor Jacob Koningsberger (see Chap. 4). In 1972 Evert became chairman of the board of directors of the publishing group Verenigde Nederlandse Uitgeversbedrijven (VNU). VNU: United Editor Companies published magazines. As chairman of the board of directors Evert had to take care of the unity of the merged publishers De Spaarnestad in Haarlem and Cebeco in 's-Hertogenbosch. *'He did an excellent job'*, said his successor Joep Brentjes [100]. After his retirement in 1981, he embarked on all kinds of activities.

---

[2]The Military Authority dealt with the day to day government of the liberated parts of the Netherlands in 1944 and 1945 on behalf of the government in London. It was dissolved in 1946. nl.wikipedia.org.

After a few years he became almost blind from one day to the next, but this handicap did not prevent him from engaging in many activities [101]. Among other things, he was chairman of the King Wilhelmina Fund and in development cooperation in the Baptist Church, and he was a member of the World Wildlife Fund. He was also a lover of trains and model railways [102]. Evert received numerous awards: he was Knight in the Order of the Dutch Lion, Knight in the Order of the Legion of Honour in France, and Officer in the Order of the Crown in Belgium. In 2000 his second wife Machteld Koningsberger died. In later years he shared his life with Ginette Sassen-Stoffels. He died in December 2007 [103].

Diet passed her final exams in 1941 and had already met Rolf Wiggelendam, a clubmate of Evert's, in 1942. The fun of studying was then completely finished. In September 1944 she went to Roermond, where her fiancée lived. She could only return to Bilthoven in January 1945 after the evacuation by the Germans from Roermond. After the liberation she married Wiggelendam in July 1945 and they had six children within eight years. From 1953 she helped in Wiggelendam's general practice. After the death of their eldest daughter in 1964, Wiggelendam became depressed. He died after many difficult years in 1991. Diet died in a nursing home in 2003 [104].

Nico's other sister To passed her final exams in 1942 and went to Groningen to study French. After the liberation she resumed her studies and graduated in Groningen in 1946. She continued her studies in Amsterdam (which was closer to Bilthoven) and passed her doctoral exam there in 1949. She started teaching (French) at Marnix College in Ede in 1950. She did that until she was sixty. She was the founder of the Alliance Française in Ede. To died in a nursing home in 2003 [105].

Herbert passed his final school exam in 1943. From October 1944 to May 1945 Herbert was in bed sick with pleurisy. After the liberation he soon recovered, and went on to study civil engineering in Delft in September 1945. In the winter of 1945–1946 he fell ill again for a long time and decided that an engineering college was perhaps not the best thing for him. He thus took up medicine in Utrecht in September 1946. Herbert was not exactly a fast worker: in 1949, his father Auke was angered by the fact that he hadn't yet passed his bachelor's examination, but in the end he only graduated in 1956. That same year he married Marretje Keyser. In 1957 he settled down as a general practitioner in Gorinchem, and he remained so until 1991. Herbert and Auke Reitze remained best friends until Herbert's death in 2010 [106].

Auke Reitze entered the sixth grade of the high school (gymnasium) in 1945. The Minister of Education took the decision that sixth-form students no longer had to take an exam and so the diploma was handed to him on a plate. In September 1945 Auke went on to study law in Utrecht. He graduated in 1949. After that he had to do his military service for two years and he became a reserve officer in the infantry and secretary of the War Council in the town of 's Hertogenbosch. He then became a lawyer in The Hague. Auke did not like life as a lawyer and in 1957 he returned to Utrecht University. In the same year he married Joop (Johanna) Baartman. Auke did research in Utrecht from 1957 to 1965. In 1965 he obtained his Ph.D. cum laude in Utrecht. In 1965 he was appointed professor of civil law in Leiden. From 1975 until 1993 he was a counsellor in the Dutch Supreme Court and until 1997 he was an extraordinary counsellor [107]. Joop died in 2011 and Auke passed away in 2016.

Nico and Deli's relationship with the Bloembergens in the Netherlands and their children

Nico's youngest brother Auke made the following remark about Nico's contacts with the family in the Netherlands: *Therefore here is an honorary salute to Deli and Nico, who have done a lot for the perpetuation of the family relationship by letting us know by letter, by visiting us regularly in the Netherlands, by celebrating both their silver and their golden wedding in the Netherlands; and by opening their house hospitably for visiting brothers and sisters, nephews and nieces, and cousins* [108].

And Auke Reitze: "*Nico was completely monomaniac about physics, which has improved. All other studies were worth nothing, according to his daughter Antonia. She did population sciences and he didn't like that. Law, that was also nothing. Now he has a certain admiration for me. But he thinks of Evert as a bit of a failure. Perhaps he could have done more. If my brothers hadn't done doctorates, I think I wouldn't have done it.*" [109].

When Nico and Deli's children were small, they played with children from the neighborhood. Later they went on a journey together. When the children were still at school, they took a car trip to California, because Bloembergen was doing a sabbatical year at Berkeley. On the way they camped, so the children got a lasting impression of the American landscape [110].

The Bloembergen children were not interested in studying science. According to Bloembergen: "*They saw how hard I worked and still had a modest salary. Antonia earned more than I did in one of her first jobs.*" [111]

Juliana Bloembergen recounts: *Evert* [Nico's brother, RH] *always visited us with his children. When visiting us from Holland, the others always left their children at home, but not Evert. He spoke three languages fluently, making jokes in different languages. He impressed us. He and my father had a very different relationship. There was a lot of competition between them. The younger brothers, like my father, liked to win. The brothers shared that.*" She continued with an anecdote about the Bloembergen brothers, which she describes as typical: "*My father was visiting Holland some fifteen or twenty years ago. He got together with his two younger brothers, Herbert and Auke, and they were going on a sailing trip. They stopped at a bar to play pool. They played badly, and my father said: 'If I'd had a little more practice, I would have won', and Herbert said: "If I had a little more physics, I would have two Nobel prizes". That was Herbert's sense of humor.*" [112]. Bloembergen also wanted to win when playing games like bowling with the children.

Clara Shoemaker, born Brink, who had introduced Bloembergen and Deli to each other and who had been a bridesmaid at their wedding, also lived in the United States, initially near the Bloembergens in Lexington. Clara was married to David Shoemaker (1920–1995), a chemist just like herself. The Bloembergens and Shoemakers saw each other regularly in Lexington at that time. Clara was a quiet person: "*She was a good housewife, quite Dutch. I don't know how long she kept working. When they had a little house in New Hampshire* [not far from Lexington, RH], *we went skiing together. We also went hiking in New Hampshire and Vermont.*" [113]. Later, the Shoemakers moved to Oregon.

Antonia tells us: '*When I was growing up in New England, with Dutch immigrant parents, ice skating was taken for granted as a regular winter activity. During lengthy stretches of very cold weather, the larger ponds and lakes would freeze over, and my father would pull out his ancient skates for big strides around the periphery. The long pieces of wood with a blade inserted and leather straps fastened over his boots attracted attention, much to the embarrassment of his kids.*' [114].

Antonia holds a Master of Arts degree in political science and demography. She currently works in market research and lives in California.

Brink attended Lexington High School and completed his education in 1971. He then obtained an MBA. Brink was vice-president of Unified Food Purchasing Company and currently lives in Kentucky [115].

Juliana went to the Public School in Lexington: "*one of the top schools in the country.*" She did chemistry, because she heard her brother Brink talk about it: "*I intended to major in chemistry, but I did not do well. I liked literature, so I ended up majoring in French.*" [116]. In 1976 she completed her training at the Radcliffe College in Harvard with a degree in Roman Languages & Literatures. Later, like her brother, she obtained an MBA, attending the Wharton Business School, part of the University of Pennsylvania. Juliana is a businesswoman and lives in Connecticut.

Travel and the question of population
Traveling is all part of a scientist's life, curious to know what the world looks like outside his own laboratory. Conferences and visits to colleagues are normal practice. Bloembergen also received many invitations as a guest professor or employee.

Why was travel so important for Bloembergen? Marika Griesel asked him in an interview: *You have been traveling far and wide. Why is it important for you to be so global, so international?* Bloembergen replied: *Harvard University attracts many foreign students, and I myself was an immigrant. So, I had many students from all over the world: it is nice to visit them later, and see how they are doing in their respective careers. We always enjoyed having an international community in the laboratory.* [117]

Deli also found traveling an interesting activity. In Bloembergen's own words: "*Fortunately, we share an interest in travel and meeting people in foreign countries.*" Deli therefore tagged along during Bloembergen's five sabbaticals. Bloembergen initially did not want to travel to Germany because of the Second World War: "*My emotional attitude towards Germany had been frozen in a hostile state since I left Holland after the German occupation. My relatives and friends had much more relaxed feelings as the years went by and I decided* [in 1980, RH] *that I should get rid of my historically prejudices.*" In 1980 Bloembergen and Deli visited Munich on invitation: "*Travel opportunities still abounded after my retirement from the Harvard Faculty in 1990. We kept in touch with colleagues and friends around the world. We remained interested in different cultural manifestations, including architecture, art, literature, music, food, and drink.*"

We are ever more convinced that human population growth presents the most serious challenge to humankind. We share an interest in Planned Parenthood activities, in educational opportunities for girls and women, and in world-wide population politics [118].

In 1996 Bloembergen wrote: *Opportunities to travel continue unabated. Since my retirement* [thus the period 1990–1996, RH], *my wife and I have made trips to the Netherlands, Indonesia, France, Japan, Mexico, Hong Kong, the People's Republic of China, Sweden, Germany, Italy, and Taiwan. In addition, I travelled without her to Korea, Egypt, and Greece* [119].

In their final years up until 2013 the Bloembergens travelled to Germany, the Netherlands, France, Peru, South Africa, Hawaii, and Spain.

Scientific heritage

'Deep' or 'broad'?

Can Bloembergen be characterized as a 'deep' or a 'broad' researcher? His research itself can certainly be called 'deep' and 'narrow', covering NMR and microwave and optical spectroscopy, areas that overlap.

He used these studies himself in two ways to broaden his horizon.

Firstly, there were the applications, sometimes leading to patents and cooperation with companies. And it cannot be emphasized enough that applications require a completely different attitude than the analysis and solution of physics problems. Applications are about making something workable, that is to say, applying what may at first seem to be of purely theoretical interest. For example, knowing the structure of the metal iron does not mean that it follows how to make nails from iron. An example of how Bloembergen saw applications is the chemical shift. Bloembergen realized that the nuclear magnetic resonance of atoms is influenced by their neighbors. If those neighbors are chemically different atoms, that is a property that can be used to recognize that particular complex, that is, to identify the chemical.

Secondly, there was the application of physics, or more generally natural science to assess the scientific merits of things like the Star Wars project. Physics and natural science were never far away during his retirement either.

Not to mention the activities for the Erasmus sponsorship or Bloembergen's interest in population issues.

Bloembergen recounted: *I have been lucky, that the two topics of the fields that interested me have both led to very important applications. In the case of nuclear magnetic resonance, the data in my thesis concerning the influence of temperature and viscosity on magnetic resonance relaxation of protons in water form the basis for taking MRI pictures. Regarding the second topic, nonlinear optics, only laser sources have high enough intensities. I was really interested in what one can do with lasers. Lasers are heavily used in surgery and in optical communication systems. And these optical communications systems, fiber optics, make the world very small. We can now e-mail anybody anywhere in the world. We can surf the World Wide Web. All this information flows over large distances under the Pacific and Atlantic Ocean.* [120].

On engineering, physics, and mathematics

Many boys have tried to understand the moving parts of a mechanical alarm clock. Why do they do that? What is so interesting about it? The fascinating movement of the escape wheel, the anchor escapement going back and forth, and the barely visible turning of the other wheels is irresistible to many.[3] The escapement causes the ticking of the clock. The sudden awareness of the concept underlying the functioning of the alarm clock gives a deep satisfaction, a sense of beauty for the technique that is difficult to describe. Machines like steam engines or internal combustion engines can also cause this feeling. In his essay *Erkönig op de motorfiets (Elf king on the motor cycle)*, inspired by the work of Robert M. Pfirsig *Zen and the Art of Motorcycle Maintenance* [121], the Dutch writer Rudy Kousbroek reflected on the machine, wondering: [122] *What is beautiful about a machine?* He concluded that the beauty of a machine lies in its function. In *Sauriërs (Saurians)*, Kousbroek states: *Of all the hidden places in this world that make the heart of nature lovers beat faster, lush places, as the English would say, a scrapyard is undoubtedly the most unforgettable [...] When making a journey of discovery, no jungle is as breathtaking as a jungle of rusting machines* [123].

Technology is largely applied physics. And in physics, similar considerations of beauty play an important role. But, of course, beauty is naturally subjective: *Beauty is in the eye of the beholder.*

Murray Gell-Mann, who won the Nobel Prize in Physics in 1969 for his research in particle physics, put it this way: "*What is especially striking and remarkable is that in fundamental physics a beautiful or elegant theory is more likely to be right that a theory that is inelegant.*" But what is it that is beautiful or elegant here? "*A theory appears to be beautiful or elegant (or simple, if you prefer) when it can be expressed concisely in terms of mathematics we already have.*" [124].

Paul Dirac, who won the Nobel Prize in Physics in 1933 for his research in atomic theory, held the same opinion about beauty in physics: *Like many theorists, he had been moved by the sheer sensual pleasure of working with Einstein's theories of relativity and Maxwell's theory. For him and his colleagues, theories were just as beautiful as Mozart's Jupiter Symphony, a Rembrandt self-portrait or a Milton sonnet* [125].

How does Bloembergen's work fit into this?

In an interview in Lindau in 2004 Bloembergen said: *I am fascinated about the very curious correspondence between mathematics and physical phenomena and that mathematics can describe so many phenomena with such accuracy.* [126] Earlier, in 1981, Bloembergen stated the same thing in slightly different terms: *What I've enjoyed since is the interaction of theory and experiment, the fascinating versatility of mathematical frameworks to correspond with embodiments in the physical world.* [127].

Bloembergen referred there to an article by Eugene Wigner (1902–1995), a Hungarian-American theoretical physicist and mathematician. In 1960, he published

---

[3]The escape wheel is a wheel coupled to a spiral spring, which rotates back and forth with an accurately determined frequency.

a classic article entitled *The Unreasonable Effectiveness of Mathematics in the Natural Sciences* [128]. Wigner makes a nice analogy with the way things go in physics: *We are in a position similar to that of a man who was provided with a bunch of keys and who, having to open several doors in succession, always hit on the right key on the first or second trial. He became skeptical concerning the uniqueness of the coordination between keys and doors.*

Wigner continues: *The first point is the enormous usefulness of mathematics in the natural sciences is something bordering on the mysterious and that there is no rational explanation for it. Second, it is just this uncanny usefulness of mathematical concepts that raises the question of the uniqueness of our physical theories.*

But: *Every empirical law has the disquieting quality that one does not know its limitations. We have seen that there are regularities in the events in the world around us which can be formulated in terms of mathematical concepts with an uncanny accuracy.*

Wigner ends with: *The miracle of the appropriateness of the language of mathematics for the formulation of the laws of physics is a wonderful gift, which we neither understand nor deserve. We should be grateful for it and hope that it will remain valid in future research and that it will extend, for better or for worse, to our pleasure, even though perhaps also to our bafflement, to wide branches of learning.*

This fascinated Bloembergen, and became a driving force for him, this mysterious relationship between mathematics and the real (physical) world, which had nothing religious for Bloembergen. That was something he was quite clear about. In an interview, in answer to the question *Are you religious?* Bloembergen answered *No.* To the question: *Are you sure of that?* Bloembergen replied: *I am sure of that, because I was brought up with a religious background. I abandoned it.* [129]

Nico Bloembergen passed away on 5 September 2017 with cardiorespiratory failure. He was 97 and had been living since 2000 in a retirement community in Tucson. Deli Brink survived him. Extensive obituaries were published in the US national journals, e.g., in The Washington Post [130], The New York Times [131], and The Boston Globe [132].

References

1. en.wikipedia.org
2. Levenson MD, Mazyr E, Oershan PS and Shen YR. Resonances. World Scientific, Singapore (1990)
3. Bloembergen N. Encounters in Magnetic Resonances. World Scientific, Singapore (1996)
4. Bloembergen N. Encounters in nonlinear optics. World Scientific, Singapore (1996)
5. Bloembergen N. Encounters in Magnetic Resonances. World Scientific, Singapore (1996)
6. mazur.harvard.edu/news.php?start = 96
7. Bloembergen AR. Het gezin van Rie en Auke Bloembergen 1917–1956. Eigen Beheer, Wassenaar (2003)
8. Interview Rob Herber with N and D Bloembergen-Brink, 7 December 2006
9. www.grandtimes.com/Academy_Village.html
10. uanews.org/node/4313

11. Telephone conversation Rob Herber with N Bloembergen, July 2010
12. Bolduc BJ. Death of a Harvard Man. The Harvard Crimson, 2 March 2009
13. Interview Rob Herber with F Spaepen, 13 June 2009
14. Interview Rob Herber with N Bloembergen, 7 December 2006
15. Interview Rob Herber with F Spaepen, 13 June 2009
16. www.thenaf.org/
17. Interview Rob Herber with F Spaepen, 13 June 2009
18. People. Time Magazine 9 June 1967
19. www.harvard.edu
20. Interview Rob Herber with F Spaepen, 13 June 2009
21. www.netherlandclub.com
22. Interview Joan Bromberg with B Lax, 15 May 1986. Niels Bohr Library & Archives, American Institute of Physics, College Park, MD USA, www.aip.org/history/ohilist/3735.html
23. Bloembergen N. (ed) Encounters in Magnetic Resonances. World Scientific, Singapore (1996)
24. Interview Rob Herber with N Bloembergen, 7 December 2006
25. Interview Rob Herber with N Bloembergen, 7 December 2006
26. Interview Rob Herber with N Bloembergen, 7 December 2006
27. Harvard University Hollis Catalogue, Accession Number 12840, Box 1B Patents and Reprints 1940–1975
28. en.wikipedia.org
29. Harvard University Hollis Catalogue, Accession Number 12840, Box 1B Patents and Reprints 1940–1975
30. Interview Rob Herber with N Bloembergen, 7 December 2006
31. en.wikipedia.org
32. Fuchs ERH. Cloning DARPA Successfully. Issues in Science and Technology. www.issues.org/26.1/fuchs.html
33. Anonymous. Phys Today Jan: 93–94 (1989)
34. www.aps.org/org/index.cfm
35. Bloembergen N. Encounters in Magnetic Resonances. World Scientific, Singapore (1996)
36. Bloembergen N. Meeting Charles H Townes. In: Chiao RY. Amazing light: a volume dedicated to Charles Hard Townes on his 80th birthday. Springer, New York (1996)
37. Interview Rob Herber with P Pershan, 18 September 2007
38. Interview Rob Herber with E Mazur, 18 September 2007
39. Interview Rob Herber with N Bloembergen and D Bloembergen-Brink, 7 December 2006
40. KNAW. Bijzonder bijeenkomst der Afdeling Natuurkunde, Amsterdam, 30 September 1978 (Royal Netherlands Academy of Science. Special meeting of the Department of Physics, Amsterdam, September 30, 1978)
41. Veltman M. Levensbericht (Life message) A. Pais. In: Levensberichten en herdenkingen (Life messages and commemorations). Huygens Institute, Royal Netherlands Academy of Arts and Sciences, Amsterdam (2002)
42. Interview Rob Herber with N Bloembergen, 6 December 2006
43. Veltman M. Levensbericht (Life message) A. Pais. In: Levensberichten en herdenkingen (Life messages and commemorations). Huygens Institute, Royal Netherlands Academy of Arts and Sciences, Amsterdam (2002)
44. Interview Rob Herber with N Bloembergen, 19 September 2009
45. Interview Rob Herber with N Bloembergen, 7 December 2006
46. Pais A. Subtle is the Lord. Oxford University Press, Oxford (2005)
47. Bloembergen N. Note on the internal secondary emission and the influence of surface states. Physica 11:343–344, 1945
48. Heerden PJ van. The Crystal Counter. A New Instrument in Nuclear Physics. Proefschrift, Rijksuniversiteit Utrecht, Utrecht, 1945
49. Interview Rob Herber with N Bloembergen, 6 December 2006
50. Benton S. Holography Reinvented. www.radicaleye.com (1998)
51. Benton S. Holography Reinvented. www.radicaleye.com (1998)

52. Heerden. PJ. The foundation of physics. With a proposal for a fundamental theory of physics. Wistik (1976)
53. Interview Rob Herber with N Bloembergen, 6 December 2006
54. Calmthout M van. Pieter, Cor en de versnellermafia (Pieter, Cor and the accelerator mafia). De Volkskrant 28 June 1996
55. Interview Rob Herber with N Bloembergen, 6 December 2006
56. en.wikipedia.org
57. US Supreme Court. Daubert v Merrell Dow Pharmaceuticals Inc, 509 US (1993)
58. en.wikipedia.org
59. en.wikipedia.com
60. en.wikipedia.org
61. www.nirs.org/radwaste/scullvalley/skullvalley.htm
62. Letter from H Bethe and S Weinberg, Federation of American Scientists, to Tom Daschle, Trent Lott, J Dennis Hastert and Richard Gephardt. 12 November 2001
63. www.nobel.se
64. Letter from the Association of American Universities. Nobel Laureates' Letter to President Bush, 21 February 2001
65. Public Interest Report. J Fed Am Scientists 55(2) 1–12 (2002)
66. Nobel laureates oppose war against Iraq. 29 January 2003. Baez J and Schroer. www.math. colummbia.edu/~woit/wordpress/?p=361
67. go.ucsusa.org/RSI_list/index.php
68. Newsletter ChildRight June 2005
69. nl.wikipedia.org
70. Interview Rob Herber with P Pershan, 18 September 2007
71. Interview Rob Herber with J Bloembergen, 20 September 2009
72. Interview Rob Herber with N Bloembergen and D Bloembergen-Brink, 7 December 2006
73. Interview Rob Herber with N Bloembergen, 6 December 2006
74. Interview Rob Herber with P Pershan, 18 September 2007
75. Interview Rob Herber with N Bloembergen, 7 December 2006
76. Interview Edward Goldwyn with N Bloembergen, 2004. http://vega.org.uk/video/programme/27
77. Interview Rob Herber with N Bloembergen, 6 December 2006
78. Interview Rob Herber with N and D Bloembergen-Brink, 7 December 2006
79. Bloembergen N. Encounters in Magnetic Resonances. World Scientific, Singapore (1996)
80. en.wikipedia.org
81. http://www.lindau-nobel.org/MeetingReports.AxCMS?ActiveID=1279
82. Film Ausflüge in die Zukunft (Excursions into the future). Vera Botterbusch. Bayerische Rundfunk, München (2005)
83. Kahneman D. Thinking, Fast and Slow. Farrar, Straus and Giroux, New York (2011)
84. http://www.lindau-nobel.org/MediaContainer.AxCMS?type=lectures&meeting=105&elementID=205
85. Interview Rob Herber with E Bloembergen, 20 November 2006
86. Interview Rob Herber with N and D Bloembergen-Brink, 7 December 2006
87. Ulrich LT. Harvard's Womanless History. Harvard Magazine, November-December 1999
88. Interview Rob Herber with N and D Bloembergen-Brink, 7 December 2006
89. Interview Rob Herber with N Bloembergen, 6 December 2006
90. www.arizonasenioracademy.org/index.php4?go=home
91. Interview Rob Herber with J Bloembergen, 20 September 2009
92. Interview Rob Herber with J Bloembergen, 20 September 2009
93. Interview Rob Herber with P Pershan, 18 September 2007
94. Interview Rob Herber with E Mazur, 18 September 2007
95. Interview Rob Herber with F Spaepen, 13 June 2009
96. Bloembergen AR. Het gezin van Rie en Auke Bloembergen 1917–1956. Eigen Beheer, Wassenaar (2003)

97. Interview Rob Herber with N Bloembergen, 6 December 2006
98. Bloembergen AR. Het gezin van Rie en Auke Bloembergen 1917–1956. Eigen Beheer, Wassenaar (2003)
99. http://www.hetutrechtsarchief.nl
100. Leistra G. Energieke alleskunner (Energetic all-rounder). Evert Bloembergen 1918–2007. Elsevier 20 January 2007
101. Bloembergen AR. Het gezin van Rie en Auke Bloembergen 1917–1956. Eigen Beheer, Wassenaar (2003)
102. Leistra G. Energieke alleskunner (Energetic all-rounder). Evert Bloembergen 1918–2007. Elsevier 20 January 2007
103. Bloembergen AR. Het gezin van Rie en Auke Bloembergen 1917–1956. Eigen Beheer, Wassenaar (2003)
104. Bloembergen AR. Het gezin van Rie en Auke Bloembergen 1917–1956. Eigen Beheer, Wassenaar (2003)
105. Bloembergen AR. Het gezin van Rie en Auke Bloembergen 1917–1956. Eigen Beheer, Wassenaar (2003)
106. Bloembergen AR. Het gezin van Rie en Auke Bloembergen 1917–1956. Eigen Beheer, Wassenaar (2003)
107. Bloembergen AR. Het gezin van Rie en Auke Bloembergen 1917–1956. Eigen Beheer, Wassenaar (2003)
108. Bloembergen AR. Het gezin van Rie en Auke Bloembergen 1917–1956. Eigen Beheer, Wassenaar (2003)
109. Interview Rob Herber with A R Bloembergen 2 February 2007
110. Interview Frank Elstner with N Bloembergen. Die Stillen Stars (The Silent Stars). Nobelpreisträger privat gesehen (Nobel prizewinners seen privately). Heute: Prof. Nicolaas Bloembergen. ZDF-documentaire (1987)
111. Interview Rob Herber with N Bloembergen, 6 December 2006
112. Interview Rob Herber with J Bloembergen, 20 September 2009
113. Interview Rob Herber with J Bloembergen, 20 September 2009
114. Bloembergen A. Memories on the Ice. Letter to the Editor. New York Times, 23 Jan 2009
115. Bloembergen N. Encounters in Magnetic Resonances. World Scientific, Singapore (1996)
116. Interview Rob Herber with J Bloembergen, 20 September 2009
117. Interview Marika Griesel with N Bloembergen, Ladda Productions AB, Nobel Web, Stockholm, (2004)
118. Bloembergen N. A Discussion of Human Population Growth. Focus 7 (1) 10–14 (1997)
119. Bloembergen N. Encounters in Magnetic Resonances. World Scientific, Singapore (1996)
120. Interview Marika Griesel with N Bloembergen, Ladda Productions AB, Nobel Web, Stockholm, (2004)
121. Pirsig RM. Zen and the Art of Motorcycle Maintenance. An Inquiry into Values. HarperCollins, New York (1974)
122. Kousbroek R. Erkönig op de motorfiets (Elf king on the motor cycle) In: Einsteins Poppenhuis: essays over Filosofie (Einstein's Doll's House: Essays on Philosophy) Meulenhoff, Amsterdam (1990)
123. Kousbroek R. Het meisjeseiland (Girl's island). Augustus, Amsterdam (2011)
124. live11.ted.com/talks/lang/nl/murray_gell_mann_on_beauty_and_truth_in_physics.html (2007)
125. Farmelo G. The Strangest Man. The Hidden Life of Paul Dirac, Quantum Genius. Faber and Faber, London (2009)
126. Interview Marika Griesel with N Bloembergen, Ladda Productions AB, Nobel Web, Stockholm, (2004)
127. Cooke R. Profile in the News. Professor Found Physics Most Difficult, Challenging. Boston Globe, 20 October 1981
128. Wigner E. The Unreasonable Effectiveness of Mathematics in the Natural Sciences. Comm Pure Applied Math 13:1–14 (1960)

129. Interview Edward Goldwyn with N Bloembergen, 2004. http://vega.org.uk/video/programme/27
130. Weil M. Nicolaas Bloembergen, winner of Nobel Prize in Physics, dies at 97. www.washingtonpost.com/local/obituaries/nicolaas-bloembergen-winner-of-nobel-prize-in-physics-dies-at-97/2017/09/09/2e8af4dc-9335-11e7-8754-d478688d23b4_story.html?noredirect=on&utm_term=.e67dc017629b
131. Fleur N St. Nicolaas Bloembergen, Who Shared Nobel for Advances with Laser Light, Dies at 97. www.nytimes.com/2017/09/11/science/nicolaas-bloembergen-who-shared-nobel-for-studies-on-lasers-dies-at-97.html
132. Marquard B. Nicolaas Bloembergen, 97, former Harvard professor and Nobel prizewinner. www.bostonglobe.com/metro/obituaries/2017/09/15/nicolaas-bloembergen-former-harvard-professor-and-nobel-prize-winner/g7qaqK43PBIvYsIKe7LmGK/story.html

Consulted Archives

Historische Kring D'Oude School (Historical Circle De Bilt), De Bilt
Het Utrechts Archief (The Utrecht Archives), Utrecht
Koninklijke Bibliotheek (Royal Library), The Hague
Tresoar (Frisian Historical and Literary Center), Leeuwarden
Historisch Centrum Leeuwarden (Historic Center), Leeuwarden
Groninger Archieven (Archives of Groningen), Groningen
Centraal Bureau voor Genealogie (Central Bureau for Genealogy), The Hague
Stadsarchief Dordrecht (City Archive), Dordrecht
Universiteitsmuseum Utrecht University, Utrecht
Centrale Archiefbewaarplaats, Central Archive Utrecht University, Utrecht
Pusey Library, Harvard University, Cambridge, Massachusetts
Leiden Institute of Physics, Leiden University
Bureau van de pedel (Bureau of the Beadle), Leiden University
Noord-Hollands Archief (Archive North-Holland), Haarlem
Niels Bohr Library, American Institute of Physics, College Park, Maryland

© Springer Nature Switzerland AG 2019
R. Herber, *Nico Bloembergen*, Springer Biographies,
https://doi.org/10.1007/978-3-030-25737-8

Printed in the United States
By Bookmasters